GENETIC FUZZY SYSTEMS

EVOLUTIONARY TUNING AND LEARNING OF FUZZY KNOWLEDGE BASES

ADVANCES IN FUZZY SYSTEMS — APPLICATIONS AND THEORY

Honorary Editor: Lotfi A. Zadeh (*Univ. of California, Berkeley*)
Series Editors: Kaoru Hirota (*Tokyo Inst. of Tech.*),
George J. Klir (*Binghamton Univ.–SUNY*),
Elie Sanchez (*Neurinfo*),
Pei-Zhuang Wang (*West Texas A&M Univ.*),
Ronald R. Yager (*Iona College*)

Advances in Fuzzy Systems — Applications and Theory   Vol. 19

GENETIC FUZZY SYSTEMS
EVOLUTIONARY TUNING AND LEARNING OF FUZZY KNOWLEDGE BASES

Oscar Cordón
University of Granada

Francisco Herrera
University of Granada

Frank Hoffmann
Royal Institute of Technology, Stockholm

Luis Magdalena
Universidad Politécnica de Madrid

World Scientific
Singapore • New Jersey • London • Hong Kong

Published by

World Scientific Publishing Co. Pte. Ltd.

P O Box 128, Farrer Road, Singapore 912805

USA office: Suite 1B, 1060 Main Street, River Edge, NJ 07661

UK office: 57 Shelton Street, Covent Garden, London WC2H 9HE

British Library Cataloguing-in-Publication Data
A catalogue record for this book is available from the British Library.

GENETIC FUZZY SYSTEMS: EVOLUTIONARY TUNING AND LEARNING
OF FUZZY KNOWLEDGE BASES
(Advances in Fuzzy Systems — Applications and Theory Vol. 19)

ISBN 981-02-4016-3
ISBN 981-02-4017-1 (pbk)

Printed in Singapore by World Scientific Printers (S) Pte Ltd

Foreword

Publication of the book "Genetic Fuzzy Systems – Evolutionary Tuning and Learning of Fuzzy Knowledge Bases" written by Prof. Cordon, Prof. Herrera, Prof. Magdalena and Dr. Hoffmann is an important event. Authored by the foremost experts in its field, the book is focused on the analysis and synthesis of systems which employ a synergistic combination of fuzzy logic and evolutionary computing. Such systems are certain to play an increasingly important role within soft computing in the years ahead, as we move farther into the realm of systems with high MIQ (Machine IQ) – systems which can perform complex physical and mental tasks without human intervention.

Basically, soft computing (SC) is a coalition of methodologies which are tolerant of imprecision, uncertainty and partial truth when such tolerance serves to achieve better performance, higher autonomy, greater tractability, lower solution cost and better rapport with reality. At this juncture, the principal members of the coalition are fuzzy logic (FL), neurocomputing (NC), evolutionary computing (EC), probabilistic computing (PC), chaotic computing (CC) and machine learning (ML).

One may think that advantages of forming a coalition are too obvious to be a matter of debate. In fact, this is not the case. When the concept of SC was introduced a decade ago, it was met, for the most part, with skepticism and hostility. In science, as in other realms of human activity, there is a tendency to be nationalistic – a tendency to make an exclusive commitment to a particular methodology and argue that that methodology, and only that methodology, should be used to formulate and solve all problems. This, of course, is the well known hammer principle: if the only tool you have is a hammer, everything looks like a nail.

Summarized as briefly possible, in the partnership of FL and EC, the main contribution of FL is what may be called the calculus of fuzzy if-then rules, while that of EC is the methodology of systematized random search which is inspired by evolutionary processes in living species. In this book, FL and EC are dealt with insight, authority and high expositionary skill. This is an area in which the authors are foremost authorities, and much of what is discussed is new, original and important.

To see the author's work in a clearer perspective, I should like to make an observation regarding FL and, more particularly, the concept of a linguistic variable – a concept which plays a pivotal role in almost all applications of FL.

When the concept of a linguistic variable was introduced close to three decades ago, I was criticized by some prominent members of the scientific establishment as a proponent of a retrogressive move from computing with numbers to computing with words. This reaction was a manifestation of a very deep-seated tradition in science of respect for what is quantitative and precise and lack of respect for what is qualitative and lacking in rigor.

But what we see more clearly today is that the concept of a linguistic variable was merely a first step in the direction of enlarging the role of natural languages in scientific theories. The need for such an enlargement becomes apparent when we consider the remarkable human capability to perform a wide variety of physical and mental tasks, employing perceptions rather than measurements of time, distance, force, shape, possibility, likelihood, intent and other values of decision-relevant variables. In essence, a natural language may be viewed as a system for describing perceptions. In this perspective, a perception may be equated to a proposition or propositions which describe it in a natural language, and operations on perceptions may be reduced to operations on these descriptors. This is the key idea in the recently developed computational theory of perceptions. FL plays an essential role in this theory.

Existing scientific theories are —almost entirely— measurement-based. The importance of the computational theory of perceptions derives from the fact that it adds to measurement-based theories the capability to operate on perception-based information. The need for such capability plays an especially important role in PC, EC, NC and decision analysis. Although the author's work does not focus on the computational theory of perceptions —a theory which is still in its initial stages of development— its incisive exposition of FL lays a firm groundwork for the theory.

To far greater extent then was the case in the past, science and technology are driven by a quest for automation of tasks which require a high level of cognitive and decision-making skills.

The book by Prof. Cordón, Prof. Herrera, Prof. Magdalena and Dr. Hoffmann makes an important contribution to the attainment of this objective. It is a must reading for anyone who is concerned with the conception, design or utilization of information/intelligent systems with high MIQ.

Lotfi A. Zadeh
Berkeley, CA
June 12th, 2000

Preface

It is difficult to add any new word to describe the successful story of fuzzy logic applications to real-world problems, the achievements of fuzzy logic are recognised world-wide. One of the keys to this success is the ability of fuzzy systems to incorporate human expert knowledge. However, the absence of a well defined and systematic design methodology, and the lack of learning capabilities in these kinds of systems, present an important drawback and potential bottle-neck when designing a fuzzy system.

During the nineties, a large amount of work has been devoted to add learning capabilities to fuzzy systems. The new field of soft computing advocates the hybridisation of fuzzy logic with neural networks or evolutionary computation. Today, a large number of books offer a comprehensive treatment on the integration of fuzzy and neural techniques into neuro-fuzzy systems. On the other hand, several hundred papers, special issues of different journals and edited books are devoted to the hybridisation of fuzzy logic and evolutionary computation. As the field of genetic fuzzy systems matures and grows in visibility, the authors feel that there is a need for a coherent, systematic treatise like the one you now hold in your hands.

Our primary objective when starting to write this monograph was to create a complete, self-contained and up to date book on *how to design/improve the knowledge base of a fuzzy rule-based system (FRBS) using genetic techniques*. A book that offers both a theoretical and an applied perspective on the subject. The idea was to compile, structure, and taxonomise the different approaches found in the literature, to offer to the reader a complete panoramic of the field, a framework to place the different approaches, and a sort of conceptual tool-box to allow rapid prototyping of genetic fuzzy rule-based systems (GFRBSs) adapted to different real-world problems. With this aim, the book has been structured in three main blocks that are described in the following paragraphs.

The first block contains Chapters 1 and 2, that offer an overview of the fundamentals of the techniques to be hybridised: fuzzy logic (focusing on FRBSs) and evolutionary computation (paying special attention to genetic algorithms).

The second block (Chapters 3 to 9) represents the core of the book. Chapter 3 gives a general overview of the topic of hybridisation in soft computing and an introduction to genetic fuzzy systems. Chapter 4 is devoted to deal with the tuning of the components of the data base of an FRBS. In Chapter 5, the argument of the book goes back to genetic algorithms (leaving for a while the area of fuzzy systems) and analyses how to apply them to learning processes. This is the starting point for the three following chapters, covering the application to FRBSs of the three fundamental approaches to genetic learning: Michigan (Chapter 6), Pittsburgh (Chapter 7) and Iterative Rule Learning (Chapter 8). Finally, Chapter 9 describes other GFRBS paradigms not covered in previous chapters.

Chapter 10 leaves the field of GFRBSs to describe other kinds of genetic fuzzy systems, including sections on genetic fuzzy-neural systems, genetic fuzzy clustering and genetic fuzzy decision tress. And finally, the last chapter of the book describes several real-world problems that have been solved by means of GFRBSs, trying to show the way from theory to application.

This book originates from several research projects, PhD dissertations, special sessions organised in different International Conferences, International Journal papers, edited books and edited special issues in International Journals. Among them, we can remark three contributions: the lectures *A General Study on Genetic Fuzzy Systems* (Cordón and Herrera, 1995) and *Genetic Algorithms in Fuzzy Control Systems* (Velasco and Magdalena, 1995), included in the Short Course on "Genetic Algorithms in Engineering and Computer Science" (EUROGEN'95), held in Las Palmas de Gran Canaria, December, 1995, and the lecture notes of the tutorial *Genetic Fuzzy Systems* (Herrera and Magdalena, 1997), organised on the occasion of the Seventh IFSA World Congress, and held in Prague, June 24, 1997. From those starting points, several years of, we hope, fruitful work culminated in the writing of this book.

Acknowledgements

First, the authors would like to thank Professor Lotfi A. Zadeh for writing the foreword to this book. In particular, O. Cordón and F. Hoffmann would like to express their gratitude towards him for the hospitality, support and inspiration they encountered during their stays with the Berkeley Initiative in Soft Computing (BISC). We will keep the wonderful time at Berkeley in keen memory.

Most of the ideas (and some of the applications) contained in this book are the results of several (public and private funded) research projects.

Part of O. Cordón and F. Herrera's research has been funded by the European Commission under the project GENESYS (Fuzzy Controllers and Smart Tuning Techniques for Energy Efficiency and Overall Performance of HVAC Systems in Buildings, JOULE contract: JOE-CT98-0090), and by the Spanish Commission for Science and Technology (CICYT) under the projects TIC96-0778 and PB98-1319.

The research of F. Hoffmann was supported by the Army Research Office (DAAH 04-96-1-0341), the Office of Naval Research (N00014-97-1-0946), the NUTEK project "Architectures for Autonomous Systems" NUTEK-1k1p-99-06200 and the EU TMR Network VIRGO.

The research of L. Magdalena has been partially funded by the European Commission under the projects MIX (Modular Integration of Connectionist and Symbolic Processing in Knowledge-Based Systems, ESPRIT 9119), and ADVOCATE (Advanced on Board Diagnosis and Control of Semi-Autonomous Mobile Systems, ESPRIT 28584), and by the Spanish CICYT under the projects ROB89-0474, ROB90-0174, TAP93-0971-E, TAP94-0115, TEL96-1399-C02-02, and TIC97-1343C02-01.

Finally, the authors would like to thank the people that accompanied them along the GFRBS research topic. O. Cordón and F. Herrera would like to recognise the work of R. Alcalá, J. Casillas, M.J. del Jesus, A. González, M. Lozano, A. Peregrín, R. Pérez, L. Sánchez, P. Villar and I. Zwir. F. Hoffmann is thankful to Prof. G. Pfister, O. Nelles and to Prof. Shankar Sastry and the entire BEAR team at the University of California-Berkeley. L. Magdalena expresses his recognition to F. Monasterio and J.R. Velasco.

Oscar Cordón, Francisco Herrera, Frank Hoffmann and Luis Magdalena
Distributed around Europe
February 13th, 2001

Contents

Chapter 1

Fuzzy Rule-Based Systems

Nowadays, one of the most important areas of application of fuzzy set theory as developed by Zadeh (1965) are fuzzy rule-based systems (FRBSs). These kinds of systems constitute an extension to classical rule-based systems, because they deal with *"IF-THEN"* rules whose antecedents and consequents are composed of fuzzy logic statements, instead of classical logic ones.

In a broad sense, an FRBS is a rule-based system where fuzzy logic (FL) is used as a tool for representing different forms of knowledge about the problem at hand, as well as for modelling the interactions and relationships that exist between its variables. Due to this property, they have been successfully applied to a wide range of problems in different domains for which uncertainty and vagueness emerge in different ways (Bardossy and Duckstein, 1995; Chi, Yan, and Pham, 1996; Hirota, 1993; Leondes, 2000; Pedrycz, 1996; Yager and Zadeh, 1992).

This chapter introduces the basic aspects of FRBSs. The different types of FRBSs, their composition and functioning are described and the design tasks that are needed to develop in order to obtain them are analysed. Moreover, a separate section is devoted to study the most common applications of FRBSs, mainly fuzzy modelling (Pedrycz, 1996), fuzzy control (Driankov, Hellendoorn, and Reinfrank, 1993) and fuzzy classification (Chi, Yan, and Pham, 1996). However, we do not focus our discussion on the foundations of FL, which are treated more thoroughly in textbooks like those by Klir and Yuan (1995) and Zimmermann (1996).

1

1.1 Framework: Fuzzy Logic and Fuzzy Systems

Rule-based systems (production rule systems) have been successfully used to model human problem-solving activity and adaptive behaviour, where the classical way to represent human knowledge is the use of *"IF-THEN"* rules. The fulfilment of the rule antecedent gives rise to the execution of the consequent, i.e., one action is performed. Conventional approaches to knowledge representation are based on bivalent logic, which is associated with a serious shortcoming: their inability to deal with the issue of uncertainty and imprecision. As a consequence, conventional approaches do not provide an adequate model for this mode of reasoning familiar to humans and most commonsense reasoning falls into this category.

FL may be viewed as an extension to classical logic systems, providing an effective conceptual framework for dealing with the problem of knowledge representation in an environment of uncertainty and imprecision. FL, as its name suggests, is a form of logic whose underlying modes of reasoning are approximate rather than exact. Its importance arises from the fact that most modes of human reasoning —and especially commonsense reasoning— are approximate in nature. FL is concerned in the main with imprecision and approximate reasoning (Zadeh, 1973). The area of fuzzy set theory originated from the pioneer contributions of Zadeh (1965, 1973) who established the foundations of FL systems.

A fuzzy system (FS) is any FL-based system, which either uses FL as the basis for the representation of different forms of knowledge, or to model the interactions and relationships among the system variables. FL and fuzzy sets position modelling into a novel and broader perspective in that they provide innovative tools to cope with complex and ill-defined systems in which, due to the complexity or the imprecision, classical tools are unsuccessful.

In particular, the application of FL to rule-based systems leads to FRBSs. These systems consider *"IF-THEN"* rules whose antecedents and consequents are composed of fuzzy statements (fuzzy rules), thus presenting the two following essential advantages over classical rule-based systems:

- the key features of knowledge captured by fuzzy sets involve handling of uncertainty, and
- inference methods become more robust and flexible with the approximate reasoning methods employed within FL.

Knowledge representation is enhanced with the use of linguistic variables and their linguistic values, that are defined by context-dependent fuzzy sets whose meanings are specified by gradual membership functions (Zadeh, 1975). On the other hand, FL inference methods such as generalised modus ponens, tollens, etc., form the bases of approximate reasoning with pattern matching scores of similarity (Zadeh, 1973). Hence, FL provides a unique computational base for inference in rule-based systems.

The following two sections will describe in depth the two existing types of FRBSs for engineering problems (i.e., dealing with real-valued inputs and outputs): the *Mamdani* and *Takagi–Sugeno–Kang FRBSs*. In each of them, both the system structure and the inference engine design will be analysed.

1.2 Mamdani Fuzzy Rule-Based Systems

The first type of FRBS that deals with real inputs and outputs was proposed by Mamdani (1974), who was able to augment Zadeh's initial formulation in a way that allows it to apply a FS to a control problem. These kinds of FSs are also referred to as FRBSs with fuzzifier and defuzzifier or, more commonly, as *fuzzy logic controllers* (FLCs) as proposed by the author in his pioneering paper (Mamdani and Assilian, 1975). The term FLC became popular as from the beginning control system design constituted the main application of Mamdani FRBSs.

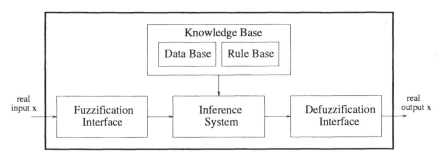

Fig. 1.1   Basic structure of a Mamdani Fuzzy Rule-Based System

The generic structure of a Mamdani FRBSs is shown in Fig. 1.1. The *knowledge base* (KB) stores the available knowledge about the problem in the form of fuzzy *"IF-THEN"* rules. The other three components compose the fuzzy inference engine, which by means of the latter rules puts into effect the inference process on the system inputs. The fuzzification interface establishes a mapping between crisp values in the input domain U and fuzzy sets defined on the same universe of discourse. On the other hand, the defuzzification interface realises the opposite operation by defining a mapping between fuzzy sets defined in the output domain V and crisp values defined in the same universe.

The next two subsections analyse in depth the two main components of a Mamdani FRBS, the KB and the inference engine.

1.2.1   *The knowledge base of Mamdani fuzzy rule-based systems*

The KB establishes the fundamental part of the Mamdani FRBS. It serves as the repository of the problem specific knowledge —that models the relationship between input and output of the underlying system— upon which the inference process reasons from an observed input to an associated output.

The most common rule structure in Mamdani FRBSs involves the use of linguistic variables (Zadeh, 1975). Hence, when dealing with multiple inputs-single output (MISO) systems, these linguistic rules possess the following form:

$$IF \ X_1 \ is \ A_1 \ and \ ... \ and \ X_n \ is \ A_n \ THEN \ Y \ is \ B,$$

with X_i and Y being input and output linguistic variables, respectively, and with A_i and B being linguistic labels with fuzzy sets associated defining their meaning.

Notice that the KB contains two different information levels, i.e., the fuzzy rule semantics (in the form of fuzzy sets) and the linguistic rules representing the expert knowledge. This conceptual distinction is reflected by the two separate entities that constitute the KB.

- A *data base* (DB), containing the linguistic term sets considered in the linguistic rules and the membership functions defining the semantics of the linguistic labels. Each linguistic variable involved in the problem will have associated a fuzzy partition of its domain representing the fuzzy set associated to each of its linguistic terms. Figure 1.2 shows an example of a fuzzy partition comprised by seven triangular-shaped fuzzy membership functions.

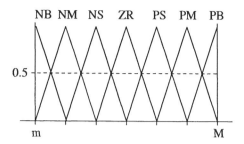

Fig. 1.2   Example of a fuzzy partition

Moreover, the DB also comprises the scaling factors or scaling functions that are used to transform between the universe of discourse in which the fuzzy sets are defined to the domain of the system input and output variables.

- A *rule base* (RB) is comprised by a collection of linguistic rules that are joined by the *also* operator. In other words, multiple rules can fire simultaneously for the same input.

It is important to notice that the RB can present several structures. The usual one is the list of rules, although a decision table (also called rule matrix) becomes an equivalent and more compact representation for the same set of linguistic rules when only a few input variables (usually one or two) are considered by the FRBS.

Consider an FRBS where two input variables, x_1 and x_2, and a single output variable, y, are involved with the following term sets associated: $\{small, medium, large\}$, $\{short, medium, long\}$ and $\{bad, medium, good\}$, respectively. The following RB composed of five linguistic rules

$R_1 : IF\ X_1\ is\ small\ and\ X_2\ is\ short\ THEN\ Y\ is\ bad,$
also
$R_2 : IF\ X_1\ is\ small\ and\ X_2\ is\ medium\ THEN\ Y\ is\ bad,$
also
$R_3 : IF\ X_1\ is\ medium\ and\ X_2\ is\ short\ THEN\ Y\ is\ medium,$
also
$R_4 : IF\ X_1\ is\ large\ and\ X_2\ is\ medium\ THEN\ Y\ is\ medium,$
also
$R_5 : IF\ X_1\ is\ large\ and\ X_2\ is\ long\ THEN\ Y\ is\ good,$

is also represented by the decision table shown in Fig. 1.3.

x_2	x_1		
	small	*medium*	*large*
short	bad	medium	
medium	bad		medium
long			good

Fig. 1.3   Example of a decision table

Before concluding this section, we should notice two aspects. On the one hand, the structure of a linguistic rule may be more generic if a connective other than the *and* operator is used to aggregate the terms in the rule antecedent. However, it has been demonstrated that the above rule structure is generic enough to subsume other possible rule representations (Wang, 1994). The above rules are therefore commonly used throughout the literature due to their simplicity and generality. On the other hand, Mamdani-type rules with a different structure can be considered, as we shall see in Sec. 1.2.6.

1.2.2   *The inference engine of Mamdani fuzzy rule-based systems*

The inference engine of a Mamdani FRBS is composed of the following three components:

- A *fuzzification interface*, that transforms the crisp input data into fuzzy values that serve as the input to the fuzzy reasoning process.
- An *inference system*, that infers from the fuzzy input to several resulting output fuzzy sets according to the information stored in the KB.
- A *defuzzification interface*, that converts the fuzzy sets obtained from the inference process into a crisp action that constitutes the global output of the FRBS.

In the following, they are briefly introduced and an example of their operation is shown.

1.2.2.1 *The fuzzification interface*

The fuzzification interface enables Mamdani-type FRBSs to handle crisp input values. Fuzzification establishes a mapping from crisp input values to fuzzy sets defined in the universe of discourse of that input. The membership function of the fuzzy set A' defined over the universe of discourse U associated to a crisp input value x_0 is computed as:

$$A' = F(x_0)$$

in which F is a fuzzification operator.

The most common choice for the fuzzification operator F is the *point wise fuzzification*, where A' is built as a singleton with support x_0, i.e., it presents the following membership function:

$$A'(x) = \begin{cases} 1, & \text{if } x = x_0 \\ 0, & \text{otherwise} \end{cases}$$

1.2.2.2 *The inference system*

The inference system is the component that derives the fuzzy outputs from the input fuzzy sets according to the relation defined through fuzzy rules. The inference scheme establishes a mapping between fuzzy sets $\mathbf{U} = \mathbf{U_1} \times \mathbf{U_2} \times \cdots \times \mathbf{U_n}$ in the input domain of $X_1, \ldots, X_n$ and fuzzy sets $\mathbf{V}$ in the output domain of Y. The fuzzy inference scheme employs the generalised modus ponens, an extension to the classical modus ponens (Zadeh, 1973):

$$\begin{array}{l} \textit{IF X is A} \quad \textit{THEN} \quad \textit{Y is B} \\ \underline{\textit{X is A'}} \\ \qquad\qquad\qquad\qquad\quad \textit{Y is B'} \end{array}$$

In order to apply the generalised modus ponens, one first has to establish the connection between the above conditional statements and the type of rule used by the FRBS. A fuzzy conditional statement of the form *"IF X is A THEN Y is B"* represents a fuzzy relation between A and B defined in $\mathbf{U} \times \mathbf{V}$. This fuzzy relation is expressed by a fuzzy set R whose membership function $\mu_R(x, y)$ is given by:

$$\mu_R(x, y) = I(\mu_A(x), \mu_B(y)), \forall x \in \mathbf{U}, y \in \mathbf{V}$$

in which $\mu_A(x)$ and $\mu_B(y)$ are the membership functions of the fuzzy sets A and B, and I is a fuzzy implication operator that models the existing fuzzy relation.

The membership function of the fuzzy set B' is obtained by applying the *compositional rule of inference* (Zadeh, 1973) in the following way: *"If R is a fuzzy relation defined in* **U** *and* **V** *and A' is a fuzzy set defined in* **U**, *then the fuzzy set B', induced by A', is obtained from the composition of R and A'"*, that is:

$$B' = A' \circ R$$

Therefore, when applied to the i-th rule of the RB,

$$R_i : IF \ X_{i1} \ is \ A_{i1} \ and \ \dots \ and \ X_{in} \ is \ A_{in} \ THEN \ Y \ is \ B_i,$$

the simplest expression of the compositional rule of inference is reduced to:

$$\mu_{B'_i}(y) = I(\mu_{A_i}(x_0), \mu_{B_i}(y))$$

where $\mu_{A_i}(x_0) = T(\mu_{A_{i1}}(x_1), \dots, \mu_{A_{in}}(x_n))$, $x_0 = (x_1, \dots, x_n)$ is the current system input, T is a fuzzy conjunctive operator (a t-norm) and I is a fuzzy implication operator. The most common choice for both of them —conjunctive and implication operators— is the minimum t-norm.

1.2.2.3   *The defuzzification interface*

The inference process in Mamdani-type FRBSs operates on the level of individual rules. Thus, the application of the compositional rule of inference to the current input using the m rules in the KB generates m output fuzzy sets B'_i. The defuzzification interface has to aggregate the information provided by the m output fuzzy sets and to obtain a crisp output value from them. This task can be done in two different ways (Bardossy and Duckstein, 1995; Cordón, Herrera, and Peregrín, 1997; Wang, 1994): *Mode A-FATI* (first aggregate, then infer) and *Mode B-FITA* (first infer, then aggregate).

Mamdani (1974) originally suggested the mode A-FATI in his first conception of FLCs. In the last few years, the Mode B-FITA is becoming more popular (Cordón, Herrera, and Peregrín, 1997; Driankov, Hellendoorn, and Reinfrank, 1993; Sugeno and Yasukawa, 1993), in particular in real-time applications which demand a fast response time.

Mode A-FATI: first aggregate, then infer

In this case, the defuzzification interface operates as follows:

- Aggregate the individual fuzzy sets B_i' into an overall fuzzy set B' by means of a fuzzy aggregation operator G (representing the *also* operator):

$$\mu_{B'}(y) = G\left\{\mu_{B_1'}(y), \mu_{B_2'}(y), \ldots, \mu_{B_n'}(y)\right\}$$

- Employ a defuzzification method, D, that transforms the fuzzy set B' into a crisp output value y_0:

$$y_0 = D(\mu_{B'}(y))$$

Usually, the aggregation operator G considered is the maximum or the minimum t-norm, and the defuzzifier D is the "centre of gravity" (CG) or the "mean of maxima" (MOM), whose expressions are shown below:

- CG:

$$y_0 = \frac{\int_Y y \cdot \mu_{B'}(y)dy}{\int_Y \mu_{B'}(y)dy}$$

- MOM:

$$y_{inf} = \inf\{z \mid \mu_{B'}(y) = \sup \mu_{B'}(y)\}$$
$$y_{sup} = \sup\{z \mid \mu_{B'}(y) = \sup \mu_{B'}(y)\}$$
$$y_0 = \frac{y_{inf} + y_{sup}}{2}$$

Mode B-FITA: first infer, then aggregate
The contribution of each fuzzy set is considered separately and the final crisp value is obtained by means of an averaging or selection operation performed on the set of crisp values derived from each of the individual fuzzy sets B_i'.

The most common choice is the CG or the "maximum value" (MV) weighted by the matching degree, whose expression is shown as follows:

$$y_0 = \frac{\sum_{i=1}^m h_i \cdot y_i}{\sum_{i=1}^m h_i}$$

with y_i being the CG or the MV of the fuzzy set inferred from rule R_i, B_i', and $h_i = \mu_A(x_0)$ being the matching between the system input x_0 and the rule antecedent.

Hence, this approach avoids aggregating the rule outputs to the final fuzzy set B', which reduces the computational burden compared to mode A-FATI defuzzification.

This defuzzification mode constitutes a different approach for the notion of the *also* operator.

1.2.3   *Example of application*

In this section, the operation mode of the inference engine will be illustrated by means of a simple example. Let us consider a problem with two input variables, *size* and *weight*, and one output variable, *quality*, with the following linguistic term sets associated:

$$D_{size} = \{small, large\} \qquad D_{weight} = \{small, large\}$$
$$D_{quality} = \{bad, medium, good\}$$

whose semantics are defined by means of the triangular-shaped fuzzy membership functions represented in Fig. 1.4.

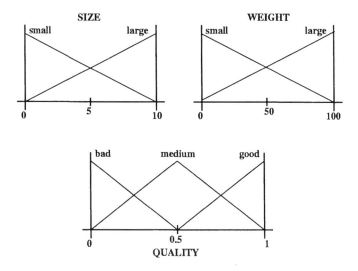

Fig. 1.4   Composition of the data base for the example

The RB for this problem is composed of the following four linguistic rules:

$R_1 : IF\ size\ is\ small\ and\ weight\ is\ small\ THEN\ quality\ is\ bad,$
also

$R_2 : IF\ size\ is\ small\ and\ weight\ is\ large\ THEN\ quality\ is\ medium,$
also

$R_3 : IF\ size\ is\ large\ and\ weight\ is\ small\ THEN\ quality\ is\ medium,$
also

$R_4 : IF\ size\ is\ large\ and\ weight\ is\ large\ THEN\ quality\ is\ good$

The inference engine considered is the classical one employed by Mamdani (1974) which considers the minimum t-norm as conjunctive and implication operators (T and I, respectively) and a Mode A-FATI defuzzification interface where the aggregation operator G is modelled by the maximum t-conorm, whilst the defuzzification method D is the CG. This fuzzy inference engine is usually called *max-min-CG*.

Let us consider that the current system output is $x_0 = (2, 25)$. This output is matched against the rule antecedents in order to determine the firing strength h_i of each rule R_i in the RB. The following results are obtained:

$$R_1 : h_1 = \min(\mu_{small}(2), \mu_{small}(25)) = \min(0.8, 0.75) = 0.75$$
$$R_2 : h_2 = \min(\mu_{small}(2), \mu_{large}(25)) = \min(0.8, 0.25) = 0.25$$
$$R_3 : h_3 = \min(\mu_{large}(2), \mu_{small}(25)) = \min(0.2, 0.75) = 0.2$$
$$R_4 : h_4 = \min(\mu_{large}(2), \mu_{large}(25)) = \min(0.2, 0.25) = 0.2$$

Then, the inference system applies the compositional rule of inference on each individual linguistic rule to obtain the inferred fuzzy sets B'_i as follows:

$$R_1 : \mu_{B'_1}(y) = \min(h_1, \mu_{B_1}(y)) = \min(0.75, \mu_{bad}(y))$$
$$R_2 : \mu_{B'_2}(y) = \min(h_2, \mu_{B_2}(y)) = \min(0.25, \mu_{medium}(y))$$
$$R_3 : \mu_{B'_3}(y) = \min(h_3, \mu_{B_3}(y)) = \min(0.2, \mu_{medium}(y))$$
$$R_4 : \mu_{B'_4}(y) = \min(h_4, \mu_{B_4}(y)) = \min(0.2, \mu_{good}(y))$$

The operation of the inference engine is graphically illustrated in Fig. 1.5 which depicts the membership functions $\mu_{B'_i}$ resulting from the inference step.

Finally, the defuzzification interface aggregates the four individual output fuzzy sets by means of the maximum t-conorm:

$$\mu_{B'}(y) = \max\left\{\mu_{B'_1}(y), \mu_{B'_2}(y), \mu_{B'_3}(y), \mu_{B'_4}(y)\right\}$$

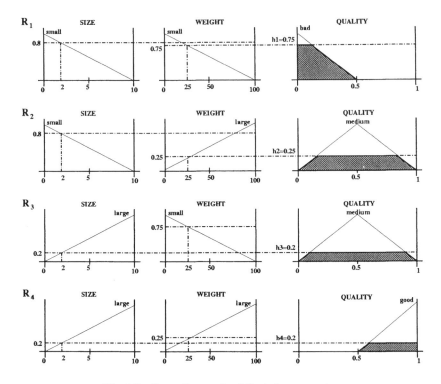

Fig. 1.5   Operation mode of the inference system

and defuzzifies the resulting aggregated fuzzy set by means of the CG strategy:

$$y_0 = \frac{\int_Y y \cdot \mu_{B'}(y) dy}{\int_Y \mu_{B'}(y) dy}$$

thus obtaining the final value $y_0 = 0.3698$. This process is graphically represented in Fig. 1.6.

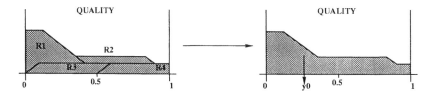

Fig. 1.6   Mode A-FATI defuzzification with the max-centre of gravity strategy

On the other hand, if a mode B-FITA defuzzification interface based on the "MV weighted by the matching" defuzzifier is selected, the final crisp output y_0 is generated by means of the expression:

$$y_0 = \frac{\sum_{i=1}^{m} h_i \cdot MV_i}{\sum_{i=1}^{m} h_i}$$

i.e.,

$$y_0 = \frac{0.75 \cdot 0.1 + 0.25 \cdot 0.5 + 0.2 \cdot 0.5 + 0.2 \cdot 0.8}{1.4} = \frac{0.46}{1.4} = 0.3286$$

as shown in Fig. 1.7.

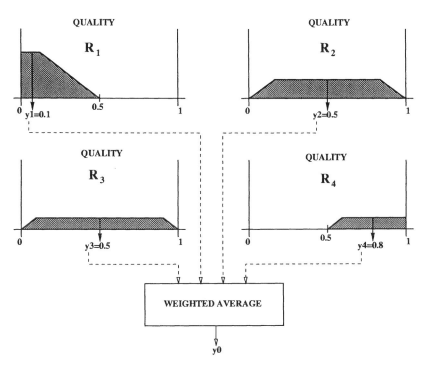

Fig. 1.7 Mode B-FITA Defuzzification with the maximum value weighted by the matching strategy

1.2.4   *Design of the inference engine*

The implementation of the inference engine of the Mamdani-type FRBS requires the following design choices (Kiszka, Kochanska, and Sliwinska, 1985):

(1) Select the operator I used in the fuzzy implication of *"IF-THEN"* rules.
    Mamdani (1974) proposed to use the minimum operator as the t-norm for implication. Since then, various other t-norms have been suggested as implication operator (Gupta and Qi, 1991), such us the algebraic product:

$$I(x, y) = x \cdot y$$

Other important family of implication operators are the fuzzy implication functions (Trillas and Valverde, 1985), being one of the most usual the Lukasiewicz's one:

$$I(x, y) = \min(1, 1 - x + y)$$

Less common implication operators such as force-implications (Dujet and Vincent, 1995; Cordón, Herrera, and Peregrín, 2000), t-conorms and operators not belonging to any of the most known implication operator families (Cao and Kandel, 1989; Cordón, Herrera, and Peregrín, 1997; Kiszka, Kochanska, and Sliwinska, 1985) have also been considered.

(2) Select the conjunctive operator T to be used in the rule antecedent. There are available different operators belonging to the t-norm family to make this choice (Gupta and Qi, 1991; Cordón, Herrera, and Peregrín, 1997).

(3) Decide the defuzzification interface operation mode to be used. If Mode A-FATI is considered, choose the aggregation operator G. A t-norm or a t-conorm are usually employed (with the minimum or the maximum, respectively, being the most common ones due to their simplicity). In (Bardossy and Duckstein, 1995), a set of more complex aggregation operators is proposed and an analysis of their properties, as well as a comparative study of their behaviour, are presented. Then, the defuzzification operator D is selected, usually within the CG and the MOM.

On the other hand, when working in mode B-FITA, the most common operators are the average, the weighted average or the selection of a characteristic value of the individual fuzzy sets, being a function of any importance degree of the rule that generated them in the inference process (Cordón, Herrera, and Peregrín, 1997). Usually, the CG and the MV are used as characteristic values, and the *area* and *height of the inferred fuzzy set* or the *matching degree* are used as importance degrees of the rule (Cordón, Herrera, and Peregrín, 1997; Hellendoorn and Thomas, 1993; Sugeno and Yasukawa, 1993). For example, the following expressions correspond respectively to the *"MV weighted by the area"* and the *"CG of the fuzzy set with largest height"*:

$$y_0 = \frac{\sum_{i=1}^m s_i \cdot MV_i}{\sum_{i=1}^m s_i}$$

with s_i and MV_i being respectively the area and the MV of the fuzzy set B_i', and

$$B_k' = \{B_i' \; s.t. \; l_i = \max_t(l_t), \quad t = 1, \ldots, m\}$$
$$y_0 = CG_k$$

with l_i and CG_i being respectively the height and the CG of B_i'.

Studies analysing the behaviour of the existing fuzzy operators for the different purposes are found in the literature. The aim of these papers is to provide some guidelines for the optimal design of the inference mechanism in the Mamdani-type FRBSs (Cao and Kandel, 1989; Cordón, Herrera, and Peregrín, 1997; Kiszka, Kochanska, and Sliwinska, 1985).

1.2.5 *Advantages and drawbacks of Mamdani-type fuzzy rule-based systems*

A Mamdani-type FRBS demonstrates several interesting features. On the one hand, it provides a natural framework to include expert knowledge in the form of linguistic rules. This knowledge can be easily combined with rules which are automatically generated from data sets that describe the relation between system input and output. On the other hand, as seen in the previous section, a Mamdani-type FRBS possesses a high degree of freedom to select the most suitable fuzzification and defuzzification interface components as well as the inference method itself. This property allows the FS designer to adapt the inference engine of the FRBS to the specific characteristics of the problem at hand.

Moreover, Mamdani-type FRBSs provide a highly flexible means to formulate knowledge, while at the same they remain interpretable. The fuzzy rules are composed of input and output linguistic variables which take values from a labelled term set that associates a meaning to each linguistic label. Therefore, each rule is a description of a condition-action statement that exhibits a clear interpretation to a human —for this reason, these kinds of systems are usually called *linguistic* or *descriptive Mamdani FRBSs*—. This property makes Mamdani FRBSs an appropriate means for applications in which the emphasis lies on model interpretability, such as fuzzy control (Driankov, Hellendoorn, and Reinfrank, 1993; Lee, 1990) and linguistic modelling (Pedrycz, 1996; Sugeno and Yasukawa, 1993).

However, although Mamdani FRBSs possess several advantages, they also come with some drawbacks. One of the problems, especially in linguistic modelling applications, is their lack of accuracy in some complex problems, which is due to the structure of the linguistic rules. Bastian (1994) and Carse, Fogarty, and Munro (1996) analysed these limitations concluding that the structure of the fuzzy linguistic *"IF-THEN"* rule is subject to certain restrictions because of the use of linguistic variables:

- There is a lack of flexibility in the FRBS due to the rigid partitioning of the input and output spaces.
- When the input variables are mutually dependent, it becomes difficult to find a proper fuzzy partition of the input space.
- The homogeneous partition of the input and output space becomes inefficient and does not scale well as the dimensionality and complexity of the input-output mapping increases.
- The size of the KB increases rapidly with the number of variables and linguistic terms in the system. In order to obtain an accurate FRBS, a fine level of granularity is needed, which requires additional linguistic terms. This increase in granularity causes the number of rules to grow, which complicates the interpretability of the system by a human. Moreover, in the vast majority of cases, it is possible to obtain an equivalent FRBS that achieves the same accuracy with a fewer number of rules whose fuzzy sets are not

restricted to a fixed input space partition.

Due to these problems, two new variants of Mamdani FRBSs have been recently proposed. Both are described in the following section.

1.2.6 *Variants of Mamdani fuzzy rule-based systems*

Both variants of linguistic Mamdani FRBSs described in this section attempt to solve the said problems by making the linguistic rule structure more flexible.

1.2.6.1 *DNF Mamdani fuzzy rule-based systems*

The first extension to Mamdani FRBSs aims at a novel rule structure, the so-called *DNF (disjunctive normal form) fuzzy rule*, which has the following form (González, Pérez, and Verdegay, 1993; González and Pérez, 1998a; Magdalena and Monasterio, 1997; Magdalena, 1997):

$$IF\ X_1\ is\ \widetilde{A_1}\ and\ \ldots\ and\ X_n\ is\ \widetilde{A_n}\ THEN\ Y\ is\ B,$$

where each input variable X_i takes as a value a set of linguistic terms $\widetilde{A_i}$, whose members are joined by a disjunctive operator, whilst the output variable remains a usual linguistic variable with a single label associated. Thus, the complete syntax for the antecedent of the rule is

$$X_1\ is\ \widetilde{A_1} = \{A_{11}\ or\ \ldots\ or\ A_{1l_1}\}\ and\ \ldots$$
$$and\ X_n\ is\ \widetilde{A_n} = \{A_{n1}\ or\ \ldots\ or\ A_{nl_n}\}$$

An example of this kind of rule is shown as follows. Let us suppose we have three input variables, X_1, X_2 and X_3, and one output variable, Y, such that the linguistic term set D_i, F associated with each one is

$$D_1 = \{A_{11}, A_{12}, A_{13}\}\quad D_2 = \{A_{21}, A_{22}, A_{23}, A_{24}, A_{25}\}$$
$$D_3 = \{A_{31}, A_{32}\}\qquad\qquad F = \{B_1, B_2, B_3\}$$

In this case, a possible DNF rule may be

$$IF\ X_1\ is\ \{A_{11}\ or\ A_{13}\}\ and\ X_2\ is\ \{A_{23}\ or\ A_{25}\}$$
$$and\ X_3\ is\ \{A_{31}\ or\ A_{32}\}\ THEN\ Y\ is\ B_2$$

1.2.6.2   *Approximate Mamdani-type fuzzy rule-based systems*

Whilst the previous DNF fuzzy rule structure does not involve an important loss in the linguistic Mamdani FRBS interpretability, the point of departure for the second extension is to obtain an FS which achieves a better accuracy at the cost of reduced interpretability. These kinds of systems are called *approximate Mamdani-type FRBSs* (Alcalá et al., 1999; Bardossy and Duckstein, 1995; Carse et al., 1996; Cordón and Herrera, 1997c; Koczy, 1996), in comparison to the conventional *descriptive* or *linguistic Mamdani FRBSs*.

The structure of an approximate FRBS is similar to the descriptive one shown in Fig. 1.1. The difference is that each rule in an approximate FRBS defines its own fuzzy sets, whereas rules in a descriptive FRBS carry a linguistic label that points to a particular fuzzy set of a linguistic partition of the underlying linguistic variable. Thus, an approximate fuzzy rule has the following form:

$$IF \ X_1 \ is \ A_1 \ and \ \ldots \ and \ X_n \ is \ A_n \ THEN \ Y \ is \ B$$

The major difference with respect to the rule structure considered in linguistic Mamdani FRBSs is the fact that the input variables X_i and the output one Y are fuzzy variables instead of linguistic variables and, thus, A_i and B are independently defined fuzzy sets that elude an intuitive linguistic interpretation. In other words, rules of approximate nature are *semantic free* whereas descriptive rules operate in the context formulated by means of the linguistic semantics.

Therefore, approximate FRBSs contain no DB that defines a semantic context in the form of linguistic variables and terms. The former separate entities DB and RB merge into a fuzzy rule base (FRB) in which each rule subsumes the definition of its underlying input and output fuzzy sets as shown in Fig. 1.8.

Approximate FRBSs demonstrate some specific advantages over linguistic FRBSs making them particularly useful for certain types of applications (Carse, Fogarty, and Munro, 1996):

- The major advantage of the approximate approach is that each rule employs its own distinct fuzzy sets resulting in additional degrees of freedom and an increase in expressiveness.
- Another important advantage is that the number of rules can be adapted to the complexity of the problem. Simple input-output

a) Descriptive Knowledge Base

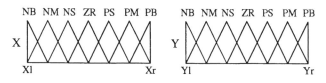

R1: If X is NB then Y is NB   R5: If X is PS then Y is PS
R2: If X is NM then Y is NM   R6: If X is PM then Y is PM
R3: If X is NS then Y is NS   R7: If X is PB then Y is PB
R4: If X is ZR then Y is ZR

b) Approximate Fuzzy Rule Base

R1: If X is △ then Y is △
R2: If X is △ then Y is △
R3: If X is △ then Y is △
R4: If X is △ then Y is △

Fig. 1.8 Comparison between a descriptive knowledge base and an approximate fuzzy rule base

relationships are modelled with a few rules, but still more rules can be added as the complexity of the problem increases. Therefore, approximate FRBSs constitute a potential remedy to the course of dimensionality that emerges when scaling to multi-dimensional systems.

These properties enable approximate FRBSs to achieve a better accuracy than linguistic FRBS in complex problem domains. However, despite their benefits, they also come with some drawbacks:

- Their main drawback compared to the descriptive FRBS is the degradation in terms of interpretability of the FRB as the fuzzy variables no longer share a unique linguistic interpretation. Still, unlike other kinds of approximate models such as neural networks that store knowledge implicitly, the knowledge in an approximate FRBS remains explicit as the system behaviour is described by local rules. Therefore, approximate FRBSs can be considered as a compromise between the apparent interpretability of descriptive FRBSs and the type of black-box behaviour, typical for non-descriptive, implicit models.

- The capability to approximate a set of training data accurately can lead to over-fitting and therefore poor generalisation to previously unseen input data.

According to their properties, fuzzy modelling (Bardossy and Duckstein, 1995) constitutes the major application of approximate FRBSs, as model accuracy is more relevant than description ability. Approximate FRBSs are usually not the first choice for linguistic modelling and fuzzy control problems. Hence, descriptive and approximate FRBSs are considered as complementary rather than competitive approaches. Depending on the problem domain and requirements on the obtained model, one should use one or the other approach. Approximate FRBSs are recommendable in case one wants to trade interpretability for improved accuracy.

1.3   Takagi–Sugeno–Kang Fuzzy Rule-Based Systems

Instead of working with linguistic rules of the kind introduced in the previous section, Takagi, Sugeno and Kang (1985, 1988) proposed a new model based on rules whose antecedent is composed of linguistic variables and the consequent is represented by a function of the input variables. The most common form of these kinds of rules is the one in which the consequent expression constitutes a linear combination of the variables involved in the antecedent:

$$IF\ X_1\ is\ A_1\ and\ \ldots\ and\ X_n\ is\ A_n$$
$$THEN\ Y = p_1 \cdot X_1 + \cdots + p_n \cdot X_n + p_0,$$

with X_i as the system input variables, Y as the output variable, and $\vec{p} = (p_0, p_1, \ldots, p_n)$ as a vector of real parameters. In regard to the A_i, they are either a direct specification of a fuzzy set (thus, X_i are fuzzy variables) or a linguistic label that points to one particular member of a fuzzy partition of a linguistic variable. These kinds of rules are usually called *TSK fuzzy rules*, in reference to their first proponents.

The output of a TSK FRBS using a KB composed of m rules is obtained as a weighted sum of the individual outputs provided by each rule, Y_i,

$i = 1, \ldots, m$, as follows:

$$\frac{\sum_{i=1}^{m} h_i \cdot Y_i}{\sum_{i=1}^{m} h_i}$$

in which $h_i = T(A_{i1}(x_1), \ldots, A_{in}(x_n))$ is the matching degree between the antecedent part of the i-th rule and the current inputs to the system, $x_0 = (x_1, \ldots, x_n)$. T stands for a conjunctive operator modelled by a t-norm. Therefore, to design the inference engine of TSK FRBSs, the designer only selects this conjunctive operator T, with the most common choices being the minimum and the algebraic product.

Takagi and Sugeno (1985) pointed out, that this type of FRBS divides the input space (uni- or multi-dimensional) in several fuzzy subspaces (uni- or multi-dimensional, in the same way) and defines a linear input-output relationship in each one of these subspaces. In the inference process, these partial relationships are combined in the said way for obtaining the global input-output relationship, taking into account the dominance of the partial relationships in their respective areas of application and the conflict emerging in the overlapping zones.

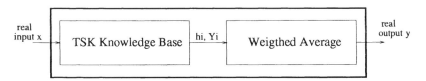

Fig. 1.9   Basic structure of a TSK Fuzzy Rule-Based System

A graphical representation of this second kind of FRBS is shown in Fig. 1.9. TSK FRBSs have been successfully applied to a large variety of practical problems. The main advantage of these systems is that they present a set of compact system equations that allows the parameters p_i to be estimated by means of classical methods, which facilitates the design process. However, the main drawback associated to TSK FRBSs is the form of the rule consequents, which does not provide a natural framework for representing expert knowledge that is afflicted with uncertainty. Still, it becomes possible to integrate expert knowledge in these FRBSs by slightly modifying the rule consequent: for each linguistic rule with consequent Y *is* B provided by an expert, its consequent is substituted by $Y = p_0$, with p_0 standing for the modal point of the fuzzy set associated to the label *B*. These kinds of rules are usually called *simplified TSK rules* or *zero-order TSK rules*. Nevertheless, it is not possible to utilise the full scope of FL principles in these kinds of systems (e.g., using the power of approximate reasoning by selecting different operators to perform the fuzzy inference process, as shown in Secs. 1.2.2 and 1.2.4).

However, TSK FRBSs are more difficult to interpret than Mamdani FRBSs due to two different reasons:

- The structure of the rule consequents is difficult to be understood by human experts.
- Their overall output simultaneously depends on the activation of the rule antecedents and on the function in the rule consequent that depends on the crisp inputs as well rather than being constant.

TSK FSs are used in fuzzy modelling (Pedrycz, 1996; Takagi and Sugeno, 1985) as well as control problems (Palm, Driankov, and Hellendoorn, 1997; Takagi and Sugeno, 1985).

1.4   Generation of the Fuzzy Rule Set

The accuracy of an FRBS directly depends on two aspects, the way in which it implements the fuzzy inference process and the composition of the fuzzy rule set (*for the remainder of this book, we use the term fuzzy rule set to refer to the collection of fuzzy rules, regardless of the type and structure of the fuzzy rules composing it*). Therefore, the FRBS design process includes two main tasks:

(1) *Conception of the inference engine*, that is, the choice of the different fuzzy operators that are employed by the inference process, task which was analysed in Sec. 1.2.4.
(2) *Generation of the fuzzy rule set* in order to formulate and describe the knowledge that is specific to the problem domain. This design task will be analysed in more detail in this section.

1.4.1   *Design tasks for obtaining the fuzzy rule set*

In the same way that in the design of the inference engine, the generation of the fuzzy rule set (KB or FRB) requires some design tasks, which vary depending on the type of FRBS considered. The main tasks are the following ones:

(1) Selection of the relevant input and output variables to the system from the set of all the possible ones.

(2) When dealing with an FRBS of descriptive nature, either a linguistic or a TSK FRBS using linguistic variables in the rule antecedents, definition of the DB structure containing the semantics of the terms that the input linguistic variables, in the latter case, or the input and output ones, in the former case, can take as a value. Thus, this includes some design subtasks:

- definition of the scale factors,
- choice of the possible term sets for each linguistic variable, which allows us to determine the desired granularity in the system,
- choice of the type of membership function to be used: triangular, trapezoidal, Gaussian or exponential-shaped, mainly (Driankov, Hellendoorn, and Reinfrank, 1993). The latter two have the advantage of achieving a smoother transition, whilst the former two are computationally simpler.
  Different studies have been made analysing the influence of the form of these functions in the accuracy of the FRBS (Baglio et al., 1993; Chang et al., 1991), although Delgado, Vila, and Voxman (1998) enunciated that trapezoidal-shaped functions might adequately approximate the remaining ones, presenting the advantage of their simplicity as well.
- definition of the membership function of the specific fuzzy set associated to each linguistic label.

As regards the approximate FRBSs, the only task that has to be developed is to select the type of membership functions that will be used in the fuzzy rules in the FRB.

(3) Derivation of the linguistic, approximate or TSK rules that will form part of the fuzzy rule set of the system. For this task, one has to determine the number of rules, as well as their composition, by defining the antecedent and consequent parts.

In the remainder of this section, we assume that the input and output variables have been determined in advance. The selection of the variables relevant to the problem is done by a human expert or by means of statistical methods, based on analysing the correlation existing between the variables available, or combinatorial methods, that analyse the influence of the subsets composed of the different combinations of variables (Bardossy and Duckstein, 1995).

1.4.2 *Kinds of information available to define the fuzzy rule set*

The problem-specific domain knowledge plays an important role in the fuzzy rule set derivation process and careful consideration can substantially improve the performance of the FRBS. In a typical application of FRBSs, i.e., modelling, control and classification, two types of information are available to the FRBS designer: numerical and linguistic. The former is usually obtained from observing the system, whilst a human expert provides the latter. Hence, there are two main ways for deriving the fuzzy rule set of a FRBS (Wang, 1994):

(1) *Derivation from experts.* In this first method, the composition of the KB is made by means of the expert information available. The human expert specifies the linguistic labels associated to each linguistic variable, the structure of the rules in the RB, and the meaning of each label. This method is the simplest one to be applied when the expert is able to express his knowledge in the form of linguistic rules. Of course, it can only be used directly when working with linguistic FRBSs although, as mentioned previously, it is possible to transform linguistic rules into simplified TSK rules to include them in an FS of the latter type.

(2) *Derivation from automatic learning methods based on the existing numerical information.* Due to the difficulties associated with the derivation of the KB from experts, researchers developed numerous inductive learning methods over the last few years for the different types of FRBSs. These design techniques are as diverse as: *ad hoc* data-driven generation methods (Bardossy and Duckstein, 1995; Casillas et al., 2000; Chi et al., 1996; Cordón and Herrera, 2000; Ishibuchi et al., 1992; Nozaki et al., 1997; Wang and Mendel, 1992b), variants of the least squares method (Bardossy and Duckstein, 1995; Takagi and Sugeno, 1985), descent methods (Nomura

et al., 1991; Nomura et al., 1992a), methods hybridising the latter two ones (Jang, 1993), neural networks (Nauck et al., 1997; Shann and Fu, 1995; Takagi and Hayashi, 1991; Takagi et al., 1992), clustering techniques (Delgado et al., 1997; Yoshinari et al., 1993) and evolutionary algorithms (Cordón and Herrera, 1995; Cordón, Herrera, and Lozano, 1997), among others. Design methods in which an evolutionary algorithm learns the fuzzy rule set are called *genetic fuzzy rule-based systems* (Cordón and Herrera, 1995), which are the central theme of this book.

When dealing with linguistic FRBSs, instead of defining the DB and RB in isolation, both parts of the KB can be designed simultaneously in case numerical and linguistic information about the problem is available. In fact, many authors (Bardossy and Duckstein, 1995; Mendel, 1995; Wang, 1994) consider this as a major advantage of FRBSs as these are the only systems able to combine both linguistic and numerical information in a seamless way. The initial FRBS is generated from expert knowledge specified in the form of linguistic rules. Subsequently, the numerical information, usually provided as input-output data pairs of the underlying system, is utilised by the automatic learning or tuning phase in order to refine the FRBS.

However, the management of the different kinds of information available and the design task for which this information may be used depend on the type of FRBS considered, as we shall analyse in the next subsections for each of the three existing FRBS types.

1.4.3 *Generation of linguistic rules*

Both information kinds can be used to design linguistic FRBSs. Although the KB derivation from linguistic information has been successfully used in many problems, the human expert is often not able to express his knowledge about the problem in the form of linguistic rules or, simply, there is no expert available that has sufficient knowledge about the problem domain. In other cases, the expert is able to provide some ideas with respect to the composition of the RB, i.e., to define some rules. The only information that he may afford about the DB is related to the universes of discourse where the problem variables are identified and their associated linguistic labels, but he is unable to define the exact membership functions defining the semantics of these labels.

In the area of FLCs, the latter problem is solved by defining a primary fuzzy partition for each variable by means of a normalisation process (Driankov et al., 1993; Harris et al., 1993; Lee, 1990). This process involves discretising the variable domain, partitioning it into a number of intervals equal to the number of linguistic labels considered and associating a label name and a fuzzy set defining its meaning to each interval. Since there is no knowledge available about the form that these fuzzy sets may have, one usually defines some uniform fuzzy partitions in which the fuzzy sets are symmetrical and of identical shape. The problem is that this approach often results in a sub-optimal performance of the final FRBS (Lee, 1990). In order to address this problem, different tuning and learning processes for the automatic generation of the DB *a posteriori* and *a priori*, respectively, have been proposed. In Chapters 4 and 9, several of these proposals based on evolutionary algorithms will be described.

In case the expert information is insufficient, the KB can be derived from numerical information. In these cases, an initial DB definition is obtained by means of a normalisation process, that generates uniform fuzzy partitions of the input and output spaces with the desired level of granularity. This preliminary definition may then be refined after the inductive RB generation process.

Hence, there are different possibilities to define the KB of a linguistic FRBS according to the information available (Bardossy and Duckstein, 1995):

(1) The expert is able to define the entire KB, i.e., he specifies the linguistic labels associated to each variable, the fuzzy sets defining the semantic of each label and the linguistic rules composing the RB.

(2) The expert is able to provide the entire or part of the said information, but there is additional numerical information available that is used to improve or complete the expert knowledge. There are different possibilities:

- Inductive methods can be used to tune the semantics of the labels, i.e, to define the DB starting from a preliminary estimate of the fuzzy sets —either provided by the expert or obtained by uniformly partitioning the input and output spaces— and then to apply an automatic tuning method that refines the original DB (Cordón and Herrera, 1997c; Herrera, Lozano, and Verdegay, 1995; Jang, 1993; Karr, 1991). We review the application of evolutionary algorithms to tune fuzzy rule sets in Chapter 4.

- In some cases the expert is able to identify the possible states of the system in a linguistic fashion. However, he is not able to indicate the correct output that is associated to these situations. The inductive methods are used in this case to complete the fuzzy rule set definition by generating the unknown consequents (Bardossy and Duckstein, 1995).

- Finally, the expert may specify a subset of rules that form an incomplete KB. In this case, the numerical information may be used to complement the missing rules of the KB (Bardossy and Duckstein, 1995).

(3) The expert is only able to identify the relevant variables and the linguistic terms associated to them, as well as their meaning, but is unable to stipulate the rules. In this case, numerical information is used to identify the rules in the RB.

(4) In case no expert knowledge is available at all, an automatic learning process based on the numerical information must be considered to generate the entire KB.

1.4.4   *Generation of approximate Mamdani-type fuzzy rules*

Approximate FRBSs operate differently than linguistic ones since the FRB is not directly defined based on expert knowledge. However, although an automatic learning method based on numerical information must be used to obtain the FRB definition, it is possible to incorporate expert knowledge in the design process.

There are three different ways to utilise expert knowledge:

(1) The knowledge is used in order to define the entire preliminary descriptive KB and afterwards the linguistic rules are globally adjusted. The KB is transformed into an approximate FRB taking into account the numerical information available. Section 4.3.3.2 introduces a genetic tuning process of this type (Herrera, Lozano, and Verdegay, 1995).

(2) The linguistic rules provided by the expert are directly included in the FRB being generated by the automatic method based on

numerical information. This method requires a simple transformation that substitutes the descriptive fuzzy sets by approximate ones. The automatic design process is capable to adjust the definition of the fuzzy sets of these rules as well (Cordón and Herrera, 1997b; Herrera, Lozano, and Verdegay, 1998a), therefore resulting in an approximate system that demonstrates the best performance. Chapter 8 presents a genetic fuzzy rule-based system with such a characteristic.

In order to understand the third approach one has to distinguish among the different kinds of approximate fuzzy rule learning methods (Alcalá et al., 1999; Cordón and Herrera, 1996b; Cordón and Herrera, 2001). Depending on whether the learning process imposes restrictions on the fuzzy sets in the generation of each approximate fuzzy rule or not, one distinguishes between methods that perform a *constrained learning* and those that perform an *unconstrained learning*:

- A learning process is *constrained* when the fuzzy sets are restricted to certain fixed intervals during the learning stage. Within this kind, two subgroups can be differentiated, *hard* and *soft constrained* learning processes:

  - In *hard constrained* learning processes, variation intervals determine the possible range of each membership function definition point during the learning process.
  - In *soft constrained* learning process, the only restriction imposed on the membership functions locations and shapes is that they lie within a fixed interval.

- In case fuzzy sets are subject to no restriction, such that their membership function can assume arbitrary values along the entire variable domain, the learning process is called *unconstrained*.

The different said restrictions to the learning process are graphically shown in Fig. 1.10. Intuitively, it becomes clear that they are related, and unconstrained learning can be considered as a special case of constrained learning in which the interval of performance associated to each fuzzy set corresponds to the entire domain. In the same way, soft constrained learning can be considered as a particular case of hard constrained learning,

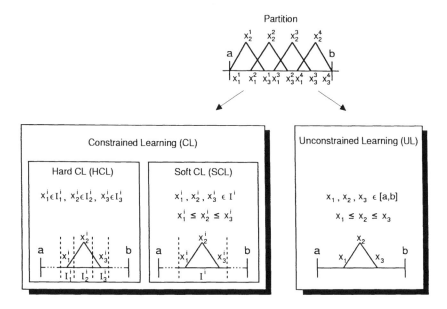

Fig. 1.10   Constrained and unconstrained approximate fuzzy rule learning

where the variation intervals of the membership function points are reduced to a common interval.

Notice that the kind of learning developed by the design process establishes a balance between the search space size and the accuracy that the approximate FRBS can achieve. Both aspects are directly related such that the dimension of the search space increases with the attainable accuracy. It was shown (Alcalá et al., 1999; Cordón and Herrera, 2001) that FRBSs generated from a constrained learning process perform well in problems of intermediate complexity, whilst FRBSs designed from an unconstrained one present a better accuracy when dealing with very complex problems.

(3) The third way to integrate expert knowledge in the FRB derivation is to work with a constrained learning process and to define the variation intervals from preliminary fuzzy partitions of the input and output variables obtained from the expert information (see Fig. 1.10) (Alcalá et al., 1999). Moreover, linguistic rules derived by considering the said fuzzy partitions can also be individually considered to specify the definition intervals as well as to initialise

the learning process. In Chapter 8, different genetic fuzzy rule-based systems performing in both said ways are presented.

1.4.5 *Generation of TSK fuzzy rules*

Finally, TSK FRBSs make use of the available information in a similar way to approximate Mamdani-type ones. Although the fuzzy rule set may be directly obtained from expert knowledge by transforming the linguistic rules provided by the expert in simplified TSK rules, this transformation often results in poor system performance. Moreover, choosing a TSK FS is not a good idea in case the only information available is the one afforded by an expert. As mentioned by Takagi and Sugeno (1985), TSK FRBSs are supposed to be designed based on numerical information in case there is no or only incomplete expert knowledge. Still, expert information remains useful in order to define the labels of the system input variables, to specify their meaning and to identify the possible system states in advance. However, the TSK rule consequent parameters are generated by means of an automated learning process alone.

1.4.6 *Basic properties of fuzzy rule sets*

The performance of an FRBS directly depends on two aspects: the way in which the fuzzy inference process is developed and the composition of the fuzzy rule set. Certain generic properties of fuzzy rule sets are beneficial in order to improve their approximation accuracy (Bardossy and Duckstein, 1995; Driankov et al., 1993; Harris et al., 1993; Lee, 1990; Pedrycz, 1989). In the following, we introduce the most relevant ones.

1.4.6.1 *Completeness of a fuzzy rule set*

A FRBS should obey the *completeness* property, namely that for each conceivable system input, it infers to a corresponding output.

For an arbitrary input $x_0 \in U$, at least one of the fuzzy rules has to trigger. Moreover, the global system output, i.e., the fuzzy set obtained by combining all the individual rule outputs, has to be non empty (Bardossy and Duckstein, 1995; Driankov, Hellendoorn, and Reinfrank, 1993). As may be seen, this definition refers to FRBSs whose defuzzification interface works in mode A-FATI since these kinds of systems are the ones in which the aggregation of some individual non empty fuzzy sets may result in an empty global output.

Following the previous analysis, the completeness property may be expressed as follows with $\sigma \in (0,1]$ bounded away from zero:

$$\forall x_0 \in \mathbf{U}, \quad Height(S(x_0)) \geq \sigma$$

with $S(x_0)$ being the global fuzzy set obtained as system output while $Height(\cdot)$ stands for the height of a fuzzy set. This property is called *σ-completeness* and it is more restrictive than the general definition introduced in the previous paragraph since it includes a degree of satisfaction in the completeness.

The σ-completeness property is very useful in practice due to the fact that it can be considered in the fuzzy rule set generation process. Fuzzy rules have to be added or modified in case no rule triggers to a degree larger than σ for a certain input state.

1.4.6.2 *Consistency of a fuzzy rule set*

A generic set of *"IF-THEN"* rules is *consistent* if it does not contain contradictory rules. This concept is clear in knowledge-based systems based on classical logic rules but it is more difficult to grasp in the case of fuzzy rule sets. In fact, there are many different interpretations of this property (Driankov, Hellendoorn, and Reinfrank, 1993).

A fuzzy rule set is inconsistent if it has rules with the same antecedent and different consequent, but this definition itself has some important inconsistencies. For example, Driankov, Hellendoorn, and Reinfrank (1993) questioned if two descriptive rules with the same antecedent and correlative linguistic labels in the consequents may be considered to be inconsistent. Thus, the authors proposed an alternative definition where two rules are only considered to be inconsistent when having the same antecedent and mutually exclusive consequents.

The study of this property is an open research issue and it seems that the best way to deal with it in fuzzy rule sets involves relaxing its definition by introducing degrees of its fulfilment. Harris, Moore, and Brown (1993) and Pedrycz (1989) took as a base the idea of considering consistent fuzzy rules with similar variations between the fuzzy sets defining their antecedents and the ones defining their consequents, and proposed an *index of inconsistency* to measure this aspect.

Finally, González and Pérez (1998a) proposed a different relaxation for the consistency property based on the concepts of *negative* and *positive examples*, which is also used by Herrera, Lozano, and Verdegay (1998a). An example is considered positive for a fuzzy rule when it matches with its antecedent and consequent, and it will be considered a negative example when it matches with its antecedent and not with its consequent. A fuzzy rule is considered inconsistent in principle when it has negative examples associated to it.

However, a fuzzy rule that covers a very small number of negative examples and a very large number of positive examples might not be considered inconsistent *per se*. Therefore, González and Pérez (1998a) introduced the *k-consistency* property which is less strict as the previous definition as it only considers the cardinality of the positive and negative example sets of a rule. Given a parameter $k \in [0, 1]$, a fuzzy rule is k-consistent when its associated number of negative examples is less than or equal to a percentage $100 \cdot k$ of the number of positive examples.

This definition is further analysed in Chapter 8.

1.4.6.3 *Low complexity of a fuzzy rule set*

This property is concerned with the number of fuzzy rules composing the base (Lee, 1990). Intuitively, one would prefer a fuzzy rule set with a fewer number of rules, as it improves comprehensibility and decreases the computational load of the inference process. The relevance of this property is related to the kind of application for which the FRBS is designed. *Low complexity* becomes important in control problems, as the main requirements are usually the process speed and the simplicity of the base, rather than the response accuracy. It is equally beneficial in linguistic modelling applications, as the main goal is to obtain a human-readable description of the real system. A linguistic model with only a few number of rules is more easily interpretable by a human. FRBSs composed of a small number of fuzzy rules are called *compact FRBSs*.

1.4.6.4 *Redundancy of a fuzzy rule set*

The *redundancy* refers to the property of FRBSs that a system state may be covered by more than one rule as the fuzzy sets in the rule antecedents overlap (Bardossy and Duckstein, 1995). However, this does not mean that rules whose premises are completely covered by other rules are redundant as they still affect the system output.

The existence of redundant rules may cause a degradation in the performance of the global FRBS. Therefore, it is important to take a closer look at redundancy in order to remove unnecessary, detrimental rules from the fuzzy rule set. The idea is to evaluate the utility of a rule by analysing its impact on the global system behaviour.

Bardossy and Duckstein (1995) presented a measure of overlap allowing redundant rules to be identified. A criterion based on this measure of overlap is then used to decide whether a rule should be discarded from the rule set. Different authors (Cordón and Herrera, 1997c; Cordón, del Jesús, and Herrera, 1998; Herrera, Lozano, and Verdegay, 1998a; Ishibuchi, Nozaki, Yamamoto, and Tanaka, 1995) proposed genetic processes for removing redundant rules from fuzzy rule sets in classification and modelling problems. Several of them based on genetic algorithms will be presented in Chapters 8 and 9. Another approach not based on this technique is also briefly described in Chapter 8.

1.5 Applying Fuzzy Rule-Based Systems

The major applications of FRBSs are fuzzy modelling, fuzzy control and fuzzy classification. This section takes a closer look at these three application areas and analyses the benefits of using FRBSs in these domains. Moreover, some representative applications in each group are mentioned.

1.5.1 *Fuzzy modelling*

1.5.1.1 *Benefits of using fuzzy rule-based systems for modelling*

The question *"why do we use FRBSs for modelling systems?"* is at large equivalent to ask: *"why do we use FRBSs for approximating continuous functions instead of more classical techniques, as regression ones, for performing this task?"* (Bardossy and Duckstein, 1995). Usually, the relationship between a group of variables is defined by a parameterised function whose parameters are estimated by means of a regression method, for example least squares, in a way that the error between the model and the data is minimised. Traditional methods usually employ a static function parameterisation as the parameter solutions had to be computed manually. The computational demand made it impossible to use functions which change their structural form during the identification process. This standard way of performing regression might not always lead to good approximations and, moreover, the parameters obtained often lack a simple interpretation. The fact that a function can be expressed in analytical form does not necessarily mean that a human is able to visualise it or to intuitively understand its behaviour. For example, even for a two-dimensional input space, it becomes difficult to imagine a third-order polynomial.

The increase in computer power over the last few years made it possible to use more flexible and adaptable representations for modelling complex systems. Fuzzy rules, even the approximate ones, offer the advantage that they provide a clear interpretation to the user, as their scope is local rather than global. An FRBS forms a collection of fuzzy rules, each one representing a local model that is easily interpretable and analysable. This local type of representation allows an improved approximation accuracy to be obtained, but at the same time increases the number of parameters to be estimated in the overall model.

FRBSs are able to represent continuous functions in a more robust manner, as modifications to a single rule only effect the approximation in a local, isolated region. This is not the case for global approximation schemes for which changing a single parameter globally alters the function.

Under certain conditions, FRBSs are *universal approximators*, in the sense that they are able to approximate any function to the desired degree (Buckley, 1993; Castro, 1995; Castro and Delgado, 1996; Kosko, 1992; Wang, 1992). Although there are classical techniques, based on polynomials and splines, that are universal approximators as well, they are not frequently used nowadays.

For modelling, FRBSs offer a significant advantage over neural networks (Wang, 1994). A suitable initial guess of the model parameters can substantially improve the convergence rate of the training algorithm since the parameters involved in FRBSs have a clear real-world sense. Thus, it is possible to find an adequate set of initial parameters for the FRBS based as the fuzzy framework allows an expert to formulate his knowledge in the form of local rules. This prior knowledge provides a guideline to obtain a preliminary definition of the membership functions and fuzzy rule set, which is subsequently improved according to the numerical data available.

In the case of neural networks, the connection between the network parameters and the input-output mapping is global and highly non-linear. Therefore, it becomes very difficult to utilise prior domain knowledge for the initialisation and one usually resorts to initialise the neural network weights at random. This benefit of FRBSs becomes of particular relevance in case only a limited amount of data but a large amount of expert knowledge about the system is available.

The very same advantage not only applies to the model design phase but also when the user wants to analyse and interpret the model obtained from the data. In case of descriptive FRBSs, the generated model is a set of linguistic rules that has a large degree of interpretability (a linguistic model). Approximate or TSK FRBSs provide better model accuracy, but even though the rules are still local, interpretability becomes more difficult as the rule parameters are no longer linguistic but numerical. Nevertheless, even approximate FRBSs are still more interpretable than neural networks that use an entirely implicit representation of knowledge. The number of rules in an FS that trigger simultaneously is limited, whereas the neural activation in response to an specific input is distributed over the entire neural network. In order to overcome the problem of global activation, so-called *radial-basis function networks* have been proposed which can be shown to be equivalent to TSK FRBSs (Jang and Sun, 1993).

1.5.1.2 *Relationship between fuzzy modelling and system identification*

There exists a direct relationship between the areas of fuzzy modelling and system identification. System identification falls in the realm of system theory and has the objective to estimate a dynamical model based on input-output data directly generated from the underlying process. Concepts in system identification are directly transferable to the domain of fuzzy modelling. The identification process involves two separate steps (Pedrycz, 1996; Sugeno and Yasukawa, 1993): the *structure identification* and the *parameter identification*. In order to identify a system, one first determines its structure and then estimates the parameters of the model in a way that it matches the numerical data. Each of the two identification process steps has to deal with particular issues:

(1) *Structure identification* consists of determining the input variables which serve as candidates to the model input. From the input candidates, one selects a subset of variables that have the largest influence on the system output. Finally, the model structure is established by a parameterised function that maps input to output variables.

(2) *Parameter identification* involves estimating the parameters of the functional model obtained in the structure identification step.

Often the input and output variables to the model are known *a priori*, such that the identification and selection of relevant inputs becomes obsolete. In some cases, the model structure is known in advance as well and the system identification task reduces to mere parameter estimation.

There exists an analogy between the said identification tasks and the derivation of a fuzzy rule set (Sugeno and Yasukawa, 1993; Pedrycz, 1996):

(1) The task of identifying the relevant input variables is identical for structure identification and FRBS design. The second step, namely establishing the structural relationship among input and output, is associated to derive the fuzzy input and output space partitions and fuzzy rules.

(2) In FRBS design, the parameter identification step involves determining the parameters of the fuzzy membership functions, and in addition for TSK FRBSs, the real-valued parameters in the rule consequents.

In summary, the structure identification step corresponds to the derivation of the RB, whereas the parameter estimation step is equivalent to defining the DB in case of linguistic or TSK FRBSs (Sugeno and Yasukawa, 1993). The distinction between the two steps is less clear in case of approximate FRBS.

Finally, notice that fuzzy modelling offers the additional advantage that the structure and parameter identification steps are separable so that it is possible for example to first define the RB (considering a preliminary definition for the DB) and to adjust its semantics afterwards. This property facilitates the identification process in fuzzy or linguistic model design. The interested reader is referred to (Cordón and Herrera, 1997b; Sugeno and Yasukawa, 1993) for fuzzy model identification methods following this approach.

1.5.1.3 *Some applications of fuzzy modelling*

Representative applications in the field of system fuzzy or linguistic modelling are:

- *Applications to the field of economics*: Applications in this domain have been developed for predicting the exchange rate in order to support decision making in currency trading (Tano, 1995; Yuize et al., 1991), for stock market trading in combination with neural networks (Ye and Gu, 1994), for forecasting financial time series (Benachenhou, 1994), and for detecting anomalous behaviour in health-care provider claims (Cox, 1995).

- *Applications to weather forecasting*: Modelling of the daily mean temperature (Bardossy and Duckstein, 1995).
- *Applications to water demand forecasting and soil water diffusion* (Bardossy and Duckstein, 1995).
- *Application to the generation of rules for sustainable reservoir operation* (Bardossy and Duckstein, 1995).
- *Applications to medicine*: Bardossy and Duckstein (1995) presented a medical diagnosis application in which the fuzzy model acts as a combined classifier that integrates multiple diagnosis provided by different medical doctors. Lee and Takagi (1996) introduced an application to the diagnosis of a teeth disease that will be analysed in Sec. 11.2.3.
- *Application to rice taste evaluation*: This problem introduced by Ishibuchi et al. (1994) will be described in detail in Sec. 11.2.2.
- *Applications to electrical engineering*: Various linguistic and fuzzy models have been designed by Cordón, Herrera, and Sánchez (1997, 1998, 1999) to solve two different electrical distribution problems, the estimation of the amount of low voltage line installed in villages and the computing of the maintenance costs of town medium voltage lines. They both will be analysed in depth in Secs. 11.2.1.1 and 11.2.1.2, respectively.

1.5.2 *Fuzzy control*

1.5.2.1 *Advantages of fuzzy logic controllers*

What are the benefits of fuzzy control systems compared to conventional approaches from control system theory? The current interest on FL emanates to a large extent from the successful application of FLCs in consumer products and industrial systems. This recent success can be understood from a theoretical and a practical perspective (Wang, 1994). The theoretical justifications for using fuzzy control are:

- As a generic rule, a good engineering technique must be able to make efficient use of all the information available. There are two sources for information in control: sensors, providing numerical measures for the system variables, and human experts, providing linguistic descriptions about the system and control instructions. FLCs, as FRBSs, constitute one of the few tools able to include both kinds of information, so they may be applied when there is a lack of one of them.

- In general, fuzzy control is a model-free approach, i.e., it does not depend on a model of the system being controlled, the same that several classical control schemes such as PID control. Model-free approaches make the controller design easier, since obtaining a mathematical model of the system is sometimes a very complex task. Therefore, the need and importance of model-free approaches continues to increase in the realm of control engineering.
- FLCs are universal approximators and are therefore suitable for non-linear control system design.

Whilst the previous reasons involve generality and rigour, aspects always asked from a theoretical point of view, the practical significance is usually focused on analysing the possibilities of potential applications for a technique. The practical reasons for the increasing visibility and utilisation of FLCs are the following:

- Fuzzy control is very easy to understand. Since FLCs emulate the human control strategy, they are easy to understand for people that are non specialists in control. This has caused its application to increase in comparison to classical control techniques based on a rigorous mathematical framework.
- FLC hardware implementation is easy and quick, and it allows a large degree of parallelisation to be used as well.
- Developing FLCs is cheap. From a practical point of view, development costs are a key aspect for obtaining a business success. Fuzzy control is easy to understand and may be learnt in a very short time, so "software costs" are low. Since it is easy to implement, "hardware costs" are low as well. All these reasons make fuzzy control a technique with an attractive balance between performance and cost.

1.5.2.2 *Differences between the design of fuzzy logic controllers and fuzzy models*

Fuzzy control operates in real time and involves a closed loop process which observes the current values of the state variables and, based on this observation, produces a control action that either keeps or drives the system to the desired output. Therefore, the design of FLCs faces some fundamentally different challenges than the design of linguistic or fuzzy models:

(1) *Types of FRBSs considered*: In control, linguistic Mamdani FRBSs are most common followed by TSK FSs. In fuzzy modelling, approximate Mamdani FRBSs are the usual choice.

(2) *Definition of the membership functions of the fuzzy sets involved in the fuzzy rule set*: In fuzzy control, the initial membership functions are usually derived by interviewing an expert, or by using an equidistant uniform partition. The original membership functions are refined upon observing the closed loop behaviour of the entire process. This adaptation is either done automatically or a "trial and error" approach —which is never used in fuzzy modelling— is employed.

(3) *Generation of the fuzzy rule set*: The linguistic rules composing the RB of an FLC may be obtained in different ways (Bardossy and Duckstein, 1995; Lee, 1990):

- From expert information provided by the human system operator.
- From the control actions applied by the operator while controlling the process.
- From a mathematical or fuzzy model of the system being controlled.
- From an automated learning process.

In fuzzy control, is it usually easier to generate training data of input-output pairs, than in the case of fuzzy modelling for which the number of training examples is often limited. This additional training data can be used to refine the FLC during operation based on the performance it demonstrates on the closed loop system. In fuzzy control, implementation cost and processing speed are in most cases more important than accuracy, so that FRBSs for control usually have a fewer number of rules than those employed in fuzzy modelling.

(4) *Design of the inference engine*: The choice of a particular inference engine is more important in fuzzy modelling than in fuzzy control as its major influence is approximation accuracy. As FLCs usually employ a small KB anyhow, the effect of using a specific inference mechanism is mainly irrelevant to the performance of the controller. As processing speed and implementation costs dominate the design choice in fuzzy control, a simple inference scheme is preferred toward a mathematical more profound but also complex one.

1.5.2.3   *Some applications of fuzzy control*

The first industrial application of FLCs was the Danish cement kiln at F. L. Smith in 1979 (Umbers and King, 1980). As there is vast number of fuzzy control applications in industrial automation and consumer products, we restrict ourselves to a few representative examples: the fuzzy control of a water treatment plant, of a refuse incineration plant, of the fermentation of the Japanese sake, of elevators, of a highway tunnel ventilation, of a fully automatic washing machine, of the flight of a space shuttle, of nuclear reactors, of a system for automated train driving, of the non linear communication channel equalisation in high speed data transmissions and applications in behaviour-based robotics (see Sec. 11.4.1) (Bardossy and Duckstein, 1995; Berenji, 1992; Bonissone, 1994; Bonarini and Basso, 1997; Hirota, 1993; Hoffmann and Pfister, 1997; Lee, 1990; Saffiotti et al., 1995).

1.5.3   *Fuzzy classification*

1.5.3.1   *Advantages of using fuzzy rule-based systems for classification*

The objective of pattern classification is to assign a class C_j from a set of alternative classifications $C = \{C_1, \ldots, C_k\}$ to an object based on its feature vector $\{x_1, \ldots, x_n\} \in \mathbb{R}^n$.

The problem of designing such a classifier is therefore to find a map

$$D : \mathbb{R}^n \longrightarrow C$$

that is optimal in the sense that it maximises some desired performance measure $\delta(D)$. The learning algorithm starts with a set of correctly classified training examples with the objective to find a classifier that assigns class labels in a way that minimises the classification error across the entire feature space. The classifier performance is evaluated on a test set of previously unseen examples to obtain an estimate of the true classification error. The classifier may be implemented from examples as an FS, a neural network, a decision tree or a Bayesian network. A fuzzy rule-based classification system (FRBCS) is a classifier that uses fuzzy rules to assign class labels to objects (Bardossy and Duckstein, 1995; Bezdek and Pal, 1992; Chi, Yan, and Pham, 1996; Kuncheva, 2000).

Humans possess the remarkable ability to recognise objects despite the presence of uncertain and incomplete information. FRBCSs provide a suitable means to deal with this kind of noisy, imprecise or incomplete information which in many classification problems is rather the norm than the exception. They make a strong effort to reconcile the empirical precision of traditional engineering techniques with the interpretability of artificial intelligence.

Classification systems may be divided into two main groups depending on the way in which they used: classification systems which are supposed to work autonomously, and those that are intended to be tools to support decision making of a human user. In the case of autonomous classifiers, the basic objective of the design process is performance, i.e., the percentage of correct classifications. Other criteria such as comprehensibility, robustness, versatility, modifiability and coherence with previous knowledge are secondary, as they are merely relevant to increase user acceptance.

In classification problems, the fundamental role of fuzzy rules is to make the classification process transparent to the user within a formal and computer-realisable framework (Zadeh, 1977), which is composed of a linguistic FRBCS.

1.5.3.2 *Components and design of fuzzy rule-based classification systems*

As any other FRBS, an FRBCS is formed by two main components:

(1) a KB, containing the fuzzy classification rules for a specific classification problem, and
(2) a fuzzy reasoning method (FRM), that classifies a new pattern, i.e., determines the class associated to it, using the information provided by the KB.

To implement the FRBCS, we start from a set of pre-classified examples and we must choose:

- the method to find or learn a set of fuzzy rules for the specific classification problem, and
- the FRM used to classify a new pattern.

The basic structure of a FRBCSs and the design process are shown in Fig. 1.11.

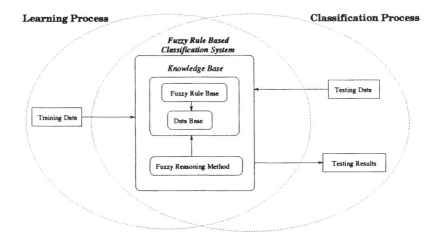

Fig. 1.11   Basic structure of a fuzzy rule-based classification system

During the last couple of years, researchers proposed several methods for generating fuzzy classification rules from numerical data pairs based on different techniques (Chi, Yan, and Pham, 1996; Cordón, del Jesús, and Herrera, 1998; González and Pérez, 1998a; Ishibuchi, Nozaki, and Tanaka, 1992; Ishibuchi, Nozaki, Yamamoto, and Tanaka, 1994). In addition, different proposals for selecting the FRM are found in the literature (Bardossy and Duckstein, 1995; Cordón, del Jesús, and Herrera, 1999b; Ishibuchi, Morisawa, and Nakashima, 1996; Mandal, Murthy, and Pal, 1992). In the following, we analyse the composition of the KB in an FRBCS, introducing the different types of rules that may be considered, and the most common types of FRMs.

The Knowledge Base

The KB is composed of an DB and an RB. The former component has the usual form previously described in this chapter but only contains information about the input variables. As usual, the granularity and form of the input space partition has a major influence on the system classification capacity (Ishibuchi, Nozaki, and Tanaka, 1993).

As regards the RB, the following three types of fuzzy classification rules may be considered:

a) *Fuzzy rules with a class in the consequent* (Abe and Thawonmas, 1997; González and Pérez, 1998a). These kinds of rules have the following structure:

$$IF\ X_1\ is\ A_1\ and\ \ldots\ and\ X_n\ is\ A_n\ THEN\ Y\ is\ C_i,$$

in which $X_1, \ldots, X_N$ are the features, $A_1, \ldots, A_n$ are the linguistic labels that partition the universe of discourse, and C_i $(i = 1, \ldots, k)$ is the class label to which the object is assigned.

b) *Fuzzy rules with a class and a certainty degree in the consequent* (Ishibuchi, Nozaki, and Tanaka, 1992):

$$IF\ X_1\ is\ A_1\ and\ \ldots\ and\ X_n\ is\ A_n\ THEN\ Y\ is\ C_i\ with\ r,$$

in which r is the degree of certainty that an object that matches the rule antecedent belongs to class C_i. This degree of certainty can be determined by the ratio $\frac{S_i}{S}$ of the number of objects S_i in the subspace defined by the rule antecedent that belong to class C_i to the overall number S of objects in that region.

c) *Fuzzy rules with a certainty degree for all classes in the consequent* (Mandal, Murthy, and Pal, 1992):

$$IF\ X_1\ is\ A_1\ and\ \ldots\ and\ X_n\ is\ A_n\ THEN\ (r_1, \ldots, r_k),$$

in which r_i $(i = 1, \ldots, k)$ is the degree of soundness for the the prediction of class C_i for an object in the region described by the antecedent. The degrees of certainty can be determined using the same ratio considered for the type b) rules.

Notice that the type c) classification rule subsumes types a) and b) by choosing the following values for $(r_1, \ldots, r_k)$. The case

$$r_h = 1, \quad r_i = 0, \quad i \neq h, \quad i = 1, \ldots, k$$

describes a type a) rule, whereas

$$r_h = r, \quad r_i = 0, \quad i \neq h, \quad i = 1, \dots, k$$

is identical to a type b) rule.

The Fuzzy Reasoning Method

The FRM infers the class label of an object from the given feature vector and the set of fuzzy *"IF-THEN"* classification rules. One of the advantages of fuzzy reasoning is that we obtain a classification even if there only exists an approximate match between the feature vector and the rule antecedents.

The most common fuzzy inference method for fuzzy classification problems is the *maximum matching* (Ishibuchi et al., 1992; Mandal et al., 1992), which selects the class label of the rule whose antecedent matches the feature vector best. Its operation mode for an FRBCS using type c) rules is as follows.

Assume, that the FRBCS RB contains the rules $R = \{R^1, \dots, R^m\}$. For a pattern $E^t = (e_1^t, \dots, e_n^t)$, the classical FRM selects the class label of rule which best matches the feature vector and has the largest degree of rule certainty r_i. It uses the following algorithm to determine the most reliable rule:

- Let $R^i(E^t)$ denote *the degree of activation* for rule R^i. Usually, $R^i(E^t)$ is obtained by applying any conjunctive operator, such as the minimum t-norm, to the matching degrees of individual clauses ("X_j is A_j^i"):

$$R^i(E^t) = T(\mu_{A_1^i}(e_1^t), \dots, \mu_{A_n^i}(e_n^t))$$

- Let $d(R^i(E^t), r_j^i)$ denote *the degree of association of the pattern E^t with class C_j according to the rule R^i*. This degree is obtained by applying a combination operator —such as the minimum, the product or the arithmetic mean— to $R^i(E^t)$ and r_j^k.
- The *degree of association of the pattern E^t with the class C_j, $Y_j(E^t)$*, is calculated for each class ($j = 1, \dots, k$):

$$Y_j(E^t) = \max_i d(R^i(E^t), r_j^i), \quad i = 1, \dots, m$$

This degree of association specifies to what extent the feature vector E^t belongs to class C_j.

- Finally, the feature vector E^t is classified according to the class C_h that has the maximal degree of association.

$$Y_h = \max_j Y_j, \; j = 1, \ldots, k$$

One of the problems of maximum matching is that a winner rule takes all decision and the FRM does not consider the classifications by other active rules. This strategy sacrifices the advantages that stem from using an approximate rather than crisp reasoning technique. Therefore, employing a reasoning method that aggregates the information of multiple active rules in a more sophisticated way might improve the generalisation capability of the FRBCS. Works such as those by Bardossy and Duckstein (1995); Chi, Yan, and Pham (1996); Ishibuchi, Morisawa, and Nakashima (1996) show that FRMs which consider multiple rules results in superior performance of FRBCSs. The interested reader can refer to (Cordón, del Jesús, and Herrera, 1999b), where a general model of reasoning that involves all these possibilities is presented.

1.5.3.3 *Some applications of fuzzy classification*

FRBCSs have been applied to the following classification problems:

- *Applications to the segmentation of geographic and satellite map images*: Chi, Yan, and Pham (1996) worked on the task of segmenting grey scale geographic maps into foreground (characters, roads, streets, boundaries, etc.) and background regions. Binaghi, Brivio, and Rampini (1996) presented different applications to the classification of synthetic aperture radar (SAR) images in order to determine the kind of surfaces present in the images (e.g., terrain classification).
- *Character recognition*: Chi, Yan, and Pham (1996) proposed two applications in the character recognition domain. An FRBCS is applied to the recognition of printed upper-case English characters. A neural network-FRBCS hybrid classifier is used for the problem of handwritten numeral recognition.
- *Applications to medical diagnosis*: FRBCSs solved different problems belonging to this domain. The diagnosis of myocardial infarction, which will be analysed in Sec. 11.1.2, is just one example (González, Pérez, and Valenzuela, 1995).

- *Applications to weather classification*: The local or regional daily precipitation, temperature and wind depend on the type of atmospheric circulation pattern occurring over the region. Thus, classification of daily atmospheric circulation patterns plays a central role to determine these meteorological variables. Bardossy and Duckstein (1995) presented a specific application to precipitation modelling.

Chapter 2

Evolutionary Computation

The goal of this chapter is to introduce the key algorithms and theory that constitute the core of evolutionary computation. The principles and foundations of a basic genetic algorithm are described in detail. In addition, a separate section discusses extensions to the simple genetic algorithm, such as refined crossover, mutation and selection schemes. The chapter also presents real-coded and messy genetic algorithms, both of which encode genetic information in a different way. The chapter concludes with an introduction to other types of evolutionary algorithms such as evolution strategies, evolutionary programming and genetic programming.

2.1 Conceptual Foundations of Evolutionary Computation

Evolutionary algorithms (EAs) constitute a class of search and optimisation methods, which imitate the principles of natural evolution (Goldberg, 1989; Holland, 1975). Fogel (1998) compiled a collection of selected readings on its historical development. The common term *evolutionary computation* comprises techniques such as *genetic algorithms*, *evolution strategies*, *evolutionary programming* and *genetic programming*. Their principal mode of operation is based on the same generic concepts, a population of competing candidate solutions, random combination and alteration of potentially useful structures to generate new solutions and a selection mechanism to increase the proportion of better solutions. The different approaches are distinguished by the genetic structures that undergo adaptation and the genetic operators that generate new candidate solutions.

Since EAs emulate natural evolution, they adopt a biological terminology to describe their structural elements and algorithmic operations. The utilisation of biological terms for their EA counterparts is an extreme simplification inasmuch as the biological objects are much more complex. Keeping the limitations of this analogy in mind, it is helpful to become acquainted with the biological terminology used throughout the following chapters.

Each cell of a living organism embodies a strand of DNA distributed over a set of chromosomes. The cellular machinery transcribes this blueprint into proteins that are used to build the organism. A gene is a functional entity that encodes a specific feature of the individual such as hair colour. The term allele describes the value of a gene, which determines the manifestation for an attribute, e.g., blonde, black, brunette hair colour. Genotype refers to the specific combination of genes carried by a particular individual, whereas the term phenotype is related to the physical makeup of an organism. Each gene is located at a certain position within the chromosome, which is called locus.

The literature in EAs alternatively avails the terms chromosome and genotype to describe a set of genetic parameters that encode a candidate solution to the optimisation problem. In the remainder of this book, we will employ the term chromosome to describe the genetic structures undergoing adaptation. Consequently, the term gene refers to a particular functional entity of the solution, e.g., a specific parameter in a multi-dimensional optimisation problem. In genetic algorithms, solutions are often encoded as binary strings. In this context, a gene corresponds to a single bit, an allele corresponds to either a 0 or 1 and loci are string positions.

Progress in natural evolution is based on three fundamental processes: mutation, recombination and selection of genetic variants. The role of mutation is the random variation of the existing genetic material in order to generate new phenotypical traits. Recombination hybridises two different chromosomes in order to integrate the advantageous features of both parents into their offspring. Selection increases the proportion of better adapted individuals in the population. Darwin coined the term *survival of the fittest* to illustrate the selection principle that explains the adaptation of species to their environment. The term fitness describes the quality of an organism, which is synonymous with the ability of the phenotype to reproduce offspring in order to promote its genes to future generations.

EAs provide a universal optimisation technique applicable to a wide range of problem domains, such as parameter optimisation, search, combinatorial problems and automatic generation of computer programs. Unlike specialised methods designed for particular types of optimisation tasks, they require no particular knowledge about the problem structure other than the objective function itself. EAs are distinguished by their robustness, the ability to exploit accumulated information about an initial unknown search space in order to bias subsequent search into useful subspaces. They provide an efficient and effective approach to manage large, complex and poorly understood search spaces, where enumerative or heuristic search methods are inappropriate.

An EA processes a population of genetic variants from one generation to the next. A particular chromosome encodes a candidate solution of the optimisation problem. The fitness of an individual with respect to the optimisation task is described by a scalar objective function. According to Darwin's principle, highly fit individuals are more likely to be selected to reproduce offspring to the next generation. Genetic operators such as recombination and mutation are applied to the parents in order to generate new candidate solutions. As a result of this evolutionary cycle of selection, recombination and mutation, more and more suitable solutions to the optimisation problem emerge within the population.

The major components and the principal structure of a generic EA are shown in Fig. 2.1. During generation t, the EA maintains a population $P(t)$ of chromosomes. The population at time $t = 0$ is initialised at random. Each individual is evaluated by means of the said scalar objective function giving some measure of fitness. A set of parents is selected from the current population in a way that more fit individuals obtain a higher chance for reproduction. Recombination merges two chromosomes to form an offspring that incorporates features of its parents. In this way, recombination fosters the exchange of genetic information among individual members of the population. The offspring are subject to mutations, which randomly modify a gene in order to create new variants. The current population is replaced by the newly generated offspring, which forms the next generation. The evolutionary cycle of evaluation, selection, recombination, mutation and replacement continues until a termination criterion is fulfilled. The stopping condition can be either defined by the maximum number of generations or in terms of a desired fitness to be achieved by the best individual in the population.

In the following sections, we introduce different variants of EAs: *genetic algorithms, evolution strategies, evolutionary programming* and *genetic pro-*

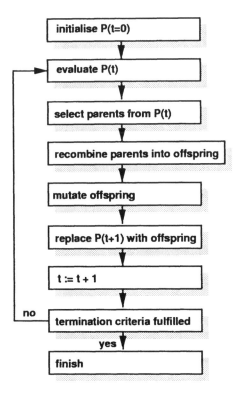

Fig. 2.1   Principal structure of an Evolutionary Algorithm

gramming. Although based on the same generic principles outlined before, each employs its particular chromosomal representation, set of genetic operators and selection and replacement scheme.

2.2   Genetic Algorithms

Genetic Algorithms (GAs) are general-purpose search algorithms that use principles inspired by natural population genetics to evolve solutions to problems. They were first proposed by Holland (1975) and are well described in text books (Bäck, 1996; Goldberg, 1989; Michalewicz, 1996; Mitchell, 1996). GAs are theoretically and empirically proven to provide a robust search in complex spaces, thereby offering a valid approach to problems requiring efficient and effective searches.

A GA operates on a population of randomly generated solutions, chromosomes often represented by binary strings. The population advances toward better solutions by applying genetic operators, such as crossover and mutation. In each generation, favourable solutions generate offspring that replace the inferior individuals. Crossover hybridises the genes of two parent chromosomes in order to exploit the search space and constitutes the main genetic operator in GAs. The purpose of mutation is to maintain the diversity of the gene pool. An evaluation or fitness function plays the role of the environment to distinguish between good and bad solutions.

2.2.1 *Main characteristics*

Solving a particular optimisation task using a GA, requires the human designer to address the five following issues.

(1) *A genetic representation of candidate solutions,*

(2) *a way to create an initial population of solutions,*

(3) *an evaluation function which describes the quality of each individual,*

(4) *genetic operators that generate new variants during reproduction, and*

(5) *values for the parameters of the GA, such as population size, number of generations and probabilities of applying genetic operators.*

In the following, the brachystochrone problem shown in Fig. 2.2 serves as an example to illustrate the operation of a simple, conventional GA. The mathematician Jakob Bernoulli originally posed the brachystochrone problem in 1696. The objective is to find an optimal planar curve between a higher starting point (x_0, y_0) and a lower destination point (x_k, y_k) in such a way that the time a movable object needs to travel along the track becomes minimal. The problem assumes a point mass which moves frictionless by means of its own gravity and has zero velocity in the beginning.

A candidate curve is approximated by a polygon defined by a set of $k + 1$ successive track-points (x_i, y_i). For simplicity, we assume a constant horizontal distance Δx among the track-points, so that only the $k - 1$ y-positions of the inner track-points $(x_1, y_1), \ldots, (x_{k-1}, y_{k-1})$ are subject to optimisation. How does the GA encode these heights $y_1, \ldots, y_{k-1}$ into a chromosome? The chromosome C in a GA is usually a binary encoded

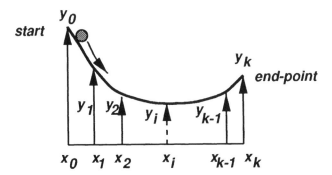

Fig. 2.2   Brachystochrone problem

string $s_0, ..., s_N$, with bit $s_i \in \{0, 1\}$. The first parameter y_1 can acquire one out of 2^n possible discrete, equidistant values in the interval $[y_{min}, y_{max}]$ encoded by the first n bits $s_0, ..., s_{n-1}$:

$$y_1 = y_{min} + \frac{y_{max} - y_{min}}{2^n - 1} \cdot \sum_{i=0}^{n-1} s_i \cdot 2^i \qquad (2.1)$$

The remaining parameters are encoded in a similar fashion by the remaining $s_n, ..., s_N$ of the string. It is important to choose n large enough, such that the discretisation provides a sufficient resolution which enables the GA to adjust the parameters y_i to a desired level of accuracy.

In our example, the polygon is composed of seven segments with six inner track-points. The outer track-points $(x_0, y_0) = (0.0, 1.0)$ and $(x_7, y_7) = (1.0, 0.0)$ remain fix and $x_i = i \cdot \Delta x = i/7$. Each of the parameters $y_1, ..., y_6$ lies in the interval $[-1, 1]$ and is encoded by six bits ($n = 6$). Therefore, the entire chromosome C for one candidate track has a length of 36 bits.

Let us consider a population P composed of chromosomes $C_1, ..., C_M$. In our case, we choose a population size of $M = 20$. The initial population is generated in a random fashion. The value of each individual bit s_i is drawn at random with equal probabilities for 0's and 1's.

The quality of a track is evaluated by a fitness function that computes the time a point mass needs to slide from the start to the end point. The overall time

$$T = \sum_{i=0}^{k-1} t_i$$

is the sum of times t_i that pass while the point mass travels the straight line from (x_i, y_i) to (x_{i+1}, y_{i+1}). The initial velocity v_i of the point mass at (x_i, y_i) can be computed from the conservation of energy

$$\frac{m \cdot v_i^2}{2} = m \cdot g \cdot (y_0 - y_i)$$

$$v_i = \sqrt{2 \cdot g \cdot (y_0 - y_i)} \qquad (2.2)$$

assuming that the velocity v_0 at the start is zero. The acceleration a_i along the track segment is constant and given by

$$a_i = \frac{g \cdot dy_i}{\sqrt{dx_i^2 + dy_i^2}} \qquad (2.3)$$

where $dx_i = x_{i+1} - x_i$ and $dy_i = y_{i+1} - y_i$. From the equations of motion the distance of the straight line equals the acceleration and velocity term.

$$\sqrt{dx^2 + dy^2} = \frac{a_i \cdot t_i^2}{2} + v_i \cdot t_i \qquad (2.4)$$

Finally, substituting Eqs. 2.2 , 2.3 into Eq. 2.4 we obtain the time

$$t_i = \frac{\sqrt{dx^2 + dy^2}}{g dy} \cdot \left(\sqrt{2g(y_0 - y_i) + g dy} - \sqrt{2g(y_0 - y_i)} \right) \quad : dy \neq 0$$

$$t_i = \frac{dx}{\sqrt{2g(y_0 - y_i)}} \quad : dy = 0$$

The scalar fitness function $f(C_i)$ evaluates the quality of a solution. In the brachystochrone problem, we want to minimise the travel time. This is achieved by a fitness function that is the inverse of travel time such that individuals which reach the end-point faster receive larger fitness.

$$f(C_i) = \frac{1}{T(C_i)}$$

Table 2.1 shows the chromosome C_i, track-points heights $y_1, \ldots, y_6$, travel time $T(C_i)$ and fitness value $f(C_i)$ for some members of the initial population. Figure 2.3 depicts the tracks corresponding to these individuals.

In a first step, the selection mechanism computes a selection probability according to which the sampling algorithm generates an intermediate population P' of parents. For each chromosome C_i in P, we calculate its

Table 2.1   Rank, chromosome C_i, track-points and travel time T for some individuals of the initial generation

#	chromosome C_i	$T(C_i)$	$f(C_i)$
	$y_1, ..., y_6$		
1	101011 010101 001010 010010 011010 010011	0.85	1.18
	0.37 -0.33 -0.68 -0.43 -0.17 -0.40		
2	100111 100110 010100 001001 000110 100001	0.87	1.15
	0.24 0.21 -0.37 -0.71 -0.81 0.05		
10	101100 101101 001100 101010 010111 101111	1.31	0.76
	0.40 0.43 -0.62 0.33 -0.27 0.49		
20	111111 011010 011001 100110 101001 100011	2.80	0.36
	1.00 -0.17 -0.21 0.21 0.30 0.11		

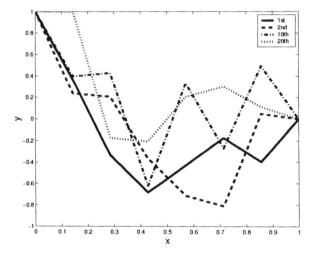

Fig. 2.3   Tracks of the individuals ranked 1st, 2nd, 10th and 20th in the initial population

selection probability

$$P_s(C_i) = \frac{f(C_i)}{\sum_{j=1}^{M} f(C_j)} \tag{2.5}$$

in a way that its chance to reproduce offspring is proportional to the fitness $f(C_i)$. The current population P is mapped onto a roulette wheel, such that the slot size of each chromosome C_i corresponds to its selection probability $P_s(C_i)$ as shown in Fig. 2.4. The intermediate population P' is generated from identical copies of chromosomes sampled by spinning the roulette wheel M times.

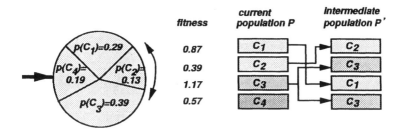

Fig. 2.4   Roulette wheel sampling of the intermediate population P' from P

In order to generate new variants, chromosomes in the parent population P' are subject to recombination and mutation. In GAs, *crossover* is considered the most important genetic operator to improve the quality of solutions. The idea is to recombine useful features of parent chromosomes in a way that allows the new offspring to benefit from favourable bit segments of both parents. Crossover explores the most promising regions of search space in a more purposeful and efficient way than a purely random global search. The interaction of crossover and selection merely generates new variants in those regions which already demonstrated to contain well adapted solutions, i.e., it performs an appropriate exploitation of search space.

Instead of every parent chromosome, only a fraction of P' selected at random according to the crossover probability P_c is subject to recombination. Typical values for the crossover rate are $P_c \in [0.6, 0.95]$. In the traditional *one-point crossover*, both parent chromosomes A and B are aligned with each other and cut at a common randomly chosen crossover position as shown in Fig. 2.5. The parents swap the segments located to the right of the crossover point resulting in two new offspring A' and B'.

In GAs, mutation plays the role of a background operator which arbitrarily inverts one or more bits of a parent chromosome to increase the structural variability of the population. Mutation provides a means to restore lost or unexplored genetic material and prevent the GA from premature convergence to sub-optimal solutions. In this sense, mutation performs

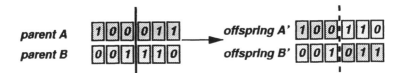

Fig. 2.5   One-point crossover

an exploration of the search space. Each bit of every chromosome in the population might become subject to an alteration according to a probability defined by a mutation rate P_m. The mutation probability per bit is usually small, typical values for P_m lie in the interval $[0.001, 0.02]$. Therefore, the mutated chromosome is usually very similar to its parents. This does not necessarily imply that the phenotypes are similar as well. In case of the standard encoding in Eq. 2.1, a mutation in the most significant bit changes the real number roughly by half the range of the interval $[y_{min}, y_{max}]$.

Finally, the current population $P(t)$ is replaced by the offspring chromosomes which constitute the next generation $P(t+1)$. As a result of this evolutionary cycle of selection, recombination and mutation, improved solutions emerge within the population. Figure 2.6 shows the population average and minimal travel time T over the course of 200 generations. The best

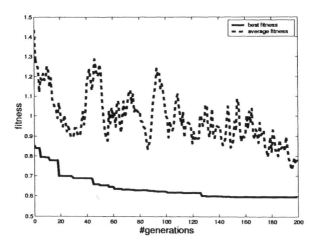

Fig. 2.6   Population average and best travel time T over 200 generations

individual emerges after 148 generations with a travel time $T = 0.5975s$.

Figure 2.7 depicts the tracks of the best individual in the population after different numbers of generations.

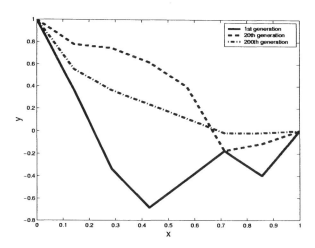

Fig. 2.7 Track-points of the best individual in the 1st (-), 20th (--) and 200th (-.) generation

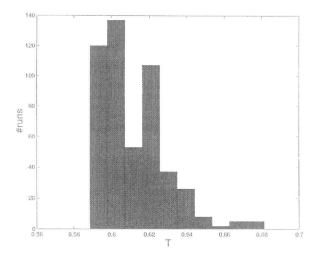

Fig. 2.8 Histogram of best travel time T for 500 runs of the GA

Due to the randomness of initialisation and genetic operators, the GA behaviour varies over different runs. Nevertheless, the GA almost always finds a reasonably good solution close to the minimal travel time. Figure 2.8 shows a histogram of the best individual performance over 500 reruns of the GA. The minimal travel time over all runs was $T = 0.5886s$, the worst run still found a solution with $T = 0.6813s$. This example demonstrates the robustness of evolutionary optimisation even for the simplest GA.

2.2.2 *Schema theorem*

A *schema* is a particular subset among the set of all possible binary strings described by a template composed of the symbols $0, 1, \star$ where the asterisk corresponds to a *don't care* term. For example the schema $1\star\star0$ corresponds to the set of strings of length four with an 1 at the first and a 0 at the last position. A string that fits a template is called an *instance* of a schema, for example 1010 is an instance of the schema $1\star\star0$. In a slight abuse of notation, one often uses the term schema to denote the subset as well as the template defining the subset itself. The order $o(h)$ of a schema h is the number of non-asterisk symbols, for example $1\star\star0$ is of order $o(1 \star \star 0) = 2$. The defining length $\delta(h)$ of a schema h is the distance between the outermost non-asterisk symbols, for example $1 \star 0\star$ has a defining length of $\delta(1 \star 0\star) = 2$.

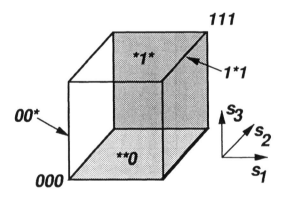

Fig. 2.9   Concept of a schema

Assume our chromosome is composed of three bits only. The solution space can be represented by a cube, whose axes correspond to the bits

s_1, s_2, s_3 as shown in Fig. 2.9. The corners are labelled by bit strings, with 000 located at the origin and 111 at the upper, right, rear corner. The bottom plane contains all bit strings that end with a 0 or in other words match the schema $\star \star 0$, the schema $00\star$ covers two instances 000 and 001 located at the left, front vertical edge. In the general case of an n-dimensional hypercube, a schema of order m corresponds to a $n - m$-dimensional hyper-plane.

The schema theorem describes how the frequency of schema instances changes considering the effects of selection, crossover and mutation. Let $m(h, t)$ describe the number of instances of schema h present in the population at generation t. The fitness $f(h)$ of a schema h is computed as the average fitness of its instances present in the current population. According to the schema theorem (Holland, 1975; Goldberg, 1989), the expected number $E(m(h, t + 1))$ of instances of schema h in the next generation is

$$E(m(h, t+1)) \geq m(h, t) \cdot \frac{f(h)}{\bar{f}} \cdot \left[1 - P_c \cdot \frac{\delta(h)}{l-1} \right] \cdot [1 - P_m]^{o(h)} \qquad (2.6)$$

in which $\bar{f}$ is the average fitness of the population, l is the length of the chromosome, $\delta(h)$ is the defining length of schema h, $o(h)$ is the order of schema h and P_c, P_m are the probabilities of applying the crossover and mutation operator. The schema theorem states that the frequency of schemata with fitness higher than the average, short defining length and low order increases in the next generation. The term $[1 - P_c \cdot \frac{\delta(h)}{l-1}]$ describes the probability that one-point crossover does not disrupt a schema. Similarly, the term $[1 - P_m]^{o(h)}$ reflects the possible disruptive effects of mutations on a schema. The selection term $f(h)/\bar{f}$ biases the sampling step toward schemata with above average fitness. Evaluating a population of chromosomes implicitly estimates the fitness of all schemata that match these instances. The observation that a GA evaluates a large number of schemata simultaneously present in the population is called *implicit parallelism* (Holland, 1975).

The schema theorem merely describes the disruptive effects of crossover and mutation but does not explain the source of evolutionary progress in a GA. According to the *building block hypothesis*, the power of GAs stems from crossover which recombines instances of low-order, highly fit schemata, so-called building blocks, to generate higher-order schemata that are even better adapted to the optimisation problem (Goldberg, 1989).

2.2.3 *Extensions to the simple genetic algorithm*

Even if the simple GA described in the previous subsection performs considerably well for a broad range of optimisation problems, it suffers from a number of drawbacks, such as premature convergence, biased search and inadequate mapping from genotype to phenotype space. This subsection introduces extensions to the genetic representations, selection schemes and genetic operators designated to improve the behaviour of GAs.

2.2.3.1 *Genetic encoding of solutions*

The success of a GA strongly depends on a proper genetic representation of candidate solutions, which matches the metric of the optimisation space. In other words, chromosomes encoding similar phenotypes are supposed to be separated by a short hamming distance. Unfortunately, this is not the case for the standard base two encoding in Eq. 2.1. The bit strings 1000 and 0111 differ in every single bit, therefore having maximal hamming distance of four, but encode the two adjacent integers 8 and 7. This adverse encoding may prevent the GA from converging to the global optimum in case it is located at a so-called *Hamming cliff*, like the one above.

Today, most GAs employ a *Gray code* which guarantees that adjacent integers are represented by bit strings which only differ in a single bit and therefore avoid *Hamming cliffs*. Given a binary string $s_1, ..., s_n$ coding integers in the standard way, the conversion formula for obtaining the corresponding Gray-coded string $g_1, ..., g_n$ is the following:

$$g_k = \begin{cases} s_1 & \text{if } k = 1 \\ s_{k+1} \oplus s_k & \text{if } k > 1 \end{cases}$$

where $\oplus$ denotes addition module 2 (i.e. $0 \oplus 0 = 0, 1 \oplus 0 = 1, 0 \oplus 1 = 1, 1 \oplus 1 = 0$). The inverse conversion is defined by means of the following expression:

$$s_k = \sum_{j=1}^{k} \oplus g_j$$

where the sum is done in module 2.

Table 2.2 compares the coding of integers 0 to 7 using binary and Gray codes. Notice that strings encoding adjacent integers have a minimal hamming distance of 1. Nevertheless, even for a Gray code, a single mutation may still result in a large change of the integer value depending on the position of the reversed bit. In general, mutation in binary-coded GAs lacks the ability to strictly focus the search on solutions that are located close to its parents in phenotype space.

Table 2.2   Comparison of binary and Gray codes

Integer	0	1	2	3	4	5	6	7
base 2	000	001	010	011	100	101	110	111
Gray	000	001	011	010	110	111	101	100

2.2.3.2 *Fitness scaling*

To compute the selection probability $P_s(C_i)$ in Eq. 2.5, the roulette-wheel selection mechanism requires positive fitness values $f(C_i)$. Positive fitness might not be guaranteed for any kind of objective functions. Furthermore, fitness-proportionate selection might result in premature convergence in early generations, in case a "super-individual" of comparatively high fitness dominates the entire population. In later generations, the selection pressure might decrease too much, since most individuals roughly achieve the same fitness value. In order to overcome these limitations, researchers proposed a variety of *fitness scaling* methods. The raw objective function values $f(C_i)$ are transformed into a scaled fitness $\hat{f}(C_i)$ that becomes ultimately used for selection. In *linear scaling*, the fitness is transformed according to

$$\hat{f}(C_i) = c_0 \cdot f(C_i) + c_1$$

where the parameters c_0, c_1 are either static or dynamically adjusted based on the raw fitness distribution of the current population. For example, c_0, c_1 are computed such that the expected number of offspring for the best individual becomes $\alpha \approx 1.5 \ldots 3.0$ and an individual with average fitness $\bar{f}$ produces one offspring on average.

$$\hat{f}_{max} = \alpha \cdot \hat{\bar{f}}$$
$$c_0 \cdot f_{max} + c_1 = \alpha \cdot (c_0 \cdot \bar{f} + c_1)$$

This determines the ratio

$$\frac{c_0}{c_1} = \frac{\alpha - 1}{f_{max} - \bar{f}}$$

which can be easily fulfilled by choosing

$$c_0 = \alpha - 1$$
$$c_1 = f_{max} - \bar{f}$$

as long as it guaranteed that the minimal transformed fitness $\hat{f}_{min}$ remains positive.

Sigma truncation scales the fitness by using the standard deviation σ of the objective function values in the current population.

$$\hat{f}(C_i) = \frac{f(C_i) - \bar{f} - \sigma}{\sigma}$$

A selection mechanism is called *elitist* if the population best fitness monotonously improves from one generation to the next. This is usually achieved by inserting a copy of the best individual into the next generation. In some selection schemes, the offspring compete with their parents which also guarantees the survival of the best individual. An elitist strategy is not advisable in case of non-stationary or probabilistic objective functions, unless the fitness of the copy is re-evaluated for every generation.

2.2.3.3 *Selection and replacement schemes*

Researchers proposed different sampling methods for the selection of parents. The *stochastic universal sampling* avoids discrepancies between the expected number $P_s(C_i) \cdot M$ and the actual number of copies an individual receives during selection (Baker, 1987). The realised number of copies of any chromosome C_i is bounded by the integer floor and ceiling of its expected number of copies. The algorithm uses a roulette wheel similar to the one depicted in Fig. 2.4. Rather than spinning the wheel M times using a single pointer, the parents are determined by a single spin of a wheel that has M equally spaced pointers. The number of copies for a chromosome C_i corresponds with the number of pointers that point to the segment associated with C_i.

Ranking selection is based on the ranking of the population according to their fitness rather than using the absolute fitness value as a selection criteria. A group of K candidates is chosen randomly with equal probability for each member in the population P. From these candidates, only the chromosome with the highest ranking is selected for reproduction, whereas the others are discarded. This tournament is repeated M times until all the slots in the intermediate parent population P' are filled.

Replacement can be regarded as the complementary operator to selection in that it determines the chromosomes to be extinguished. The standard GA described in the previous subsection employs a *generational replacement* scheme, in which the entire population is replaced by the new generation of offspring. A *steady-state* GA operates on overlapping populations in which only a subset of the current population is replaced in each generation. In an *incremental* GA, a newly generated offspring replaces either one of its parents, a random member, the currently worst member of the population or the individual that is most similar to it.

2.2.3.4 *Niching genetic algorithms*

GAs present the drawback that when dealing with multi-modal functions with multiple peaks of unequal value, they tend to converge to the highest peak of the fitness landscape. This characteristic prevents them from adequate individual sampling over peaks in other regions. This phenomenon is called *genetic drift* (Deb and Goldberg, 1989) and constitutes an undesirable behaviour for those kinds of problems in which one is interested in the location of other function optima as well.

In (Deb and Goldberg, 1989; Goldberg, 1989), the concepts of *niche, species* and *individual fitness sharing* are introduced to GAs in order to overcome this behaviour. Following the principal guideline of evolutionary computation, they attempt to translate concepts from natural evolution to the area of GAs. In biology, a niche is viewed as an environment with certain characteristics that impose a specific challenge on the organisms inhabiting it. A species is a collection of individuals having similar features in common. Separate niches induce the formation of stable sub-populations of organisms by forcing similar individuals to share the available resources.

The population is divided in different sub-populations (species) according to the similarity of the individuals. These sub-populations form niches in the genotype as well as in the phenotype space, thus establishing *genotypic* and *phenotypic sharing*. Each niche provides a certain amount of resources available to its inhabitants. Similar to nature, individuals belonging to the same niche share the associated payoff among them. A *sharing function* determines the degree of sharing among individuals according to their closeness measured by some metric either defined in the genotype or the phenotype spaces.

2.2.3.5 *Recombination*

One-point crossover operator has the drawback of a positional bias, in that genes located at both ends of the chromosomes are disrupted more frequently than those in the centre. *Two-point crossover* depicted in Fig. 2.10 avoids this positional asymmetry in cutting the chromosome at two locations rather than one and swapping the middle segments in the offspring.

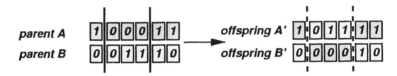

Fig. 2.10   Two-point crossover

A generalisation to two-point-crossover allows n crossover locations in which the offspring inherits subsequent segments from both parents in alternating order. *Uniform crossover* takes n-point-crossover to the extreme by reducing the swapped segments to single bits which are independently drawn from either one of the parents.

2.2.4 *Real-coded genetic algorithms*

Fixed-length and binary-coded strings for the representation of candidate solutions historically tend to dominate in research and applications of GAs. This trend results mainly from a number of theoretical results such as Holland's *schema theorem* obtained for binary representations (Holland, 1975; Goldberg, 1991). More recently, authors proposed the use of non-binary representations in GAs more adequate for a certain class of optimisation problems (Wright, 1991; Eshelman and Schaffer, 1993; Herrera, Lozano, and Verdegay, 1998b). In fact, *evolution strategies*, which are described in subsection 2.3.1, constitute an entire class of EAs that operate on real-coded chromosomes. The similar encoding of solutions in evolution strategies and *real-coded GAs* (RCGAs) in particular facilitates the exchange of genetic operators between both domains (Mühlenbein and Schlierkamp-Voosen, 1993).

Using a *real number representation* seems particularly natural for optimisation problems in the continuous domain. The chromosome in RCGAs is a vector of floating point numbers, $C_i = (x_1^i, \ldots, x_n^i)$, whose size n equals the dimension of the continuous search space, or in other words each gene x_j^i corresponds to a particular parameter of the problem.

RCGAs offer the advantage that the continuous parameters can gradually adapt to the fitness landscape over the entire search space whereas parameter values in binary implementations are limited to a certain interval and resolution. RCGAs blur the distinction between genotype and phenotype, since in many problems the real-number vector already embodies a solution in a natural way.

2.2.4.1 *Recombination in real-coded genetic algorithms*

As a result of the modified genetic representation RCGAs require customisation of the standard mutation and crossover operators. In *flat crossover*, an offspring $C' = (x_1', \ldots, x_i', \ldots, x_n')$ is generated by

$$x_i' = \lambda_i \cdot x_i^1 + (1 - \lambda_i) \cdot x_i^2$$

from parents $C_1 = (x_1^1, \ldots, x_n^1), C_2 = (x_1^2, \ldots, x_n^2)$ where λ_i is a random number drawn uniformly from the interval $[0, 1]$.

BLX-α crossover is an extension to flat crossover that allows the offspring gene x_i' to be located outside the interval $[x_i^1, x_i^2]$ (Herrera, Lozano, and Verdegay, 1998b). x_i' is a randomly, uniformly chosen number from the interval $[x_{min} - I \cdot \alpha, x_{max} + I \cdot \alpha]$, where $x_{max} = \max(x_i^1, x_i^2)$, $x_{min} = \min(x_i^1, x_i^2)$, $I = x_{max} - x_{min}$. The parameter α is chosen by the user and determines the possible range for the offspring gene x_i'. For $\alpha = 0$, BLX-α crossover becomes identical to flat crossover.

Simple crossover is equivalent to the conventional one-point crossover in swapping the parent tail segments starting from a randomly chosen crossover site $k \in \{1, 2, \ldots, n-1\}$

$$
\begin{aligned}
C' &= (x_1^1, x_2^1, \ldots, x_k^1, x_{k+1}^2, \ldots, x_n^2) \\
C'' &= (x_1^2, x_2^2, \ldots, x_k^2, x_{k+1}^1, \ldots, x_n^1)
\end{aligned}
$$

Discrete crossover is the counterpart to uniform crossover in binary codings, each gene x_i' is randomly chosen either from x_i^1 or x_i^2.

2.2.4.2   *Mutation in real-coded genetic algorithms*

Let us assume that the allelic value x_i of the i-th gene ranges over the domain $[a_i, b_i]$. In *random mutation*, the mutated gene x_i' is drawn randomly, uniformly from the interval $[a_i, b_i]$. In *non-uniform mutation*, the mutation step-size decreases with increasing number of generations (Michalewicz, 1996). Assume t is the current generation and t_{max} is the maximum number of generations,

$$x_i' = \begin{cases} x_i + \Delta(t, b_i - x_i), & \text{if } \tau = 0 \\ x_i - \Delta(t, x_i - a_i), & \text{if } \tau = 1 \end{cases}$$

where τ represents an unbiased coin flip $p(\tau = 0) = p(\tau = 1) = 0.5$, and

$$\Delta(t, x) = x \cdot \left(1 - \lambda^{\left(1 - \frac{t}{t_{max}} \right)^b} \right)$$

defines the mutation step, where λ is a random number from the interval $[0, 1]$. The function Δ computes a value in the range $[0, x]$ such that the probability of returning a number close to zero increases as the algorithm advances. The parameter b determines the impact of time on the probability distribution of Δ over $[0, x]$. Large values of b decrease the likelihood of large mutations in a smaller number of generations.

This subsection only presented a subset of the genetic operators proposed for RCGAs. For a complete list of crossover and mutation operators we refer the interested reader to (Herrera, Lozano, and Verdegay, 1998b).

2.2.5   *Messy genetic algorithms*

The GAs described so far encode candidate solutions by chromosomes of fixed length in which the locus of a gene determines its functionality. Messy GAs relax the fixed-locus assumption and operate with chromosomes of flexible length in which genes can be arranged in any order (Goldberg, 1989). In a messy GA, each gene, in addition to the bit representing its value, contains an additional integer defining its functionality. For example, the standard binary string **1001** is transformed into a string of pairs **(1,1) (2,0) (3,0) (4,1)**, where the first element serves as a label that uniquely defines the gene functionality. The second element represents the usual allelic value of the gene. Notice that the messy coding scheme no longer requires a particular order of genes. For example, the chromosome **(4,1)**

(3,0) (1,1) (2,0) with the shuffled genes also lends itself to represent the same bit string **1001**.

The schema theorem states that the frequency of schemata with above average fitness, low order and short defining length increases in the next generation. Assume that the genes labelled **1** and **4** in the above example are tightly coupled in a way that instances of the building block **1⋆⋆1** have a high fitness compared to other chromosomes in the population. According to the schema theorem in Eq. 2.6, the large defining length $\delta(1\star\star1) = 3$ limits the frequency increase of the schema since every crossover operation disrupts the schema. A chromosome in which the genes **1** and **4** are adjacent, is less prone to disruption by crossover. Messy GAs are able to form better building blocks since the position independent coding allows them to permute the order of genes. Assume the genes in the example are arranged in the order **(3,0) (2,0) (4,1) (1,1)**. The defining length of the permuted schema ⋆ ⋆ 11 decreases because genes **1** and **4** become adjacent.

2.2.5.1 *Under- and over-specification*

The more flexible coding scheme of messy GAs comes with the prize of incomplete or ambiguous chromosomes. *Under-specification* occurs whenever an incomplete chromosome lacks a particular gene. In the 4-bit example from above, the string **(2,0) (4,1) (1,1)** is incomplete because it does not contain a gene with label **3**. Over-specification constitutes a problem whenever the string contains multiple, ambiguous alleles for the same gene. The string **(2,0) (4,1) (3,0) (2,1) (1,1)** embodies two genes with label **2** having ambiguous alleles **0** and **1**. Messy GAs require some additional processing when mapping genotypes to phenotypes in order to handle under and over-specification.

Over-specification is resolved by a "first come, first serve" rule, which only considers the leftmost gene of multiple occurrence. In the above example, the coding scheme only expresses the leftmost gene **(2,0)** whereas it discards the second gene **(2,1)**. Under-specification turns out to be a more difficult problem. Messy GAs can deal with under-specification under the assumption that a partial string can at least be partially evaluated by the fitness function. As will be shown in later chapters, this prerequisite holds for the task to evolve a KB using a messy GA. The fact that a fuzzy rule set is a collection of rules facilitates it's encoding by a messy genotype. Many optimisation problems require all parameters to compute the objective function. Goldberg (1989) proposed *competitive templates* in order to determine the fitness of a partial chromosome. Unspecified bits are filled in from a template string, a chromosome that constitutes a local optimum of the objective function.

2.2.5.2 *Genetic operators*

Messy GAs employ a slightly modified crossover, composed of two steps *cut* and *splice* as depicted in Fig. 2.11.

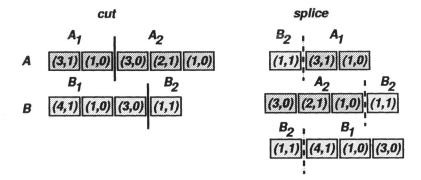

Fig. 2.11   Schematics of cut and splice operators

The parent strings A and B are cut at a uniformly, randomly chosen position, resulting in four segments A_1, A_2, B_1 and B_2 available for recombination. Notice that unlike in position dependent coding, the location of both parent cuts does not necessarily coincide. The splice operation randomly selects two out of these four segments and concatenates them in a random order. Fig. 2.11 shows three possible offspring $B_2 - A_1$, $A_2 - B_2$ and $B_2 - B_1$ resulting from a splice operation. Notice that in case the segments originate from the same parent, such as in the offspring $B_2 - B_1$, cut and splice mimic an inversion-like operation.

The mutation operator only affects the allele but does not alter the label of a gene. According to a small per bit mutation probability, an allele is flipped from **1** to **0** or *vice versa*.

2.3 Other Evolutionary Algorithms

2.3.1 *Evolution strategies*

Rechenberg and Schwefel developed evolution strategies (ESs) in the 1960s (Rechenberg, 1973; Schwefel, 1995), independently from the work of Holland (1975) on GAs at the same time. Both methods share common features of EAs, a population of genetic structures, selection to exploit promising regions of the search space and recombination and mutation to generate new variants. The major difference between ESs and GAs exists in the genetic representation of candidate solutions. The chromosome in an ES consists of a vector of real numbers, whereas GAs usually process a population of binary strings[*].

In ESs, mutation plays a central role, whereas in GAs mutation is considered to be a secondary operator merely useful to prevent premature convergence of the population. The opposite holds for recombination. The first ES did not even employ recombination at all. ESs are distinguished by self-adaptation of additional strategy parameters, which enables them to adapt the evolutionary optimisation process to the structure of the fitness landscape. The selection operator in ESs is entirely deterministic in the sense that only the μ best individuals serve as parents. Due to these characteristics, ESs are usually preferable for problems in the domain of continuous optimisation whereas GAs are an approved method for problems of discrete or combinatorial nature.

ESs operate with chromosomes composed of two vectors of real numbers. The vector $\vec{x} \in \mathbb{R}^n$ contains the object variables subject to evaluation by the objective function. Additional strategy parameters are stored in the vector $\vec{\sigma} \in \mathbb{R}^n_+$. The parameter σ_i defines the standard deviation of random mutations applied to the corresponding object variable x_i. Mutation adds a normally distributed noise $N_i(0, \sigma_i^2)$, with variance σ_i^2 and zero mean, to the object variable x_i

$$x'_i = x_i + N_i(0, \sigma_i^2) \tag{2.7}$$

Notice that the probability distribution for the offspring $\vec{x'}$ is constant across the surface of a hyper-ellipsoid with axes radii $r_i = \alpha \cdot \sigma_i^2$ centred at $\vec{x}$.

---

[*]More recently, researchers investigate real number representations for GAs like those presented in Sec. 2.2.4

The strategy parameters $\vec{\sigma}$ are subject to optimisation as well and therefore enable the search process to adapt itself to the topology of the fitness landscape. The strategy parameters are adapted first, and the new values are used for the mutation of the object parameters. The σ_i are mutated using a logarithmic normal distribution

$$\sigma_i' = \sigma_i \exp\left(\tau' \cdot N(0,1) + \tau \cdot N_i(0,1)\right) \tag{2.8}$$

where $\exp\left(\tau' \cdot N(0,1)\right)$ is a global factor increasing or decreasing the overall mutability and $\exp\left(\tau \cdot N_i(0,1)\right)$ locally adapts the individual step-sizes σ_i. The random variable $N(0,1)$ is only sampled once for the entire chromosome, whereas the local mutations $N_i(0,1)$ are uncorrelated for different σ_i. The logarithmic normal distribution guarantees that the mutation standard deviations σ_i remain positive. The mutation is unbiased in the sense that increases and decreases of σ_i by the same factor occur with identical probability.

Schwefel (1995) proposed to choose the learning rates

$$\tau' \sim 1/\sqrt{2 \cdot n}$$
$$\tau \sim 1/\sqrt{2 \cdot \sqrt{n}}$$

for a chromosome of size n.

In the course of evolution, the self-adaptive process favours strategy parameters, which potentially result in mutations of optimal magnitude and direction. Due to the mechanism of self-adaptation, an exogenous control of step-sizes, utilised by standard mathematical optimisation methods, becomes obsolete.

In the standard ES, mutations are uncorrelated between different object variables x_i (Eq. 2.7), or in other words the mutation hyper-ellipsoids are axis-parallel. In order to allow general, correlated mutations, the random variables are drawn from a generalised normal distribution with covariance matrix $\overleftrightarrow{C}$

$$\vec{x}' = \vec{x} + \vec{N}(\vec{0}, \overleftrightarrow{C})$$

The $n \times n$ covariance matrix $\overleftrightarrow{C}$ can be parameterised by a vector of length $n \cdot (n+1)/2$, composed of the n standard deviations σ_i and $n \cdot (n-1)/2$ rotations by angles α_i. Since the covariance matrix implicitly defines the rotation angles α_i, it is also called *angle vector*. The angle vector defines linearly correlated mutations between the object variables x_i. The standard deviations σ_i are mutated according to Eq. 2.8. Since the rotation angles are not restricted to $\mathbb{R}_+$, they are mutated similar to the object variables by

$$\alpha'_i = \alpha_i + N(0, \beta^2)$$

with the parameter $\beta \approx 0.0873 \approx 5°$. The α'_i are mapped to the interval $[-\pi, \pi)$, whenever mutation lefts the feasible range.

While mutation is the major genetic operator in ESs, recombination can be helpful to improve the search. There are different ways to perform the recombination of one or more parents to obtain a preliminary offspring that will then become subject to mutation. For example, in intermediate recombination, an offspring inherits the median of its parents A, B

$$x_i = \frac{x_i^A + x_i^B}{2}$$

The discrete recombination resembles uniform crossover in GAs, where each parameter is randomly picked from either one of the parents

$$x_i = \begin{cases} x_i^A & \text{with p=1/2} \\ x_i^B & \text{with p=1/2} \end{cases}$$

Usually, intermediate recombination is chosen for the strategy parameters, while discrete recombination is preferred for the object variables.

The first ES did not include the notion of a population, but rather employed a very simple selection and replacement scheme. It maintains a single parent which generates one mutated offspring. If the offspring has a better fitness, it becomes the new parent in the next generation, otherwise it is discarded and the original parent survives. This simple selection and replacement scheme is called a (1+1)-strategy. The $(\mu+1)$-strategy introduces the notion of a population in which μ parents generate one offspring at a time, which then replaces the current worst member of the population. This selection scheme was generalised to the multi-membered $(\mu+\lambda)$- and (μ,λ)-strategies. In a $(\mu+\lambda)$-strategy, μ parents generate λ offspring. The μ best individuals out of the union of parents and offspring form the parents of the next generation. In the (μ,λ)-strategy, which became the standard selection scheme in ESs, no parents survive and only the μ best offspring constitute the new population. In this case, λ must be greater than μ in order for the offspring to outnumber the parents.

Notice that these selection schemes are entirely deterministic and are solely based on the fitness ranking of individuals. Therefore, fitness scaling like in roulette wheel selection becomes obsolete. A $(\mu+\lambda)$-strategy is elitist, whereas (μ,λ)-selection is not.

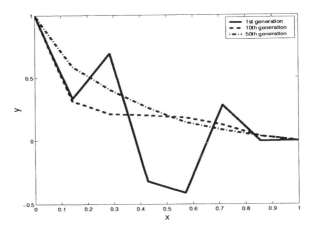

Fig. 2.12   Track-points of the best individual in the 1st (-), 10th (--) and 50th (-.) generation

In the following, we apply a $(4, 20)$-strategy to solve the brachystochrone problem (Sec. 2.2.1). The chromosome vector of object variables x_i encodes the height y_i of the inner track-points. The ESs employs the basic mutation operator in Eq. 2.7 as the sole genetic operator. Fig. 2.12 depicts the tracks of the current best individual after 1, 10 and 50 generations.

The graphs in Fig. 2.13 shows the population average and minimal travel time over the course of 100 generations. Notice that the ESs finds a solution that is already very close to the optimum within less than 20 generations. The travel time for the best configuration of track-points was $T = 0.5883s$ which is slightly better than the overall best time $T = 0.5886$ found in 500 reruns of a GA. The ES is able to generate a better solution since the resolution of the real-number variables is only limited by the floating-point precision of the machine which runs the algorithm, whereas in a GA the resolution depends on the number of bits used in the coding. The ES converges very reliably to the optimal solution, 98% of the runs find a track with a travel time that exceeds the optimum by less than $0.002s$.

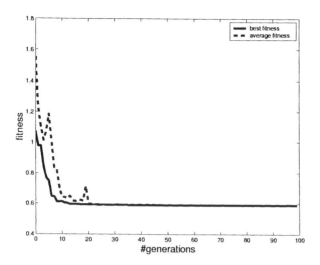

Fig. 2.13 Population average and best travel time T over 100 generations

2.3.2 *Evolutionary programming*

Evolutionary programming (EP) originated from the work of L.J. Fogel in the 60s (Fogel, 1962) and was further extended by his son D.B. Fogel in the late 80s (Fogel, 1991). In its current status, EP shares a number of characteristic features with ES, such as a real-number representation, normally distributed mutations and adaptation of mutation variances.

The chromosome is of the form $\vec{a} = (\vec{x}, \vec{\nu})$, in which $\vec{x} \in \mathbb{R}^n$ is a vector of real-number optimisation parameters and $\vec{\nu} \in \mathbb{R}^n_+$ is a vector of variances. EP uses a scaling function δ to map objective function values $f(\vec{x})$ to fitness values

$$\Phi(\vec{a}) = \delta(f(\vec{x}))$$

Originally, EP mutation was of the form

$$x'_i = x_i + \sqrt{\beta_i \cdot \Phi(\vec{a}) + \gamma_i} \cdot N_i(0, 1)$$

in which $N_i(0, 1)$ is drawn from a normal distribution with variance one and zero mean. The constants β_i and γ_i are exogenous parameters that must be manually adapted to a particular optimisation task. This approach faces the difficulty that either the scaling function δ or the linear transformation between fitness and variances has to be tuned in a such a way that the variances $\sqrt{\beta_i \cdot \Phi(\vec{a}) + \gamma_i}$ go to zero as $f(x)$ reaches its minimum.

More recently, Fogel (1992) proposed a meta-EP which addresses the issue of mutation step-size adaptation in a way similar to ESs

$$x_i' = x_i + \sqrt{\nu_i} \cdot N_i(0,1)$$
$$\nu_i' = \nu_i + \sqrt{\chi \cdot \nu_i} \cdot N_i(0,1)$$

The control parameter χ is adjusted in a way that guarantees the ν_i to remain positive. Despite the strong similarity, there is a minor difference in the mutation operator between EP and ESs. An ES uses a logarithmic normal distribution to mutate the variances (see Eq. 2.8) that automatically maintains positive variances, whereas EP has to verify this property explicitly after each mutation. EP generates new variants solely by means of mutation and does not employ any recombination operator.

EP employs a combination of tournament and ranking selection in which the fitness of an individual is compared with q other randomly picked competitors taken from the union of parents and offspring. The score $w_i \in 0, \ldots, q$ of an individual is computed as the number of competitors with lower fitness $\Phi(\vec{a})$. The parents and offspring are ranked according to their score and the μ best individuals form the parents of the next generation. This selection scheme is an elitist strategy since the best individual achieves the maximum score q and therefore becomes always selected. For a large tournament size q, this selection mechanism has almost the same effect as a $(\mu + \mu)$-ES (Bäck, 1996).

2.3.3 *Genetic programming*

Genetic programming (GP) is concerned with the automatic generation of computer programs. Proposed in the early 90's by John Koza (1992), this new field has rapidly grown in visibility as a promising approach to adapt programs to particular tasks by means of an EA (Koza, 1994; Banzhaf, Nordin, Keller, and Francone, 1998). GP improves a population of computer programs by testing their performance on some training examples. GP has been applied to a remarkable variety of different domains, such as symbolic regression, electronic circuit design, data mining, biochemistry, robot control, optimisation, pattern recognition, planning and evolving game-playing strategies.

Most GP systems employ a tree structure to encode the program which performs the computation to achieve a particular task. Recently, authors proposed linear and graph-based programs representations for GP (Nordin, 1994). In the following we describe the most common approach in GP. For a general treatise on GP, the interested reader is referred to (Banzhaf, Nordin, Keller, and Francone, 1998).

The trees in GP are composed of primitives taken from the sets of *function* and *terminal* symbols. Terminal symbols refer to nodes at the leaves of the tree and provide a value, for example an input variable, a constant or a zero-argument node. Function symbols correspond to non-leaf nodes in the tree and compute or process an argument passed from its children nodes. Commonly used functions are arithmetic and logic operators, conditional statements and loop statements.

Figure 2.14 shows an example of a parse tree, with a terminal set $T = \{x_1, x_2\}$ and a function set $F = \{+, -, *, /\}$. The tree is parsed from left to right in a depth first order resulting in the expression $+(x_1 (*(x_1 (-x_2 x_1))))$ in prefix notation.

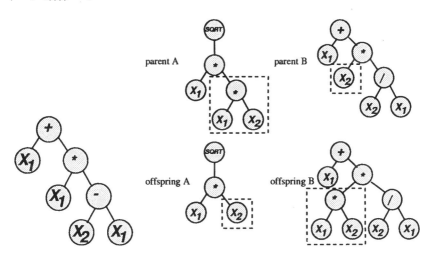

Fig. 2.14   Left: genetic programming tree, right: tree-based crossover

The first generation of programs is seeded with a variety of tree structures. A program tree is generated at random by selecting a symbol from the function and the terminal set for each node. There are two different methods to grow a tree. In the full method, a tree grows only using function symbols as nodes until each branch reaches a maximal depth and ends with a terminal symbol. In the grow method, function and terminal symbols are chosen randomly as nodes at each level of depth, such that the trees can grow in an irregular fashion.

Tree-based crossover swaps a sub-tree of one parent with a sub-tree of the other as shown in Fig. 2.14. Two parent trees are chosen based on the selection scheme, for example fitness-proportional selection. A random sub-tree is selected in each parent denoted by the dashed rectangle. The selected sub-trees are swapped over the parents resulting in two new offspring. In self-crossover, sub-trees are exchanged within a single individual.

The mutation operator in tree-based GP replaces an existing sub-tree with a newly generated random sub-tree. Point mutation exchanges a single either function or terminal node against a random node of the same type, for example the arithmetic function + is replaced by *. Other GP mutation operators substitute terminals for sub-trees, or *vice versa*, sub-trees for terminals.

The GP designer can utilise his domain-specific knowledge of the problem to define the function and terminal set. In the block stacking problem depicted in Fig. 2.15, the task is to move the blocks from an initial configuration into the correct order such that the stack forms the word UNIVERSAL.

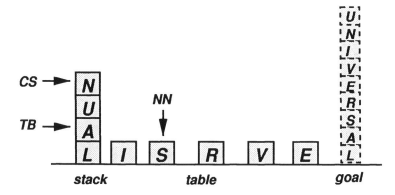

Fig. 2.15   Block stacking problem: the task is to build the stack such that the blocks form the word UNIVERSAL

The terminal set includes sensor information about the current status of the stack:

- CS : refers to the current block on top of the stack

- TB : refers to the top correct block such that it and all blocks below are in the correct order
- NN : refers to the next block needed above TB

The function set contains the following actions, logic operators and loop structures. Each function has a return value, that is passed to its parent node while evaluating the tree program.

- MS(*arg1*) : move block *arg1* from the table to the top of the stack, returns *arg1*
- MT(*arg1*) : move the block on top of the stack to the table, if block *arg1* is anywhere in the stack, returns *arg1*
- DU(*arg1,arg2*) : do until, execute the first argument until the second argument becomes true
- EQ(*arg1,arg2*) : returns true if the first and second argument refer to the same block

Each program is tested on a set of initial stack configurations of different complexity. In some test cases, the initial stack is already correct and the remaining task becomes to put the blocks that are on the table onto the stack in the right order. The fitness is the number of test cases for which the stack is correct after the execution of the genetic program.

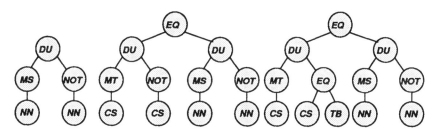

Fig. 2.16   Programs for the block-stacking problem evolved by GP

The program tree in the left of Fig. 2.16 —DU (MS NN) (NOT NN)— constitutes a partial solution. It puts the blocks from the table on the stack in the right order. It generates the valid solution UNIVERSAL, if the initial stack is already ordered correctly, e.g. SAL. The tree in the centre of Fig. 2.16 —EQ (DU (MT CS) (NOT CS)) (DU (MS NN) (NOT NN))— is a complete solution which produces the correct stack from any possible initial configuration. The left sub-tree first disassembles the stack until all blocks are located on the table. Then, it employs the previous partial solution to rebuild the stack in proper order. The EQ function at the root connects the two sub-trees which are executed sequentially. The program EQ (DU (MT CS) (EQ (CS TB))) (DU (MS NN) (NOT NN)) —on the right in Fig. 2.16— is almost identical to the previous one. It constitutes the optimal solution in the sense that it requires the smallest numbers of stack operations. Instead of unmounting the stack all the way, the program only removes blocks until the topmost correct block TB is also the top block CS of the stack, or in other words until the remaining stack is correct.

Chapter 3

Introduction to Genetic Fuzzy Systems

The previous chapters introduced FSs and GAs, the complementary constituents of *genetic fuzzy systems*. This chapter analyses the main principles and operation of genetic fuzzy systems, which are part of a larger class of hybrid methodologies that combine FL, neural networks, probabilistic reasoning and GAs in one way or another. The term soft computing has been coined for this family of robust, intelligent systems, that in contrast to precise, traditional modes of computation, are able to deal with vague and partial knowledge.

3.1 Soft Computing

The term soft computing refers to a family of computing techniques comprising four different partners: FL, evolutionary computation (EC), neural networks (NNs) and probabilistic reasoning (PR). The term *soft computing* distinguishes these techniques from *hard* computing that is considered less flexible and computationally demanding. The key point of the transition from hard to soft computing is the observation that the computational effort required by conventional computing techniques sometimes not only makes a problem intractable, but is also unnecessary as in many applications precision can be sacrificed in order to accomplish more economical, less complex and more feasible solutions. Imprecision results from our limited capability to resolve detail and encompasses the notions of partial, vague, noisy and incomplete information about the real world. In other words, it becomes not only difficult or even impossible, but also inappropriate to apply hard computing techniques when dealing with situations in which the required

79

information is not available, the behaviour of the considered system is not completely known or the measures of the underlying variables are noisy.

Soft computing techniques are meant to operate in an environment that is subject to uncertainty and imprecision. According to Zadeh (1997), the guiding principle of soft computing is:

> exploit the tolerance for imprecision, uncertainty, partial
> truth, and approximation to achieve tractability, robust-
> ness, low solution cost and better rapport with reality.

All four methodologies that constitute the realm of soft computing have been conceptualised and developed over the past forty years. Each method offers its own advantages and brings certain weaknesses. Although they share some common characteristics, they are considered complementary as desirable features lacking in one approach are present in another. Consequently, after a first stage in which they were applied in isolation, the last decade witnessed an increasing interest on hybrid systems obtained by symbiotically combining the four components of soft computing. Figure 3.1 shows some hybrid systems positioned in the corresponding intersection of soft computing techniques.

3.2 Hybridisation in Soft Computing

The four constituents of soft computing, being inspired from such diverse domains as logic, biology, physiology or statistics, underwent a first stage of independent development in their respective community of practitioners. Their success progressively attracted people in the other fields, who aimed for different approaches to hybridise the individual methods.

This section describes some of these attempts such as neuro-fuzzy and neuro-genetic systems, while postponing the subject of genetic fuzzy systems to the remainder of this chapter.

3.2.1   *Fuzzy logic and neural networks*

Neuro-fuzzy systems and fuzzy-neural networks constitute the first appearance of hybridisation in soft computing, in that they incorporate elements from FL and NNs. The following arguments substantiate the interest and large research effort devoted to this area:

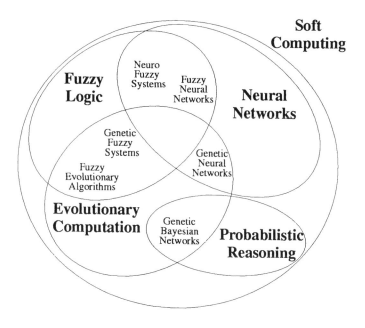

Fig. 3.1 Hybridisation in soft computing

- Neural-like systems fail to reflect a type of imprecision that is associated with the lack of sharp transition from the occurrence of an event to its non-occurrence. Fuzziness is then introduced in the model of the neuron to deal with this imprecision (Lee and Lee, 1975).

- The marriage of FL and NNs has a sound technical basis, because these two approaches generally attack the design of intelligent systems from quite different angles. NNs are low-level computational algorithms dealing with sensor data, and FL often deals with reasoning in a higher level (Bezdek, 1992).

- Neuro-fuzzy systems pursue the goal to design new architectures in which enhanced learning and improved knowledge representational capabilities go hand in hand (Pedrycz, 1992).

- Incorporating FL into the NN enables a system to deal with cognitive uncertainties in a way similar to humans. This involves an increase in required computation that can be readily offset by using fuzzy-neural systems, having the potential of parallel computation with high flexibility (Gupta and Rao, 1993).

- Commercial applications of FL obtained large success over the past years. This success resulted in products that solve tasks in a more intelligent and thereby efficient manner. The increased complexity of technical systems is handled by NNs that speed up the computation of complex rule-based systems (Takagi, 1994).

These ideas originate from two observations:

(1) FSs are neither capable of learning, adaptation or parallel computation, whereas these characteristics are clearly attributed to NNs.
(2) NNs lack flexibility, human interaction or knowledge representation, which lies at the core of FL.

Although the reasons to merge fuzzy and neural techniques into neuro-fuzzy systems are somehow specific, the general arguments for combining different methods apply to other kinds of hybrid systems as well.

Depending on which component dominates in a hybrid system, one roughly distinguishes between neuro-fuzzy systems and fuzzy-neural networks.

3.2.1.1 *Neuro-fuzzy systems*

In neuro-fuzzy systems, the dominant component is the FS, which is therefore regarded as an FS with the capability of neural adaptation (Nauck, Klawoon, and Kruse, 1997).

One type of neuro-fuzzy systems applies the learning techniques to the membership function shapes by considering the usual fuzzy inference system, as seen in Sec. 1.2.2. A learning or tuning process is targeted at parameterised fuzzy sets in the rule antecedents and consequents. Ichihashi and Tokunaga (1993) and Nomura, Hayashi, and Wakami (1992a) proposed learning schemes based on gradient descent and adapting Gaussian membership functions and triangular fuzzy sets, respectively. The approach of Wang and Mendel (1992a) utilised Gaussian membership functions subject to an orthogonal least squares learning algorithm.

In some cases, FL plays the role of an interface between an expert that possesses the knowledge to solve a certain problem, and an NN that then processes and utilises this knowledge. The extracted knowledge is either used to train an NN or is transformed into a network of appropriate neural topology and weights. The systems are designed to directly incorporate the knowledge described by a fuzzy rule set with linear equations as consequents (Jang, 1992), or with consequents that are fuzzy expressions (Narazaki and Ralescu, 1993; Nauck and Kruse, 1993). These systems can refine the rules obtained from experts by applying learning to the NN. Usually, expert knowledge is viewed as a preliminary knowledge, whereas learning by the network is regarded as a subsequent tuning and optimisation process. The above approaches represent FSs that are enhanced by analytical learning and/or parallel processing abilities.

3.2.1.2 *Fuzzy-neural networks*

Fuzzy-neural networks are hybrid systems that more resemble an NN rather than an FS. In this case, the notion of imprecision described by means of fuzzy sets is applied to pattern recognition neural techniques as a meaningful way to reflect vaguely defined categories. Often, the imprecision of definitions is related to the complexity of categories, that can be addressed and resolved by using concepts from FL. Bezdek et al. (1992); Carpenter et al. (1992); Keller and Hunt (1985); Kuo et al. (1993); Simpson (1992, 1993); Zhang et al. (1994), proposed different NNs that utilise fuzziness at the level of inputs, outputs and internal states, to relax properties, conditions or restrictions of the problem at hand.

Another class of fuzzy-neural networks is composed of NNs that implement logical operators in their nodes, which enables an NN to process symbolic information (Hayashi, Czogala, and Buckley, 1992; Pedrycz, 1992; Kwan and Cai, 1994). Learning methods such as gradient descent applied to logical functions may cause convergence problems as the derivative of logical functions is no longer continuous (Pedrycz, 1991). The idea to augment NNs with fuzzy operators was studied by Gupta (1992).

In some cases, the fuzzy-neural system could be described as a rule-like NN, i.e., a network that topologically and conceptually is structured as a rule-based system with *"IF-THEN"* clauses. In these kinds of systems, a part of the NN (that plays the role of IF part) generates an activation value or a truth value for each THEN part. This value defines the membership function of the actual input to the domain of each rule, expressing these domains as fuzzy sets having the input space as universe of discourse. The IF part of the rules is obtained by applying (fuzzy) clustering techniques, using the generated clusters to tune (by applying a supervised learning method) the elements of the net that represent the IF part. Different blocks of the connectionist structure play the role of THEN parts, generating the output related with the corresponding rule. The output of the whole system is obtained by evaluating the outputs from each THEN part jointly with the truth value of the corresponding rule. The THEN parts of the rules are obtained through a process of supervised learning based on input-output data and on an optimisation function. The consequents (THEN part) may have a fuzzy structure or a numerical structure (Mamdani or TSK types), both cases were analysed by Horikawa, Furuhashi, and Uchikawa (1992). Other example of those kinds of systems is NARA (Neural networks designed on Approximate Reasoning Architecture) (Takagi and Hayashi, 1991; Takagi et al., 1992).

3.2.2 *Neural networks and evolutionary computation*

Recently, researches in soft computing analysed the combination of NNs with EC techniques. This case of hybridisation seems quite intuitive, since animal and human brains are designed by means of natural evolution. Two main applications of GAs to NNs have been proposed:

- to train the weights of the NN, and
- to learn the topology of the network.

3.2.2.1 *Genetic algorithms for training neural networks*

A GA is used to optimise the parameters of the learning method that adapts the synaptic weights of the neural net. Harp, Samad, and Guha (1989) proposed a GA to determine the optimal learning and momentum rates of the NN.

GAs perform a coarsely granulated, global search whereas the back-propagation constitutes a finer granulated local hill-climbing method. This observation motivates a search strategy in which the GA obtains a set of suitable initial weights which is then fine tuned by a gradient descent method (Kitano, 1990; McInerney and Dhawan, 1993).

The application of GAs for obtaining the weights of a NN became merely obsolete for two reasons. First of all, gradient methods in training feed-forward NNs became more efficient which reduces the need for evolutionary-based weight optimisation. The second reason for the declining interest in genetic optimisation of neural weights is the so-called *com-*

peting conventions problem (Withley, 1995). This problem emerges from the non-unique translation from phenotype to genotype. There are a large number of equivalent NNs (generated by permutations of neurons) that, although they produce an identical output, are generated from completely different chromosomes. Given a NN, if we simultaneously exchange the first and second neurons of the hidden layer and their connection weights, the obtained NN is functionally equivalent to the previous one, although its chromosome is entirely different. In fact, the more complex the topology of the network, the larger the number of permuted chromosomes that generate an equivalent NN. As the chromosomes in the population do not share a common *convention* on the role of hidden neurons, crossover in almost all cases result in chromosomes that demonstrate inferior performance compared to its parents. Different solutions to cope with this problem, as the one by Montana and Davis (1989), have been proposed.

Weiland (1991) and Whitley, Dominic, and Das (1991, 1993) used EAs to train NNs for reinforcement learning problems.

3.2.2.2 *Genetic algorithms for learning the topology of the network*

More closer to the scenario of natural evolution are approaches in which GAs evolve the topology of the NN.

Assuming a bounded number of hidden neurons, GAs are applied to find the network structure that minimises the number of training cycles (Miller, Todd, and Hedge, 1989; Whitley, Starkweather, and Bogart, 1990).

Angeline, Saunders, and Pollack (1994) applied an EA solely based on selection and mutation to simultaneously learn the weights and topology of a recurrent NN.

3.2.3 *Genetic algorithms and probabilistic reasoning*

Fusion of GAs with PR into hybrid systems has not been largely visible until now. GAs have been used to find the optimal structure of a Bayesian network for a given database of training cases (Larrañaga, Kuijpers, Murga, and Yurramendi, 1996; Larrañaga, Poza, Yurramendi, Murga, and Kuijpers, 1996). Recently, Bayesian networks for modelling GAs have been proposed that go by the name of Bayesian optimisation algorithms (Pelikan, Goldberg, and Lobo, 2000; Pelikan and Goldberg, 2000; Pelikan, Goldberg, and Cantu-Paz, 2000).

3.3   Integration of Evolutionary Algorithms and Fuzzy Logic

The principles and operation of EC and fuzzy computation have been broadly described in the two introductory chapters. At this point, we discuss those characteristics that are relevant if one attempts a synergistic combination of both approaches.

EAs provide robust search capabilities —global and local— in complex spaces. FSs present robust and flexible inference methods in domains subject to imprecision and uncertainty. The linguistic representation of knowledge allows a human to interact with an FS in an intuitive, seamless manner. Two main hybrid approaches have been analysed:

- fuzzy evolutionary algorithms, and
- genetic fuzzy systems (GFSs).

Fuzzy evolutionary algorithms are EAs whose inherent parameters such as fitness function and stopping criterion are fuzzified, thus taking advantage of a tolerance for imprecision in order to save computational resources. FSs are either used as a tool for dynamic control of the algorithm parameters or certain components —specially the crossover operator— of the EA are designed using FL techniques.

On the other hand, in a GFS, a GA evolves an FS by: tuning fuzzy membership functions, learning fuzzy rules or adapting the context.

3.3.1   *Fuzzy evolutionary algorithms*

A fuzzy evolutionary algorithm is an EA that uses FL techniques to adapt its parameters or operators to improve its performance. One distinguishes between:

- adaptive GAs that adapt control parameters such as mutation and crossover rates, population size, selective pressure, and
- GAs with fuzzified versions of the genetic operators.

A complete review of fuzzy genetic algorithms can be found in Herrera and Lozano (1998).

3.3.1.1 *Adaptation of genetic algorithm control parameters*

FLCs are used to dynamically compute optimal values for the GA parameters such as crossover and mutation rates. The objective of this self-adaptation is to adjust the GA optimisation process itself to the fitness landscape of the underlying problem. A GA expert formulates fuzzy control rules to adapt the GA parameters, which achieve an optimal balance between exploitation and exploration through the GA execution. The main idea is to use an FLC whose inputs are a combination of GA performance measures —regarding population diversity indices, among others (for an study of these kinds of measures, the interested reader can refer to (Herrera and Lozano, 1996))— and current control parameters from which it computes new settings for the control parameters subsequently used by the GA.

The behaviour of a GA resulting from the interaction between the genetic operators is fairly complex. Therefore, finding fuzzy rules that actually improve the performance of the GA online is not straightforward. The RB is either derived from some heuristics (Herrera and Lozano, 1996; Streifel, Marks, Red, Choi, and Healy, 1999; Voget and Kolonko, 1998; Xu and Vukovich, 1993), by a previous learning of the fuzzy rules (Lee and Takagi, 1993a; Lee and Takagi, 1994) or by some kind of meta-learning in which a second outer GA co-evolves the proper fuzzy rules for the inner GA parameter adaptation (Herrera and Lozano, 2001).

To conclude this section, we will show two examples of the FLCs involved in these kinds of fuzzy evolutionary algorithms. On the one hand, Xu and Vukovich (1993) designed two different FLCs to adapt the crossover and mutation rates, both of which considered the same two state variables, the *generation* being currently run by the GA and the actual *population size*. The decision table of both RBs is shown in Fig. 3.2.

	Population size		
Generation	*Small*	*Medium*	*Large*
Short	Medium	Small	Small
Medium	Large	Large	Medium
Large	Very Large	Very Large	Large

Fig. 3.2 Decision table of the rule bases considered by Xu and Vukovich

On the other hand, Herrera and Lozano (1996) also considered two two-input single-output FLCs to adapt the values of two different variables:

- a parameter establishing the frequency of application of each of the different crossover operators —with different exploratory and exploitative behaviours— considered by the GA, and
- a parameter establishing the selective pressure performed by the selection operator.

The state variables considered in this case were two different diversity measures, a genotypic and a phenotypic one, both of which were considered by both FLCs. The RBs were derived by a human expert following the behaviour of encouraging exploration (by augmenting the frequency of application of these kinds of crossover operators and decreasing the selective pressure) when the population diversity is low, and encouraging exploitation (by doing the opposite actions) when the diversity is high.

3.3.1.2   *Genetic algorithm components based on fuzzy tools*

Different fuzzified versions of genetic operators have been proposed, such as fuzzy connective-based crossover (Herrera, Lozano, and Verdegay, 1996, 1997; Herrera and Lozano, 2000) and soft genetic operators (Voigt, Mühlenbein, and Cvetković, 1995). As an example, we will show the operation mode of one of the crossover operators belonging to the former family, the *max-min-arithmetical crossover*.

If $C_v^t = (c_1, ..., c_k, ..., c_H)$ and $C_w^t = (c_1', ..., c_k', ..., c_H')$ are to be crossed, the following four offspring are generated

$$C_1^{t+1} = a \cdot C_w^t + (1 - a) \cdot C_v^t$$

$$C_2^{t+1} = a \cdot C_v^t + (1 - a) \cdot C_w^t$$

$$C_3^{t+1} \text{ with } c_{3k}^{t+1} = \min\{c_k, c_k'\}$$

$$C_4^{t+1} \text{ with } c_{4k}^{t+1} = \max\{c_k, c_k'\}$$

The min and max operators considered can be substituted for any pair of dual t-norm and t-conorm, thus generating different fuzzy connective-based crossover operators. On the other hand, the parameter a can be either a constant or a variable whose value depends on the age of the population. The resulting descendents are the two best of the four aforesaid offspring.

3.4 Genetic Fuzzy Systems

A GFS is an FS that is augmented with an evolutionary learning process. The most extended GFS type is the *genetic fuzzy rule-based system* (GFRBS), where an EA is employed to learn or tune different components of an FRBS. This section will be completely dedicated to these kinds of systems. Other kinds of GFSs include genetic fuzzy clustering systems, genetic fuzzy neural systems and genetic fuzzy decision trees, which are discussed in a separate chapter (Chapter 10).

3.4.1 *Genetic fuzzy rule-based systems*

As seen in Chapter 1, the two main tasks in the FRBS design process are:

(1) the design of the inference mechanism (Sec. 1.2.4), and
(2) the generation of the fuzzy rule set (KB or FRB) (Sec. 1.4).

One of the major drawbacks of FRBSs is that they are not able to learn, but require the KB to be derived from expert knowledge. The key point is to employ an evolutionary learning process to automate the FRBS design.

The automatic definition of an FRBS can be seen as an optimisation or search problem, and GAs are a well known and widely used global search technique with the ability to explore a large search space for suitable solutions only requiring a simple scalar performance measure. In addition to its ability to find near optimal solutions in complex search spaces, the generic code structure and independent performance features of GAs make them suitable candidates to incorporate *a priori* knowledge. In the case of FRBSs, this *a priori* knowledge may be in the form of linguistic variables, fuzzy membership function parameters, fuzzy rules, number of rules, etc (see Secs. 1.4.2 and 1.4.3). These capabilities extended the use of GAs in the development of a wide range of approaches for designing FRBSs over the last few years.

Figure 3.3 illustrates this idea. It is important to notice that the genetic learning process aims at designing or optimising the KB. Consequently, a GFRBS is a design method for FRBSs which incorporates evolutionary techniques to achieve the automatic generation or modification of the entire or part of the KB.

From the view point of optimisation, the problem of finding an appropriate KB to solve an specific problem, is that of defining a parameterised KB

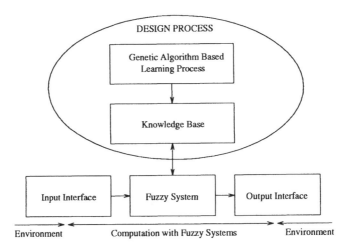

Fig. 3.3   Genetic fuzzy rule-based systems

where a set of parameters describes the fuzzy rules and fuzzy membership functions, and obtaining a suitable set of parameter values according to the optimisation criterion. The KB parameters constitute the optimisation space, the phenotype space, which has to be transformed into a suitable genetic representation, the genotype space. As seen in Chapter 2, in order for the EA to search the genotype space, it requires some mechanism to generate new variants from the currently existing candidate solutions. The objective of the search process is to maximise or minimise a fitness function that describes the desired behaviour of the system.

In summary, the genetic process is the result of the interaction between the evaluation, selection and creation of genetically encoded candidate solutions, which represent the contents of the KB of an FRBS. The following section describes the key features of a genetic learning process for FRBSs. The final section in this chapter discusses a specific aspect of GFRBSs, the so-called *cooperation versus competition problem*.

3.4.2   *Defining the phenotype space for a genetic fuzzy rule-based system*

The objective of evolutionary learning of an FRBS is to find a KB such that the resulting FS solves a given problem. This section describes possible structures and parameters of the optimisation space for GFRBSs.

The first step in designing a GFRBS is to decide which parts of the KB are subject to optimisation by the EA. The KB of an FRBS does not have an homogeneous structure but is rather the union of qualitatively different components. As described in Sec. 1.2 (and shown in Fig. 1.1), the KB of a descriptive Mamdani-type FRBS is comprised of two components: a DB, containing the definitions of the scaling factors and the membership functions of the fuzzy sets associated with the linguistic labels, and an RB, constituted by the collection of fuzzy rules.

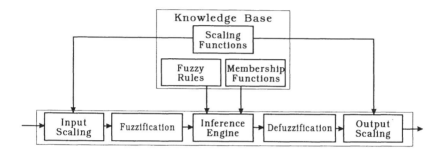

Fig. 3.4 Structured knowledge base

Figure 3.4 shows the structure of a KB integrated in a linguistic FRBS. Compared to Fig. 1.1, it contains two additional elements, the normalisation and denormalisation interfaces for scaling the input and output variables. The tuning/learning process adapts the contents of either the DB, the RB or both. The search space in which the EA operates looks different depending on which part of the KB is subject to optimisation.

It is important to distinguish between tuning and learning problems. Tuning is more concerned with optimisation of an existing FRBS, whereas learning constitutes an automated design method for fuzzy rule sets that starts from scratch. Tuning processes assume a predefined RB and have the objective to find a set of optimal parameters for the membership and/or the scaling functions. Learning processes perform a more elaborated search in the space of possible RBs or whole KBs and do not depend on a predefined set of rules.

The decision about which part of the KB to adapt depends on two conflicting objectives:

- A search space of smaller dimension results in a faster and simpler learning process, but the achievable solutions might be sub-optimal.
- A larger, complete search space that comprises the entire KB and has a finer granularity is more likely to contain optimal solutions, but the search process itself might become prohibitively inefficient and slow.

In summary, there is an obvious trade-off between the completeness and granularity of the search space and the efficiency of the search.

3.4.2.1 *Genetic tuning of the data base*

The tuning of the scaling functions and fuzzy membership functions is an important task in FRBS design. The parameterised scaling functions and membership functions are adapted by the EA according to a fitness function that specifies the design criteria in a quantitative manner.

Tuning the scaling functions

Scaling functions applied to the input and output variables of an FRBS (as shown in Fig. 3.4) normalise the universes of discourse in which the fuzzy membership functions are defined. From a control engineering point of view, these scaling functions play the role of gains associated with the variables. From a knowledge engineering point of view, they contain context information as translate relative semantics into absolute ones. Usually, the scaling functions are parameterised by a single scaling factor or a lower and upper bound in case of linear scaling and a contraction/dilatation factor in case of non-linear scaling. These parameters are adapted such that the scaled universe of discourse better matches the underlying variable range.

Tuning the membership functions

In the case of tuning membership functions, an individual represents the entire DB as its chromosome encodes the parameterised membership functions associated to the linguistic terms. Triangular membership functions are usually encoded by their left, centre and right point, whilst Gaussian membership functions by their centre and width.

The structure of the chromosome is different for FRBSs of the descriptive or the approximate type (see Sec. 1.2). When tuning the membership functions in a linguistic model, the whole fuzzy partitions are encoded into the chromosome and they are globally adapted to maintain the global semantic in the RB. On the other hand, tuning the membership functions of an approximate model is a particular instantiation of KB learning since the rules are completely defined by their membership functions instead of referring to linguistic terms in the DB.

The genetic tuning of scaling factors and membership functions is analysed in detail in Chapter 4.

3.4.2.2 *Genetic learning of the rule base*

Genetic learning of the RB assumes a predefined set of fuzzy membership functions in the DB to which the rules refer to by means of linguistic labels. The EA adapts the RB, either working with chromosomes that describe a single fuzzy rule or an entire RB. The RB is either represented by a relational matrix, a decision table or a list of rules. Each of the three possible representations for GFRBSs involves the issues of completeness, consistency and complexity of the RB (see Sec. 1.4.6).

Genetic learning of the RB only applies to descriptive FRBSs, as in the approximate approach adapting rules is equivalent to modify the membership functions.

3.4.2.3 *Genetic learning of the knowledge base*

As genetic learning of the KB deals with a heterogeneous search space, it encompasses different genetic representations such as variable length chromosomes, multi-chromosome genomes and chromosomes encoding single rules instead of a whole KB. The computational cost and efficiency of the search grow with increasing complexity of the search space. An GFRBS that encodes individual rules rather than entire KBs is one possible means to maintain a flexible, complex rule space in which the search for a solution remains feasible and efficient.

3.4.2.4 *A phenotype space of rules or rule bases/knowledge bases*

Genetic learning processes can operate in two different classes of phenotype spaces: the space of rules and the space of entire RBs or KBs. In case each chromosome represents an individual rule, the population as a whole constitutes the solution, namely the optimal set of rules. The rules that form the RB are either evolved simultaneously as in the so-called Michigan approach or are progressively added to the list as in the iterative rule learning approach. In the second, so-called Pittsburgh approach, individuals represent a complete RB or KB. The result of the evolutionary process is a population of KBs, of which the best one is selected as the final solution.

The principles of genetic learning approaches are analysed in Chapter 5, while the application of the three particular variants —Michigan, Pittsburgh and iterative approach— to the design of GFRBSs i studied in detail in Chapters 6, 7 and 8.

3.4.3   *From phenotype to genotype spaces*

The previous section discussed the general structure that is subject to adaptation, namely the components that form the KB, independent of the learning technique itself. The following two sections focus on the learning method, in particular on how the GA solves the optimisation problem. That relates to some of the key issues of evolutionary optimisation as described in the beginning of Sec. 2.2.1.

The optimisation problem corresponds to define the KB of an FRBS. The information describing a candidate KB is translated into a genetic representation to which genetic operators such as recombination and mutation are applied.

This mapping from phenotype to genotype can be realised in two different ways:

- Designing a coding function that maps to the standard binary code and apply the garden-variety genetic operators to this representation.
- Choosing a non-standard representation that closely matches the KB structure and develop specialised genetic operators that take the underlying phenotype structure into account.

Choosing a vanilla flavour type of GA offers the advantage that a large amount of existing theoretical and empirical knowledge reduce the effort of tuning the GA, e.g. parameter setting. The drawback is that a standard coding is sub-optimal as it introduces non-linearities and discontinuities that complicate the optimisation process. The simple GA can be extended such that the metrics of genotype and phenotype space become more conform (see Secs. 2.2.3 and 2.2.4). Such a representation often improves the efficiency of optimisation but requires the conception of specialised genetic operators which guarantee that the generated offspring genotype corresponds to a valid, feasible phenotype. Finding an adequate, coherent genetic representation remains a challenge in the design of any GA as it is of crucial importance to the success of evolutionary optimisation.

3.4.4 *Generating new genetic material*

As discussed in Sec. 2.2, new genetic variants are generated by means of selection, reproduction and genetic operators applied to the genotype. The choice of genetic operators depends on the phenotype (see Sec. 3.4.2) and genotype structure (see Sec. 3.4.3) and the coding function that maps one into the other.

The genetic operators and the chromosome structure have to be compatible with each other such that the GA searches the phenotype space in an efficient manner. Consequently, genetic operators, chromosome structure and the coding function have to be designed in conjunction such that they are compatible to each other.

Moreover, the definition of genetic operators not only depends on the genetic representation but is also conditioned by the type of search process used by the GFRBS. A successful search strategy has to take two different aspects into account: the exploitation of the better solutions and the exploration of unknown regions of the search space. The set of genetic operators has to achieve an adequate balance between exploration and exploitation which is crucial to the success of evolutionary learning.

3.4.5 *Evaluating the genetic material*

One distinguishes between two different groups of optimisation problems in FRBSs. The first group contains those problems in which optimisation targets the isolated behaviour of the FRBS, whereas the second one refers to those problems in which optimisation aims at the global performance of an overall system regulated by the FRBS. The first group encompasses problems such as modelling, classification, prediction and identification problems in general. In this case, the optimisation process searches for an FRBS able to reproduce the behaviour of a given target system. The most prominent example for the second class of problems is fuzzy control, with the objective to tune or learn an FRBS such that the controlled plant demonstrates a desired behaviour.

In identification problems, the performance index is usually based on error measures that characterise the difference between the desired output and the actual output of the system. In control problems, the performance index is either based on information describing the desired behaviour of the closed loop system, or describing the desired behaviour of the isolated FLC itself. The second situation is closely related to identification problems. In the first case, the performance index is based on some cost function defined over the state and control variables, rather than the error between a nominal and the output of the FRBS. Transportation problems are another example in which performance can only be defined in terms of the overall behaviour of the process regulated by the FS.

These aspects will be clarified in Chapter 11, where different applications of GFRBSs, belonging to the different said groups, are analysed.

3.4.6   *The cooperation versus competition problem*

There is one major problem that appears when designing FRBSs by means of EAs which has to be solved adequately in order to obtain accurate FRBSs from the design process. As with other system design processes, GFRBSs try to combine the main aspects of the design tool and of the system being designed —an EA/GA and an FRBS, in this case—:

- On the one hand, GFRBSs take one of the most interesting features of FRBSs into account, the interpolative reasoning they develop, which *is a consequence of the cooperation among the fuzzy rules composing the KB*. As seen in Chapter 1, the output obtained from an FRBS is not usually due to a single fuzzy rule but to the cooperative action of several fuzzy rules that have been fired because they match the system input to any degree. This characteristic plays a key role in the high performance of FRBSs.
- On the other hand, the main feature of an EA is considered, which is the *competition induced among the population members representing possible solutions to the problem being solved*, which allows us to obtain better ones. In this case, this characteristic is due to the mechanisms of the natural selection in which the EA is based, as seen in Chapter 2.

Therefore, since a GFRBS combines both said features, it works by *inducing competition to get the best possible cooperation*. This seems to be a very interesting way to solve the problem of designing an FRBS, because the different members of the population compete among them to provide a final solution (KB) presenting the best cooperation among the fuzzy rules composing it. The problem is to find the best possible way to put this into effect. This problem is referred to as *cooperation versus competition problem* (CCP) (Bonarini, 1996b).

The difficulty of solving the introduced problem directly depends on the genetic learning approach followed by the GFRBS: Michigan, Pittsburgh or iterative rule learning. The Pittsburgh approach solves the CCP by evolving a population of RBs instead of individual rules, thus implicitly evaluating the cooperation of fuzzy rules (see Chapter 7). This method is very time consuming as in every generation each member of the population is evaluated separately and is therefore not suitable for online learning tasks.

In the Michigan approach, the population consists of competing rules (see Chapter 6). However, competition is local in the sense that only rules with the same antecedent compete with each other (Bonarini, 1996b). The population is partitioned into disjoint sub-populations according to the similarity of rule antecedents. Rules within the same sub-population are subject to local competition, whereas the best rules taken from complementary sub-populations cooperate on a global level.

The iterative rule learning approach solves the CCP by dividing the learning process into two separate stages (see Chapter 8). In the generation stage, rules merely compete with each other similar to the Michigan approach. The method incrementally adds the current best rule to an intermediate fuzzy rule set, and the examples that are covered by this rule are removed from the training set. The second, so-called post-processing stage, stimulates cooperation of fuzzy rules, by removing redundant or unnecessary fuzzy rules from the previously generated fuzzy rule set or by adjusting the fuzzy membership functions.

Chapter 4

Genetic Tuning Processes

The objective of a genetic tuning process is to adapt a given fuzzy rule set such that the resulting FRBS demonstrates better performance. The following components of the KB are potential candidates for optimisation (Zheng, 1992):

- *DB components*: scaling functions* and membership function parameters, and
- *RB components*: "IF-THEN" rule consequents.

This chapter discusses the tuning of scaling functions and membership function parameters by an EA under the assumption of a predefined RB that is not subject to adaptation. First, we analyse tuning of FRBSs in general before taking a closer look at genetic tuning processes applied to the DB.

Notice that adjusting scaling and membership function parameters modifies the *context* in which the FS operates. A substantial modification of the membership functions changes the FRBS behaviour so profoundly that it eventually requires the complete reformulation of the RB. As we assume a fixed, predefined RB, the tuning process has to be constrained such that the current rule consequents remain valid for the new membership function and scaling parameters. Approaches that adjust the RB and the DB parameters simultaneously go by the name *DB learning* rather than *DB tuning* and are the subject of Chapter 9.

---

*The initial approaches used a scaling factor K (Procyk and Mamdani, 1979), via the function $f(x) = k \cdot x$.

4.1   Tuning of Fuzzy Rule-Based Systems

4.1.1   *Tuning of scaling functions*

The scaling functions map the input and output variables onto the universe of discourse over which the fuzzy sets are defined.

From a linguistic point of view, the scaling function can be interpreted as a sort of context information. While the membership functions describe the relative semantics (context-independent) of the linguistic variables, scaling and membership functions together establish the absolute semantics of the linguistic variables (context-dependent through the scaling functions).

The initial tuning approaches used a scaling factor (Procyk and Mamdani, 1979), assuming that the membership functions were scaled by the same ratio. Notice that scaling an input or output variable has a macroscopic impact as it affects every rule in the RB. A single, modified membership function has a medium size effect as it only affects a particular column or row of the rule matrix. Finally, a modified rule consequent has a small size effect that only affects one entry of the rule matrix. It is important to keep this hierarchy of parameter sensitivity in mind when considering the order in which DB and RB parameters are tuned (Zheng, 1992).

One distinguishes among linear and non-linear scaling functions, which are well known in classical control theory.

(1) *Linear context*: A linear scaling function is of the form

$$f(x) = \alpha \cdot x + \beta$$

The scaling factor α enlarges or reduces the operating range, which in turn decreases or increases the sensitivity of the controller in respect to that input variable, or the corresponding gain in case of an output variable. The parameter β shifts the operating range and plays the role of an offset to the corresponding variable.

(2) *Non-linear context*: A common non-linear scaling function for a variable that is symmetric with the respect to the origin is of the form:

$$f(x) = \text{sign}(x) \cdot |x|^{\alpha}$$

Non-linear scaling increases ($\alpha > 1$) or decreases ($\alpha < 1$) the relative sensitivity in the region around the origin and has the opposite effect at the boundaries of the operating range.

The effect of scaling functions on the performance characteristics such as overshoot, settling time, rise time of PD, PI, and PID-like FLCs has been analysed soon after fuzzy control was introduced (Procyk and Mamdani, 1979). Based on this analysis, different tuning methods to obtain the optimal scaling parameters have been proposed. Zheng (1992) described a heuristics to tune PI-like FLCs. Daugherity, Rathakrishnan, and Yen (1992) tuned, by a supervisory FLC, the input scaling factors of a PI-like FLC that regulates a gas-fired water heater. The scaling factors were adjusted according to the difference between the observed and the desired performance measures such as overshoot, rise time and oscillation. The meta fuzzy rules used for the tuning were of the form:

$$IF \ performance\text{-}measure \ is \ A_i \ THEN \ \Delta s_e is \ B_i,$$

where Δs_e is a change of the scaling factor s_e for the error e, and A_i, B_i are fuzzy sets for the corresponding performance measure or scaling factor. A method for tuning scaling factors using cross-correlation of the controller input and output was proposed by Palm (1995). Burkhardt and Bonissone (1992) defined a non-linear and discontinuous scaling function, adapting its parameters by a gradient descent method. Pedrycz, Gudwin, and Gomide (1997) proposed non-linear scaling functions implemented as NNs.

The following two subsections analyse the two linear and non-linear context approaches in more depth.

4.1.1.1 *Linear contexts*

A linear scaling function of the form

$$f(x) = \alpha \cdot x + \beta$$

is parameterised by the scaling factor α and the offset β.

Scaling allows it to define the fuzzy membership functions over a normalised universe of discourse. Figure 4.1 shows a possible fuzzy partition of the normalised variable *body height*.

This variable is normalised by applying the parameterised scaling function contained in Fig. 4.2, which incorporates the context information through the parameters *min* and *max*. It is important to notice, from Fig. 4.2, that values smaller than *min* are normalised to 0, and those larger

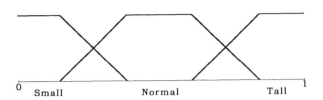

Fig. 4.1   Fuzzy partition for the normalised variable body-height

than *max* are normalised to 1. Table 4.1 proposes possible values for *min* and *max*.

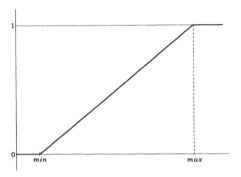

Fig. 4.2   Parameterised scaling function for the variable body-height

Table 4.1   Scaling function parameters for different contexts

Context	*min*	*max*
Pygmy tribe	1 m	1.50 m
NBA players	6 ft	7 ft
Kindergarten boys	0.60 m	1.20 m

From a control engineering perspective, the scaling functions represent context information. In some cases, this information is related to the physical properties or dimensions of the controlled system or process, including restrictions imposed due to the measurement acquisition or actuators. When designing a control system for driving a car, it is a good idea to normalise the variable *car speed* to the operating ranges [0, *maximum speed of the car*] (physical property) or [0, *speed limit of the road*] (restriction). In other cases, the scaling functions are not directly related to physical properties, but represent information affecting the overall behaviour of the controlled system. To normalise the variable *car acceleration*, a wider range of values is employed under race driving conditions than for a more comfortable operation in normal traffic. In this case, the scaling functions are mainly conditioned by the desired behaviour of the controlled system and not so much by physical considerations, although the engine power obviously imposes a physical limit on the maximal acceleration.

Therefore, a linear scaling function could be interpreted as gains associated with the variables (from a control engineering point of view) or as context information that translates relative semantics into absolute ones (from a knowledge engineering point of view). The latter perspective is the most common one in FRBSs, whether fixed, tuned or learned scaling functions are employed.

4.1.1.2 *Non-linear contexts*

The main disadvantage of linear scaling is the fixed relative distribution of the membership functions. Non-linear scaling provides a remedy to this problem as it modifies the relative distribution and changes the shape of the membership functions.

The effects of modifying contexts (scaling functions) and semantics (membership functions) of an FRBS are tightly coupled. This is reflected in the idea of considering context and semantic information as a single block for the purpose of adaptation. Assume the membership functions and scaling functions are already parameterised in one way or another. The main question is how to reduce the large number of parameters while keeping the *reachable* portion of the space of solutions as large as possible.

To analyse the interaction between scaling and membership functions, it is useful to make the following observation:

Given two fuzzy partitions $A = \{A_1, \dots, A_n\}$ and $B = \{B_1, \dots, B_n\}$ of the same cardinality, defined respectively on the intervals $X = [x_{min}, x_{max}]$ and $Y = [y_{min}, y_{max}]$, and such that their membership functions:

- are normal and convex,
- are continuous,

- sum up to one,

$$\sum_{i=1}^{n} \mu_{A_i}(x) = 1, \quad \forall x \in X$$

and

$$\sum_{i=1}^{n} \mu_{B_i}(y) = 1, \quad \forall y \in Y$$

- the grades of membership are greater than zero for at most two elements of the fuzzy partition, at any point of the universes of discourse

$$\mu_{A_i}(x) > 0,\ \mu_{A_j}(x) > 0,\ i \neq j \ \Rightarrow \mu_{A_k}(x) = 0, \quad \forall k \neq i, j$$

there is an increasing real function $f : [x_{min}, x_{max}] \in X \to Y$ such that

$$\forall x \in X,\ \mu_{A_i}(x) = \mu_{B_i}(f(x)), \quad \forall i = 1, \dots, n$$

Consequently, it is possible to argue that an FRBS using normalised membership functions (e.g., partitions composed of isosceles triangles regularly distributed, Fig. 4.3) can mimic any FRBS using partitions that satisfy the previous restrictions, by only applying an adequate family of scaling functions $\{f_{i1}, \dots, f_{iN}, f_{o1}, \dots, f_{oM}\}$ to its N input and M output variables.

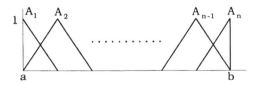

Fig. 4.3   An example of fuzzy partition

The conditions imposed are not highly restrictive for fuzzy partitions defined in a control context (and the other FRBS application fields), on the contrary they are rather common conditions. Under the above assumptions, the question of adjusting fuzzy partitions can be reduced to that of learning scaling functions. The problem is the large number of parameters that the scaling function requires to be adequately defined.

It has been previously said that using a fixed set of normalised membership functions (e.g., partitions composed of isosceles triangles regularly distributed, Fig. 4.3) with an appropriate family of parameterised non-linear scaling functions, it is possible to generate a wide range of fuzzy partitions (any fuzzy partition fulfilling a set of restrictions). Different non-linear scaling functions are proposed throughout the literature (Pedrycz, Gudwin, and Gomide, 1997; Gudwin, Gomide, and Pedrycz, 1997; Gudwin, Gomide, and Pedrycz, 1998; Magdalena, 1997).

Our goal is to define a family of scaling functions that produces the widest range of fuzzy partitions with the smallest number of parameters. A small number of parameters reduces the complexity of the search, while a wide range of possible fuzzy partitions increases the approximation accuracy of the FRBS.

A possible parameterisation of the non-linear scaling function are the four parameters $\{V_{min}, V_{max}, S \text{ and } a\}$, proposed by Magdalena and Velasco (1997):

- V_{min} and V_{max} are real numbers defining the upper and lower limits of the universe of discourse,
- a is a real number, greater than zero, that produces the non-linearity through the parameterised function $(f : [-1, 1] \rightarrow [-1, 1])$

$$f(x) = \text{sign}(x) \cdot |x|^a, \text{ with } a > 0,$$

an odd function that maintains the extremes of the interval unchanged in any case ([-1,0], [-1,1] or [0,1]). Figure 4.4 shows the function graphs for values $a = 2$ and $a = \frac{1}{4}$.
- S is a parameter in $\{-1, 0, 1\}$ to distinguish between non-linearities with symmetric shape (lower granularity for middle or for extreme values, Fig. 4.5) and asymmetric shape (lower granularity for lowest or for highest values, Fig. 4.6).

In this case, the obtained fuzzy partitions cover the range from lower granularity for middle values of the variable to lower granularity for extreme values (including homogeneous granularity) and from lower granularity for lowest values to lower granularity for highest values (including homogeneous granularity too).

The normalised fuzzy partition is shown in Fig. 4.7. The scaling or normalisation process includes three steps:

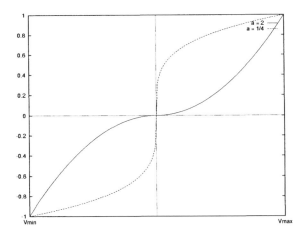

Fig. 4.4   Non-linear scaling functions

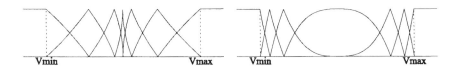

Fig. 4.5   Denormalized fuzzy partition with S=0 (a=1/2, a=3)

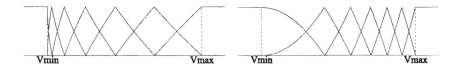

Fig. 4.6   Denormalized fuzzy partition with S=1 (a=1/2, a=2)

(1) The first step based on the parameters $\{V_{min}, V_{max}$ and $S\}$ produces a linear mapping from $[V_{min}, V_{max}]$ interval to $[-1,0]$ (when $S = -1$), to $[-1,1]$ (when $S = 0$) or to $[0,1]$ (when $S = 1$).

(2) The second step scales the intervals $[-1,0]$, $[-1,1]$ or $[0,1]$ in a non-linear way by means of the parameter a.

(3) Finally, a second linear mapping transforms the resulting interval ($[-1,0]$, $[-1,1]$ or $[0,1]$) into $[-1,1]$.

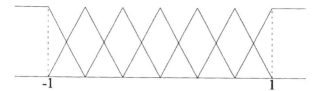

Fig. 4.7   Normalised fuzzy partition

The overall scaling transformation (for any value of S) is a non-linear mapping from the interval $[V_{min}, V_{max}]$ to the interval [-1,1].

Notice that non-linear scaling functions may substantially change the membership functions to an extent that makes it necessary to refine the RB.

In fact, these scaling function approaches can be regarded as global tuning procedures embedded in evolutionary learning processes as those presented in Chapters 7 and 9. In contrast, tuning of membership functions discussed in the following section constitutes a local adaptation process that assumes a static, predefined RB.

4.1.2   *Tuning of membership functions*

The second method of parameter adaptation tunes the membership functions to which the linguistic labels in the fuzzy rules refer. Assume a predefined RB that represents the structural knowledge about the system to be modelled expressed in a linguistic way. The performance of the FRBS can be improved by properly tuning the fuzzy sets without having to change the RB itself. Figure 4.8 shows two possible membership function distributions before and after the tuning that result in a different FRBS behaviour.

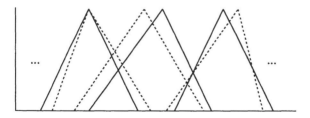

Fig. 4.8   Alternative membership function distributions

Valente de Oliveira (1999) presented a study of semantically driven conditions to constraint the optimisation process of an FRBS in such a way that the resulting membership functions can still represent human readable linguistic terms. The set of semantic properties (requirements of semantic integrity as were pointed out by Pedrycz and Valente de Oliveira (1995)) includes a moderate number of membership functions, distinguishability, normality, natural zero positioning, and coverage. Some of the semantic properties of membership functions are easily satisfied by an *a priori* selection of certain membership function features (Valente de Oliveira, 1999). This is the case for the moderate number of membership functions, the use of normalised membership functions or by choosing the core of a membership function to be positioned at zero. However, the optimisation process must monitor and explicitly maintain the coverage and distinguishability properties.

Tuning of membership functions can be an effective means of tuning an FRBS, applying different optimisation techniques such as pure gradient descent (Glorennec, 1991; Nomura, Hayashi, and Wakami, 1991; Vishnupad and Shin, 1999), a mixture of back-propagation and mean least squares estimation as in ANFIS (Jang, 1993) or NEFCLASS (with an NN acting as a simple heuristic) (Nauck and Kruse, 1997), or NN-based on gradient descent method (Shi and Mizumoto, 2000), and simulated annealing (Benítez, Castro, and Requena, 2001; Garibaldi and Ifeator, 1999; Guêly, La, and Siarry, 1999).

A further distinction is to be made between membership function tuning in Mamdani and in TSK FRBSs. TSK FSs are tuned in a different way due to parameters in the TSK rule consequents. The following sections present some examples of evolutionary tuning of membership functions in TSK and Mamdani FRBSs.

4.1.3 *Tuning of fuzzy rules*

The third tuning approach adapts the fuzzy label in the THEN-part of rules and the resulting systems are called self-organising controllers (Procyk and Mamdani, 1979).

The basic idea is to identify rules in the RB which cause poor controller performance and to replace them with better rules. Replacing a single rule consequent has a very local effect, as the control behaviour only changes in the fuzzy subspace defined by the rule antecedent.

As mentioned by Bonissone, Khedkar, and Chen (1996), tuning individual rules has a microscopic scope, compared to the macroscopic effects caused by tuning the scaling factors and the medium-size effects of tuning membership functions.

Many different approaches have been presented for the optimisation of the RB by GAs as (Chan, Xie, and Rad, 2000) among others. All of them code an RB as a chromosome and initialise the population either randomly and/or based on a predefined set of rules. These kinds of genetic processes are subsumed within the genetic learning of RB based on the Pittsburgh approach, presented in Chapter 7.

4.2   Genetic Tuning of Scaling Functions

The scaling function is defined by means of two parameters, a scaling factor and an offset. Scaling corresponds to a linear map that transforms the interval [a,b] into a normalised reference interval (e.g., [0,1]). The use of a single scaling factor maps the interval [-a,a] to a symmetric reference interval (e.g., [-1,1]). This allows us to work with normalised universes of discourse in which the fuzzy membership functions are defined. This kind of context is the most broadly applied, using fixed, tuned or learned scaling functions.

The genetic process adapts the scaling parameters such as upper and lower limits of the operating range or scaling factor for input and output variables (Bonissone, Khedkar, and Chen, 1996; Hanebeck and Schmidt, 1996; Li, Chan, Rad, and Wong, 1997; Magdalena, 1997; Ng and Li, 1994). The partial coding of the parameters of each individual variable are concatenated to form the chromosome. In some approaches, the scaling parameters only constitute a fraction of the chromosome, which in addition contains membership function parameters and the RB itself.

Bonissone, Khedkar, and Chen (1996) presented a comparative study of manual versus genetic tuning, in which a GA successfully tunes an FLC for velocity control of a freight train. The membership functions are tuned in a hierarchical way in the order of significance: scaling factors are adapted first for a coarse global tuning, followed by tuning of membership functions with a finer granulation. Both components could be tuned simultaneously (Hanebeck and Schmidt, 1996).

4.3   Genetic Tuning of Membership Functions of Mamdani Fuzzy Rule-Based Systems

The genetic tuning processes of membership functions in Mamdani-type FRBSs are different for descriptive and approximate fuzzy models (Sec. 1.2 and Fig. 1.8).

Early work in this area by C. Karr (1991, 1993) considered descriptive Mamdani-type FRBSs. The DB definition is encoded in the chromosome, which contains the concatenated parameters of the input and output fuzzy sets, which are either represented as piece-wise linear, i.e., triangular and trapezoidal, or as Gaussian membership functions.

This section analyses the genetic tuning processes according to the shape of the membership functions together with the coding representation in Sec. 4.3.1. Section 4.3.2 analyses genetic tuning processes from the perspective of the semantic, studying the constraints imposed to the genetic tuning methods. Finally, Sec. 4.3.3 describes two approaches in more detail.

The following preliminary considerations serve to clarify the discussion:

- Genetic tuning processes usually utilise the mean square error between the FRBS output and a training data set as their *fitness function*. Surmann, Kanstein, and Goser (1993) proposed to incorporate an additional entropy term into the fitness function that penalises a large number of simultaneously activated rules.
- We consider approaches that tune the membership functions independently of the learning of the RB. A simultaneous adaptation of the RB and the membership functions is possible, but the discussion of these genetic learning approaches is delayed until Chapters 6 to 9.
- Some approaches first employ a GA to generate the RB, before tuning the membership functions. This order of the tuning process is similar to that typically used by self-organising controllers (Burkhardt and Bonissone, 1992).

4.3.1   *Shape of the membership functions*

Two main groups of parameterised membership functions have been proposed and applied: piece-wise linear functions and differentiable functions.

4.3.1.1 *Piece-wise linear functions*

The most common parameterised fuzzy sets employ piece-wise linear membership functions, with the following choices (Fig. 4.9):

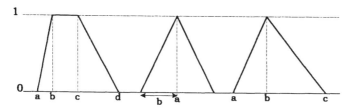

Fig. 4.9 Examples of parameterised piece-wise linear functions

- *Isosceles triangular shapes* (Fig. 4.9 centre). The triangular membership functions at the boundaries are right-angled, the inner ones are isosceles. The inner membership functions are parameterised by the centre and width of the triangle or equivalently its left and right most point, the boundary membership functions by the width of the triangle or a single point only as the other point coincides with the limits of the operating range. The set of all membership functions is either encoded using binary genes (Karr, 1991) or real coding (Park, Kandel, and Langholz, 1994), in which each tuning parameter can assume a discrete or continuous value within a given range.
- *Asymmetric triangular shapes* (Fig. 4.9 right). Cordón and Herrera (1997c) employed an RCGA to tune asymmetric triangular membership functions characterised by their left, centre and right points, which vary in predefined intervals of adjustment.

  Kinzel, Klawoon, and Kruse (1994) proposed a special coding of fuzzy sets in which fuzzy partitions were described by discrete, step-wise constant membership functions. Each gene represented the constant membership value of a fuzzy set in a certain interval.

  Glorennec (1997) employed an integer representation in which each membership function parameter can assume one out of five possible discrete values. In combination with a small population size, the approach is computationally efficient enough to be used in a robotic application.

- *Trapezoidal shapes* (Fig. 4.9 left). The left, right and the two central points parameterise the membership function. Similar to the case of triangular membership functions, binary genes (Bonissone, Khedkar, and Chen, 1996; Karr and Gentry, 1993) or real genes (Fathi-Torbaghan and Hildebrand, 1997; Herrera, Lozano, and Verdegay, 1995) are used to encode the parameters.

4.3.1.2 *Differentiable functions*

Gaussian, radial basis and sigmoidal (Fig. 4.10) are examples of parameterised differentiable membership functions. These membership functions are broadly applied in neuro-fuzzy systems (Magdalena, 1996) and GFRBSs (Fig. 4.10, left).

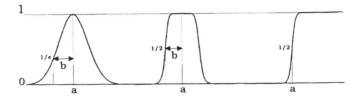

Fig. 4.10   Examples of parameterised differentiable functions

It is possible to code the centre and width of the Gaussian functions (Farag, Quintana, and Lambert-Torres, 1998; Hanebeck and Schmidt, 1996; Surmann, Kanstein, and Goser, 1993) or only the centres (Gurocak, 1999).

To translate the parameters of the function into genetic information, different approaches can be used, as a binary code (Gurocak, 1999; Surmann, Kanstein, and Goser, 1993), real-valued numbers (Hanebeck and Schmidt, 1996), or decimal-integer strings (Farag, Quintana, and Lambert-Torres, 1998).

4.3.2 *Scope of the semantics*

4.3.2.1 *Tuning of descriptive Mamdani fuzzy rule-based systems*

As seen in Chapter 1, the fuzzy rules in a descriptive FRBS use linguistic labels to refer to a common set of membership functions defined in the DB. Therefore, it can be said that the system uses a global semantic.

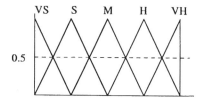

Fig. 4.11   Strong fuzzy partition with triangular membership functions

We can distinguish among different genetic tuning approaches for descriptive FSs, depending on the constraints imposed by the coding of the fuzzy partition.

- *Strong fuzzy partition* (Fig. 4.11). In a strong fuzzy partition, the membership degrees form a partition of unity:

$$\sum_{i=1}^{n} \mu_{A_i(x)} = 1, \quad \forall x \in X_A$$

  In case of triangular membership functions, two neighbouring fuzzy sets intersect half-way between their centre points at a height of 0.5. Figure 4.11 shows an example of a strong partition for a linguistic variable with five labels.

  The normalisation constraint is easily satisfied by only coding the modal points of the membership functions: one point for triangular shapes (Glorennec, 1997) and two points for trapezoidal shapes. The latter partitions the universe of discourse into intervals which alternate between core and overlapping regions (Bonissone, Khedkar, and Chen, 1996). The closest modal points of the two neighbouring fuzzy sets define the support of a fuzzy set.

- *Semantics ordering relation*. The order of the linguistic labels is maintained. Considering a_i (respectively, a_j) as the modal value of A_i (respectively, A_j), the ordering relation requires that

$$\forall i < j, \ a_i < a_j$$

  Approaches using this constraint are those proposed by Cordón and Herrera (1997c); Fathi-Torbaghan and Hildebrand (1997); Glorennec (1996); Gurocak (1999); Hanebeck and Schmidt (1996). Coding the relative distance between neighbouring fuzzy sets rather than the absolute location preserves the ordering relation. In that case, the optimisation process needs to verify that the modal points of the outer fuzzy sets do not exceed the domain boundaries.

- *Completeness.* The σ-completeness property (see Sec. 1.4.6.1) requires that for each point x, there exists a fuzzy set i that has a membership degree $\mu_{A_i}(x)$ larger than σ:

$$\forall x \in U, \;\; \exists j \text{ such that } \mu_{A_j}(x) \geq \sigma > 0$$

with (A_j) being a fuzzy set defined on the domain U of x, and σ a given completeness degree.

Cordón and Herrera (1997c) introduced a variant of the previous expression, the τ-completeness property, to be used in a genetic tuning process for Mamdani-type fuzzy rule sets, which will be introduced in Sec. 4.3.3.1.

Finally, there is a class of *non-semantic restrictive* tuning approaches that do not impose any constraints on the membership function parameters, other than their potential adjustment intervals resulting from the particular coding.

4.3.2.2 *Tuning of approximate Mamdani fuzzy rule-based systems*

Approximate FSs employ a local semantic, in which each fuzzy rule defines its own fuzzy sets rather than referring to common linguistic variables (see Sec. 1.2.6.2). Therefore, the chromosome concatenates the parameters of the individual rule membership functions.

Herrera, Lozano, and Verdegay (1995) proposed an approximate genetic tuning approach, directly tuning each rule used by the FRBS. It uses an RCGA (see Sec. 2.2.4) in which trapezoidal membership functions are encoded by the four characteristic points. Each rule is represented by the concatenation of the membership function parameters in the rule antecedent and consequent. The tuning range of each parameter directly depends on the width of the original membership function. The overall chromosome is obtained by concatenation of the rule codes.

4.3.3 *Example: genetic tuning processes of Mamdani-type fuzzy rule sets*

As we shall see in Chapter 8, MOGUL (Methodology for Obtaining GFRBSs Under the iterative rule Learning approach) (Cordón, del Jesús, Herrera, and Lozano, 1999) is a methodology composed of some design guidelines that allow different users to customise their own GFRBSs based on the iterative rule learning approach (see Sec. 5.4), in order to design FRBSs able to cope with their specific problems. In MOGUL, the user can obtain different GFRBSs by defining different evolutionary learning processes in each of the learning stages, as long as they meet the MOGUL specification.

The final learning stage of GFRBSs in MOGUL is always an evolutionary tuning process that refines the definition of the fuzzy rule set generated up to this point. The descriptive and approximate genetic tuning processes provided by MOGUL for Mamdani-type FRBSs share a common structure. The only difference between them lies in the coding scheme. While the chromosome in the approximate tuning process encodes separate membership functions for each individual rule; the chromosomes considered in the descriptive approach only encode the primary fuzzy partitions constituting the DB which are shared by the linguistic rules in the RB. Both methods assume an existing fuzzy rule set, which is subsequently refined by the evolutionary tuning process.

4.3.3.1 *Common aspects of both genetic tuning processes*

The GA employs a real-valued coding scheme and uses stochastic universal sampling as selection procedure together with an elitist scheme (Baker, 1987). The genetic operators considered for the generation of new variants are Michalewicz's non-uniform mutation (Michalewicz, 1996) (Sec. 2.2.4.2) and the max-min-arithmetical crossover (Herrera, Lozano, and Verdegay, 1995; Herrera, Lozano, and Verdegay, 1997) (Sec. 3.3.1.2).

The objective of the tuning process is to improve the approximation accuracy of the FRBS over the training data set E_p. The fitness value for a particular chromosome C_j is computed as the mean square error (MSE) between the FRBS output $S(ex^l)$ and the training data output ey^l (Herrera, Lozano, and Verdegay, 1998a):

$$E(C_j) = \frac{1}{2 \cdot |E_p|} \sum_{e_l \in E_p} (ey^l - S(ex^l))^2$$

with $S(ex^l)$ being the output value obtained from the FRBS using the fuzzy rule set coded in C_j. Notice that in this particular case the EA tries to minimise rather than maximise the fitness $E(C_j)$.

In addition to approximation accuracy, the adapted fuzzy rule set should also satisfy the completeness property. The so-called τ-completeness condition (Cordón and Herrera, 1997c) requires every example in the training set to be covered by at least one fuzzy rule to a degree greater than or equal to τ:

$$C_{R(C_j)}(e_l) = \bigcup_{j=1..T} R_j(e_l) \geq \tau, \quad \forall\, e_l \in E_p \text{ and } R_j \in R(C_j)$$

with $R_j(e_l)$ being the *compatibility degree* between the rule R_j and the example e_l: $R_j(e_l) = T(A_{j1}(ex_1^l), ..., A_{jn}(ex_n^l), B_j(ey^l))$ (where T is a t-norm), and with τ being the minimal training set completeness degree accepted for the adjusted fuzzy rule set.

Therefore, a *training set completeness degree* of $R(C_j)$ over the set of examples E_p is defined as

$$TSCD(R(C_j), E_p) = \bigcap_{e_l \in E_p} C_{R(C_j)}(e_l)$$

The fitness function for the approximation accuracy criterion is modified in a way that a fuzzy rule set that does not satisfy the completeness property is strongly penalised:

$$F(C_j) = \begin{cases} E(C_j), & \text{if } TSCD(R(C_j), E_p) \geq \tau \\ \infty, & \text{otherwise} \end{cases} \tag{4.1}$$

4.3.3.2   *The approximate genetic tuning process*

This process was proposed by Herrera, Lozano, and Verdegay (1995) and used within a learning process by Cordón and Herrera (1997c); Herrera, Lozano, and Verdegay (1998a). As mentioned earlier, each chromosome encodes a complete approximate Mamdani-type FRB. The k-th triangular-shaped membership functions is parameterised by a 3-tuple of real values, such that the i-th rule is encoded by the partial chromosome C_{ji}, $i = 1, \ldots, m = n + 1$, as follows:

$$C_{ji} = (a_{i1}, b_{i1}, c_{i1}, \ldots, a_{in}, b_{in}, c_{in}, a_i, b_i, c_i)$$

in which (a_{ik}, b_{ik}, c_{ik}) are the three characteristic points of the k-th input fuzzy set in the rule antecedent, and (a_i, b_i, c_i), those for the output fuzzy set in the rule consequent.

Therefore, the chromosome C_j for the entire FRB definition is the concatenation of rule chromosomes C_{ij}:

$$C_j = C_{j1} \ C_{j2} \ \ldots \ C_{jm}$$

The predefined FRB is converted into a prototype chromosome C_1 which is then used to initialise the chromosomes in the first generation. Each gene c_h, $h = 1 \ldots (n+1) \cdot m \cdot 3$, in C_1 either represents a left, centre or right point of a particular triangular membership function. The interval of adjustment $[c_h^l, c_h^r]$ determines the possible range of the parameter represented by the gene c_h. For a triangular fuzzy set with the left, centre and right point (c_t, c_{t+1}, c_{t+2}), the possible intervals of adjustment —shown in Fig. 4.12— are computed as:

$$c_t \in [c_t^l, c_t^r] \ = \ [c_t - \frac{c_{t+1} - c_t}{2}, c_t + \frac{c_{t+1} - c_t}{2}]$$

$$c_{t+1} \in [c_{t+1}^l, c_{t+1}^r] \ = \ [c_{t+1} - \frac{c_{t+1} - c_t}{2}, c_{t+1} + \frac{c_{t+2} - c_{t+1}}{2}]$$

$$c_{t+2} \in [c_{t+2}^l, c_{t+2}^r] \ = \ [c_{t+2} - \frac{c_{t+2} - c_{t+1}}{2}, c_{t+2} + \frac{c_{t+3} - c_{t+2}}{2}]$$

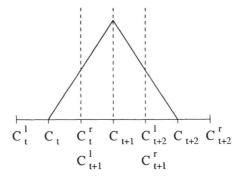

Fig. 4.12 Membership function and intervals of performance for the Mamdani tuning process

The initial population contains one exact copy of the prototype chromosome C_1, while the remaining chromosomes are initialised by uniformly drawing the values c_h at random from its interval of adjustment $[c_h^l, c_h^r]$.

4.3.3.3   *The descriptive genetic tuning process*

The descriptive genetic tuning process (Cordón, Herrera, and Lozano, 1996; Cordón and Herrera, 1997c) is a modified version of the approximate one. In this case, the preliminary fuzzy rule set is a Mamdani-type KB (that is, an initial DB definition and an RB) and each chromosome encodes a different DB definition. A primary fuzzy partition like the one shown in Fig. 4.11 is represented as an array composed of $3 \cdot N$ real values, in which N is the number of linguistic terms. The complete DB for m linguistic variables, is encoded into a fixed length real-coded chromosome C_j formed by the partial chromosomes C_{ji} of the individual fuzzy partitions.

$$C_{ji} = (a_{i1}, b_{i1}, c_{i1}, \ldots, a_{iN_i}, b_{iN_i}, c_{iN_i})$$
$$C_j = C_{j1} \ C_{j2} \ \ldots \ C_{jm}$$

The first generation is initialised in the same way as in the approximate genetic tuning process. The only difference is that the prototype chromosome C_1 is generated from the original DB definition.

4.4   Genetic Tuning of TSK Fuzzy Rule Sets

Two different sets of parameters are subject to tuning in a TSK KB: the membership function parameters in the rule antecedents and the weights of the linear function in the rule consequents. The different role of the parameters is reflected in the structure of the chromosome that encodes the TSK RB, which is usually composed of two parts, C^1 for the parameters in the antecedent and C^2 for the parameters in the consequent part of the rule.

The first part of the chromosome possesses the same structure as in a Mamdani FRBS, encoding the parameterised membership functions of either the fuzzy partitions (approximate) or fuzzy rules (descriptive).

Assume, a descriptive TSK KB with rules such as:

$$IF \ X_1 \ is \ A_1 \ and \ \ldots \ and \ X_n \ is \ A_n$$
$$THEN \ Y = p_1 \cdot X_1 + \cdots + p_n \cdot X_n + p_0$$

If trapezoidal membership functions are considered in the antecedent, the fuzzy partitions of the FRBS are encoded by 4-tuples $a_{ij}, b_{ij}, c_{ij}, d_{ij}$ as follows:

$$C_i^1 = (a_{i1}, b_{i1}, c_{i1}, d_{i1}, \ldots, a_{iN_i}, b_{iN_i}, c_{iN_i}, d_{iN_i})$$
$$C^1 = C_1^1 \, C_2^1 \, \ldots \, C_n^1$$

where N_i is the number of linguistic labels in the term set of the i-th input variable.

4.4.1 Genetic tuning of TSK rule consequent parameters

The second part of the chromosome, C^2, encodes the consequent parameters of the TSK fuzzy rules. The chromosome C^2 of length $m \cdot (n+1)$ is the concatenation of the partial chromosomes C_i^2 that encode the weights p_{ij} in the linear function of the i-th rule consequent:

$$C_i^2 = (p_{i0}, p_{i1}, \ldots, p_{in}), \quad i = 1, \ldots, m$$
$$C^2 = C_1^2 \, C_2^2 \, \ldots \, C_m^2$$

In order to find a suitable binary or real valued coding of the parameters p_{ij}, their possible range must be known in advance. Unfortunately, this information is usually not available in the case of learning or tuning the TSK rule consequent parameters.

This problem is usually solved by choosing conservatively large intervals for the parameter ranges (Lee and Takagi, 1993c; Lee and Takagi, 1996; Siarry and Guely, 1998; Setness and Roubos, 2000; de Sousa and Madrid, 2000; Wu and Yu, 2000). This remedy tends to solve the problem of parameter range, although there is no guarantee for the optimal solution to lie within the parameter ranges. Moreover, in case of a binary code, large parameter ranges carry the drawback that the number of bits per parameter has to be increased in order to obtain the same resolution.

Cordón and Herrera (1997a) proposed a novel coding scheme, called *angular coding*, that applies a non-linear transformation before encoding the TSK rule consequent parameters. Rather than directly coding the weights p_{ij}, the new parameters represent the direction of the tangent $\alpha_{ij} = \arctan p_{ij}$. The range for the parameters α_{ij} is the interval $(-\frac{\pi}{2}, \frac{\pi}{2})$, such that the weights p_{ij} can assume any real value.

As mentioned in Sec. 1.3, the partial linear relation defined by the consequent of a TSK rule determines a hyper-plane in the input space. For a single input system, a TSK rule $Y = p_1 \cdot X + p_0$ defines a straight line. The

real value p_1 is simply the tangent of the angle between this line and the
X-axis. Thus, if we code the angular value instead of the tangent itself by
means of the non-linear transformation

$$f : \mathbb{R} \rightarrow (-\frac{\pi}{2}, \frac{\pi}{2}) \quad ; \quad f(x) = \arctan(x)$$

the possible values of parameter p_1 lie inside the interval $(-\frac{\pi}{2}, \frac{\pi}{2})$. Figure
4.13 shows some examples of the transformation f.

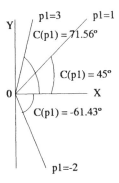

Fig. 4.13   Examples of angular coding

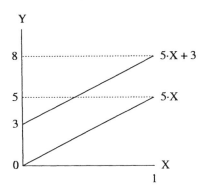

Fig. 4.14   Geometrical interpretation of the parameter p_0

The parameter p_0 determines the offset of the straight line along the Y
axis, as shown in Fig. 4.14. Since the values of the parameter p_0 may vary
between different TSK rules, no common fixed interval of definition can be
considered and using angular coding is a good choice. In this case, there

is no geometric interpretation in the coding, but the transformation allows the EA to adequately search in the solution space to learn the values of these parameters.

4.4.2   *Example: the evolutionary tuning process of MOGUL for TSK knowledge bases*

This evolutionary process was proposed by Cordón and Herrera (1999b). It is composed of a genetic local search algorithm composed of an RCGA including an $(1+1)$-ES as another genetic operator to improve the search process, guided by a fitness function based on the MSE as shown in Sec. 4.3.3.1. The remainder of this section describes the hybrid EA components that are different from the previous Mamdani tuning processes.

4.4.2.1   *Representation*

The chromosome encodes the TSK KB in the way described in Sec. 4.4 using real values to encode the membership function parameters. However, the intervals of adjustment are computed in a different way to the one shown in that section. The parameters of the triangular fuzzy sets $D_{ij} = (a_{ij}, b_{ij}, c_{ij})$, $i = 1, \ldots, n$, $j = 1, \ldots, N_i$, are allowed to vary freely in the interval $[D_{ij}^{min}, D_{ij}^{max}]$. The interval limits are computed before running the refinement process in the following way:

$$[D_{ij}^{min}, D_{ij}^{max}] = [a_{ij} - \frac{b_{ij} - a_{ij}}{2}, c_{ij} + \frac{c_{ij} - b_{ij}}{2}]$$

in which a_{ij}, b_{ij}, c_{ij} are the left, centre and right point of the original triangular membership function in the predefined DB.

Therefore, the interval of adjustment $[c_t^l, c_t^r]$ of a gene c_t in C^1 depends on the fuzzy membership function to which it is associated. If $(t \bmod 3) = 1$, then c_t is the left value of the support of a fuzzy set, which is defined by the three parameters. The variable ranges for the left c_t, centre c_{t+1} and right point c_{t+2}) of the triangular membership function are given by:

$$c_t \in [c_t^l, c_t^r] = [D^{min}, c_{t+1}]$$
$$c_{t+1} \in [c_{t+1}^l, c_{t+1}^r] = [c_t, c_{t+2}]$$
$$c_{t+2} \in [c_{t+2}^l, c_{t+2}^r] = [c_{t+1}, D^{max}]$$

During the optimisation the 3-tuple (c_t, c_{t+1}, c_{t+2}) can assume any values within the interval $[D^{min}, D^{max}]$ as long as they satisfy the order relation $c_t \leq c_{t+1} \leq c_{t+2})$. Figure 4.15 shows an example of these intervals.

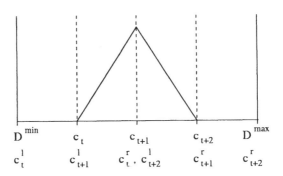

Fig. 4.15   Membership function and parameter range for the TSK genetic tuning process

The second part of the chromosome, C^2, represents the consequent parameters of the TSK rules based on the angular coding introduced in the previous section.

4.4.2.2   *Initial population*

The TSK genetic tuning process of MOGUL uses the preliminary definition of the KB to initialise the first generation. The initial population of size M is generated in three steps as follows:

(1) The preliminary definition of the KB taken as process input is encoded directly into a chromosome, denoted as C_1.

(2) The following $\frac{M}{2} - 1$ chromosomes are initiated by generating, at random, the first part, C^1, with each gene being in its respective interval of adjustment, and by encoding the preliminary definition of the rule consequent parameters in C^2.

(3) The remaining $\frac{M}{2}$ are set up by generating C^1 in the same way followed in the previous step, and by generating the values for C^2 by adding a random value distributed following a normal distribution $N(0, d)$ to the values in the C^2 part of the previous chromosomes.

4.4.2.3 *Genetic operators*

The mutation and crossover operators are the same considered in the genetic tuning processes for Mamdani-type fuzzy rule sets presented in the previous sections, i.e., Michalewicz's mutation and max-min-arithmetical crossover. On the other hand, there is a third genetic operator to be applied which consists of an *(1+1)-ES*. This optimisation technique was selected and integrated into the genetic recombination process in order to perform a local tuning of the best population individuals in each run, following the assumptions of the so-called *genetic local search* (Jog, Suh, and Gutch, 1989; Suh and Gutch, 1987). Each time a GA generation is performed, the ES will be applied over a percentage δ of the best different population individuals existing in the current genetic population.

The basis of the ES employed was briefly presented in Sec. 2.3.1. The coding scheme and the fitness function considered in the $(1+1)$-ES are the same as those used in the GA. Thus, the only changes to be performed have to be done in the generic ES mutation scheme, and the great majority of them only when working with the first part of the chromosome, C^1. In this case, the following two changes have to be put into effect:

- *Definition of multiple step sizes:* The mutation strength depends directly on the value of the parameter σ, which determines the standard deviation of the normally distributed random variable z_i. In our case, the step size σ can not be a single value because the membership functions encoded in the first part of the chromosome are defined over different universes and so require different order mutations. Therefore, a step size $\sigma_i = \sigma \cdot s_i$ for each component in C^1 is going to be used in the *(1+1)-ES*. Anyway the relations of all σ_i were fixed by the values s_i and only the common factor σ is adapted following the assumptions proposed by Bäck (1996).

- *Incremental optimisation of the individual parameters*: Usually, the different parent components are not related and the ES is used in its usual working mode in which all of them are adapted at the same time. Unfortunately, in this problem each three correlative parameters, (x_0, x_1, x_2), in C^1 define a triangular-shaped membership function, and the property $x_0 \le x_1 \le x_2$ must be verified in order to obtain meaningful fuzzy sets. Therefore, there is a need to develop an incremental optimisation of the individual parameters because the intervals of adjustment for each one of them will depend on the value of any of the others.

As mentioned in the description of the coding scheme, a global interval of adjustment (in which the three parameters defining the membership function may vary freely) is defined for each fuzzy set involved in the optimisation process. With $C_{ij} = (x_0, x_1, x_2)$ being the membership function currently adapted, the associated interval of adjustment is $[C_{ij}^{min}, C_{ij}^{max}] = [x_0 - \frac{x_1 - x_0}{2}, x_2 + \frac{x_2 - x_1}{2}]$. The incremental adaptation is based on generating the mutated fuzzy set $C_{ij}' = (x_0', x_1', x_2')$ by first adapting the modal point x_1 obtaining the mutated value x_1' defined in the interval $[x_0, x_2]$, and then adapting the left and right points x_0 and x_2 obtaining the values x_0' and x_2' defined respectively in the intervals $[C_{ij}^{min}, x_1']$ and $[x_1', C_{ij}^{max}]$. It may be clearly observed that the progressive application of this process allows us to obtain fuzzy sets freely defined in the said interval of performance.

The value of the parameter $s(x_i)$ determining the particular step sizes, $\sigma_i = \sigma \cdot s(x_i)$, is computed each time the component x_i is going to be mutated. When $i = 1$, i.e., the modal point is being adapted, and then $s(x_1)$ is equal to $\frac{Min(x_1 - x_0, x_2 - x_1)}{2}$. In the other two cases, $i = 0$ and $i = 2$, $s(x_0) = \frac{Min(x_0 - C_{ij}^{min}, x_1' - x_0)}{2}$ and $s(x_2) = \frac{Min(x_2 - x_1', C_{ij}^{max} - x_2)}{2}$. Hence, when σ takes value 1 at the first ES generation, the obtaining of a large quantity of z_i normal values performing a successful x_i mutation (i.e., the corresponding $x_i' = x_i + z_i$ with $z_i \sim N_i(0, \sigma_i'^2)$ lying in the expected interval for x_i) is ensured. If the mutated value lies outside, it is assigned the value of the interval extent closest to $x_i + z_i$.

The next algorithm summarises the application of the adaptation process on a membership function encoded in the parent. The steps to follow —being $C_{ij} = (x_0, x_1, x_2)$ the fuzzy set currently adapted— are:

(1) *Compute the step size of central point,* $s(x_1) \leftarrow \frac{Min\{x_1 - x_0, x_2 - x_1\}}{2}$.

(2) *Generate* $z_1 \sim N(0, \sigma_1^2)$ *and compute* x_1' *in the following way:*

$$x_1' \leftarrow \begin{cases} x_1 + z_1, & \text{if } x_1 + z_1 \in [x_0, x_2] \\ x_0, & \text{if } x_1 + z_1 < x_0 \\ x_2, & \text{if } x_1 + z_1 > x_2 \end{cases}$$

(3) *Adapt the remaining two points:*

(a) $s(x_0) \leftarrow \dfrac{Min\{x_0 - C_{ij}^{min}, x_1' - x_0\}}{2}$

Generate $z_0 \sim N(0, \sigma_0^2)$

$$x_0' \leftarrow \begin{cases} x_0 + z_0, & \text{if } x_0 + z_0 \in [C_i^l, x_1'] \\ C_i^l, & \text{if } x_0 + z_0 < C_i^l \\ x_1', & \text{if } x_0 + z_0 > x_1' \end{cases}$$

(b) $s(x_2) \leftarrow \dfrac{Min\{x_2 - x_1', C_{ij}^{max} - x_2\}}{2}$

Generate $z_2 \sim N(0, \sigma_2^2)$

$$x_2' \leftarrow \begin{cases} x_2 + z_2, & \text{if } x_2 + z_2 \in [x_1', C_i^r] \\ x_1', & \text{if } x_2 + z_2 < x_1' \\ C_i^r, & \text{if } x_2 + z_2 > C_i^r \end{cases}$$

When working with the second part of the chromosome, C^2, the latter problem does not appear. In this case, the different components are not related and the mutation can be performed in its usual way. The only change that has to be made is to adapt the step size to the components in C^2. As all of them are defined over the same interval of performance, $(-\frac{\pi}{2}, \frac{\pi}{2})$, they all will use the same step size $\sigma_i = \sigma \cdot s_i$.

Chapter 5

Learning with Genetic Algorithms

The previous chapter was devoted to the application of genetic or evolutionary techniques to tune the membership and/or scaling functions of FRBSs. Now, we go back to GAs and analyse how to apply them to learning issues. This chapter is the starting point for the three following chapters, covering the application to FRBSs of the three fundamental genetic learning approaches that are reviewed in the following: Michigan, Pittsburgh and iterative rule learning.

To do so, the first section shows the difference between tuning and learning and introduces the topic of learning with GAs. The remaining three sections in the chapter are devoted to introduce the three said genetic learning approaches, describing the composition of the genetic learning processes considering them in depth and presenting several specific learning systems of each kind.

5.1 Genetic Learning Processes. Introduction

As said, the previous chapter introduced the topic of genetic tuning of FRBSs. The idea of tuning (as considered in this book) is closely related to that of parameter optimisation. As a matter of fact, all the described approaches work on the basis of parameterised membership or scaling functions where the adjustment of the parameter values is the objective of the search performed by the EA.

It is difficult to make a clear distinction between tuning and learning processes, since establishing a precise border line becomes as difficult as defining the concept of learning itself. However, one of the common aspects of most learning approaches is that of generating systems which possess the ability to change their underlying structure with the objective of improving their performance or the quality of their knowledge according to some criteria. In this book, the level of sophistication of the complexity of the structural changes available to a specific learning system has been used to establish the limit between tuning and learning processes.

According to DeJong (1988), the different levels of complexity in the structural changes produced by GAs are:

> In the simplest and most straightforward approaches, GAs alter a set of parameters that control the behaviour of a pre-developed, parameterised performance program. A second, more interesting, approach involves changing more complex data structures, such as "agendas", that control the behaviour of the task subsystem. A third and even more intriguing approach involves changing the task program itself.

These three levels were established for the general case of GAs, but are easily translated to the case of rule-based applications.

Those systems described in the previous chapter could be characterised as belonging to the first level of complexity, that of parameter optimisation. On the other hand, one of the topics that DeJong includes in the third level of complexity is that of learning production-system programs, and in a more specific sense, learning the rule set of a production system. Considering this area of work (learning the rule set of a production system), DeJong (1988) analysed the existence of two main approaches:

> To anyone who has read Holland (1975), a natural way to proceed is to represent an entire rule set as a string (an individual), maintain a population of candidate rule sets, and use selection and genetic operators to produce new generations of rule sets. Historically, this was the approach taken by DeJong and his students while at the University of Pittsburgh (e.g., see Smith (1980, 1983)), which gave rise to the phrase "the Pitt approach".
>
> However, during the same period, Holland developed a model of cognition (classifier system) in which the members of the population are individual rules and a rule set is rep-

resented by the entire population (e.g., see Holland and Reitman (1978); Booker (1982)). This quickly became known
as "the Michigan approach" and initiated a friendly but
provocative series of discussions concerning the strengths
and weaknesses of the two approaches.

A different possibility is the one that is known as the iterative rule
learning approach, where a chromosome codes an individual rule and a
novel rule is adapted and added to the rule set in an iterative fashion
in every GA run. The GA provides a partial solution to the problem of
learning and, contrary to both previous approaches, it is run several times
to obtain the complete set of rules. Venturini (1993) initially proposed this
approach with the SIA system. It has been widely developed in the field of
GFRBSs as we will describe in Chapter 8.

There are other approaches to learning based on GAs. As an example,
we have the following systems:

COGIN

Greene and Smith (1992, 1993, 1994) proposed COGIN (COverage-based
Genetic INduction), a GA-based inductive system that exploits the conventions of induction from examples to provide this framework. The novelty
of COGIN lies in its use of training set coverage to simultaneously promote
competition in various classification niches within the model and constrain
overall model complexity.

REGAL

REGAL, proposed by Giordana and Saitta (1993); Neri and Giordana
(1995); Giordana and Neri (1996), is focused on the problem of learning
multi-modal concepts in first order logic, i.e., concepts requiring disjunctive
descriptions. REGAL's main features are: a task-oriented selection operator, called Universal Suffrage operator, and a distributed architecture. RE-
GAL's approach is a hybrid one, because each individual encodes a partial
solution —a tentative conjunctive description for one concept modality—
and the whole population is a redundant set of partial solutions, which
hopefully contains a general solution for the induction problem —i.e., a disjunction of conjunctive descriptions covering the entire target concept— at
the end of the run. The difference with respect to the Michigan's approach
is that the fitness of an individual does not depend upon the cooperation
with the other ones.

DOGMA

Hekanaho (1996, 1997) proposed DOGMA (Domain Oriented Genetic Machine), a genetic-based machine learning algorithm that supports two distinct levels, with accompanying operators. On the lower level, DOGMA uses Michigan-type fixed-length genetic chromosomes, which are manipulated by mutation and crossover operators. This lower level is similar to REGAL. On the higher level, the chromosomes are combined into genetic families, through special operators that merge and break families. The chromosomes represent single rules, whereas the genetic families encode rule sets that express classification theories. DOGMA incorporates special stochastic operators which induce knowledge from a given background theory.

These approaches are not represented in GFRBSs and will not be considered in the following sections, where the Michigan, Pittsburgh and iterative rule learning approaches are described.

5.2 The Michigan Approach. Classifier Systems

Classifier systems (CSs) are production rule systems that automatically generate populations of rules cooperating to accomplish a desired task. They are massively parallel, message-passing, rule-based systems that learn through credit assignment and rule discovery.

CSs typically operate in environments that exhibit one or more of the following characteristics: (i) perpetually novel events accompanied by large amounts of noisy or irrelevant data; (ii) continuous, often real-time, requirements for action; (iii) implicitly or inexactly defined goals; and (iv) sparse payoff or reinforcement obtainable only through long action sequences. CSs are designed to absorb new information continuously from such environments, devising sets of competing hypotheses (expressed as rules) without disturbing significantly capabilities already acquired.

CSs (Holland, 1976) were inspired by the Michigan approach whose foundations were laid by Holland (1975).

In contrast to traditional expert systems where rules are handcrafted by knowledge engineers, CSs use the GA as a discovery operator to generate classifiers. Each classifier is an "IF-THEN" rule, with a condition part and an action part. A message list is used to store the current environmental state and any internal message. Associated to each classifier is a numerical value called its strength. Holland and Reitman (1978) adjusted classifier strength through backwards averaging and other central methods. This led to concerns with apportioning credit in parallel systems. Early considerations, such as those of Holland and Reitman (1978), gave rise to an algorithm called the *bucket brigade algorithm* (see (Holland, 1986)) that only uses local interactions between rules to distribute credit.

Many different CS configurations exist in the literature and some aspects have been debated and are still subject of debate. The prototype organisation is composed of the three following main parts, as illustrated in Fig. 5.1:

- *The Performance System.*
- *The Credit Assignment System.*
- *The Classifier Discovery System.*

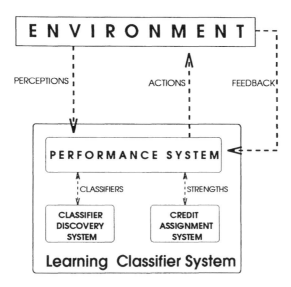

Fig. 5.1   A Classifier system

5.2.1   *The performance system*

The performance system is the part of the overall system that directly interacts with the environment. Its activities are environmental interaction and message processing. It is composed of six basic elements:

(1) *An input interface,* which consists of at least one detector that translates the current state of the environment into standard messages (external messages). Generally, all messages are required to be of a fixed length over a specified alphabet, typically k-bit binary strings.

(2) *An output interface,* which consists of at least one effector that translates some messages (action messages) into actions that modify the state of the environment.

(3) *A set of rules,* called *classifier list,* represented as strings of symbols over a three-valued alphabet $(A = \{0, 1, \#\})$ with a *condition/action* format. The condition part specifies the messages that satisfy (activate) the classifier and the action part specifies the messages (internal messages) to be sent when the classifier is satisfied. A limited number of classifiers fire in parallel in each cycle. Details on classifier coding are given in (Booker, 1990; Goldberg, 1989).

(4) *A pattern-matching system,* which identifies which classifiers are matched or satisfied (matched classifiers) in each cycle of the CS.

(5) *A message list,* which contains all current messages, i.e., those generated by the detectors and those sent by fired classifiers.

(6) *A conflict-resolution (CR) system,* which has two functions:

   (a) Determining which matched classifiers fire when the size of the message list is smaller than the number of matched classifiers.

   (b) Deciding which actions to choose in case of inconsistency in the actions proposed to effectors, e.g. "turn left" and "turn right".

The CR system acts according to some usefulness measures associated to each competing classifier, i.e., *the relevance to the current situation* or *specificity* and *the strength* or *past usefulness.* The CR system is based on an economic analogy and involves of a *bid competition* between classifiers. In this system, matched classifiers bid a certain proportion of their usefulness measures and classifier conflicts are resolved based on a probability distribution over these bids. Higher bidding classifiers are favoured to fire and post their messages. If a classifier fires, it must pay out its bid. Thus, each classifier that fires risks a certain percentage of its strength with the possibility of receiving a reward as compensation. No global information is maintained to suggest which classifiers compete with each other. Each classifier maintains only its own statistics, which are only updated when the classifier is active.

A descriptive scheme of the performance system is presented in Fig. 5.2.

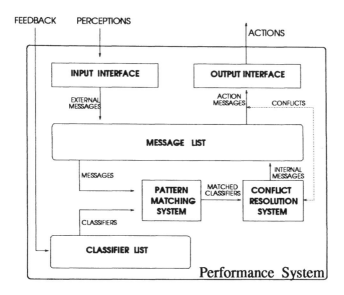

Fig. 5.2   A general performance system

Usually, the rules in the performance system are of varying usefulness and some, or even most, of them may be incorrect. Somehow the system must evaluate the rules. This activity is often called *credit assignment* or *apportionment of credit.*

5.2.2   *The credit assignment system*

The main task of the Credit Assignment (CA) system includes the activity of learning by the modification and adjustment of conflict-resolution parameters of the classifier set, i.e., their strengths. There are many different CA system schemes proposed in the literature. The two most important ones are shown as follows:

(1) *The bucket brigade algorithm* (BBA) (Booker, 1982; Booker, Goldberg, and Holland, 1989; Goldberg, 1989; Holland, 1985), which is a local learning scheme that requires a few memory and computational requirements. Usually, the BBA is linked to the CR system and set up the mechanism of a competitive economy known as CA/CR system. In this classical CS CA/CR system, each satisfied classifier C_j makes a bid Bid_j,

$$Bid_j = b \cdot Str_j \cdot Spec_j$$

where b is a small constant less than 1, called *risk factor*, Str_j is C_j's strength (initialised with the same value for all classifiers), and $Spec_j$ is C_j's specificity; i.e., a quantity expressing the classifier relevance to particular environmental situations. The probability that a bidding classifier C_j wins the competition is given by

$$\frac{Bid_j}{\sum_{C_l \in B} Bid_l}$$

where B is the set of all bidding classifiers. A winning classifier C_j reduces its strength by the amount of its bid according to

$$Str_j = Str_j - Bid_j$$

and this amount (Bid_j) is shared among those classifiers (predecessors) whose preceding activities enabled C_j to became active according to the following rule:

$$Str_i = Str_i + \frac{Bid_j}{|j|}, \quad \forall\, C_i \in \mathcal{P}_j$$

where $\mathcal{P}_j$ is the set of predecessors, that is,

$$\mathcal{P}_j = \{C_i \ : \ \exists\, T \in \mathcal{T}_j,\ \exists\, m \in T,\ (C_i \text{ sent } m)\}$$

where $\mathcal{T}_j$ is the set of all message tuples that satisfy C_j and m is a message.

Obviously, if a classifier does not bid enough to win the competition, it pays nothing. Additionally, if an external reward is received from the environment then it is equally distributed among the classifiers that sent the effector-activating messages. In this way, all classifiers that are directly or indirectly useful in achieving specific goals are rewarded although they are not active when the external reward is obtained. Some variants of BBA were proposed by Dorigo (1991); Weiβ (1991); Wilson (1987).

(2) *The profit sharing plan (PSP)* (Holland and Reitman, 1978; Grefenstette, 1988), which is a global learning scheme typically achieving a clearly better performance than the BBA. Here, the bidding and selection of the winning classifiers is done as in BBA and the external reward Ext is distributed among sequences of active classifiers, that is, among classifiers C_j that were active at least one time during an episode (where an episode is defined as the time interval between the receipts of two successive external rewards) according to rule

$$Str_j = Str_j - Bid_j + b \cdot Ext$$

Therefore, the BBA is an incremental learning algorithm according to which the strengths are updated each cycle. Against that, the PSP is an episodical learning algorithm according to which the strengths are only updated at the end of each episode. Some variants of PSP were proposed by Grefenstette (1988) and Weiβ (1992).

There exist some approaches to a synthesis of the said CA schemes, for example, the so-called RUDI system proposed by Grefenstette (1988). More recently, Weiβ (1994) proposed a new CA system approach, *the hierarchical chunking algorithm*, approaching to both the lower computational requirements of the BBA and higher performance level of the PSP, being designed to solve the local versus global dilemma.

The traditional CS CA/CR scheme is based on an economic analogy and consists of a *bid competition* between classifiers. In this scheme, matched classifiers bid a certain proportion of their strength, and rule conflicts are resolved based on a probability distribution over these bids. Higher bidding classifiers are favoured to fire and post their messages. If a classifier fires, it must pay out its bid. Thus, each classifier that fires risks a certain percentage of its strength, with the possibility of receiving rewards as compensation. The incorporation of the CA/CR mechanism into a CS divides the learning process into two phases: the adjustment of classifier strengths and the modification of the classifier set with the GA. These phases are interrelated, so that the accuracy of evaluation of rule fitness influences the GA's ability to improve the rule set.

5.2.3 *The classifier discovery system*

As previously mentioned, the learning of a CS is divided in two learning processes:

(1) the learning process developed in the *credit assignment system*,
(2) the learning process developed in the *classifier discovery system*.

In the former one, the set of classifiers is given and its use is learned by means of environmental feedback. In the latter, new and possibly useful classifiers can be created using past experience. The classifier discovery process for CSs uses GAs for its task. The system develops its learning process generating new classifiers from the classifier set by means of GAs.

Basically, a GA selects high fitness classifiers as parents forming off-spring by recombining components from the parent classifiers. The fitness of a classifier is determined by its usefulness or strength calculated with the CA system instead of a fitness function. In typical CS implementations, the GA population is a portion of the classifier set consisting of those that demonstrate high strength. In order to preserve the system performance, the GA is allowed to replace only a subset of the classifiers, i.e., a subset of the worst m classifiers is replaced by another of m new classifiers created by the application of the GA on the selected portion of the classifier population. As the GA uses classifier strength as a measure of fitness, it can be usefully applied to the set of classifiers only when the CA system has reached steady state, i.e., when a classifier strength accurately reflects its usefulness. Therefore, it is applied with a lower frequency, usually between 1000 and 10000 CA system cycles. A GA basic execution cycle is as follows:

(1) Take the classifier set as initial population P.

(2) Rank individuals of P in decreasing fitness order using the strength associated to every classifier as a measure of fitness.

(3) Choose $2 \cdot k$ individuals to be replaced among low ranked useless ones.

(4) Choose k pairs of individuals to be replicated among high ranked useful ones.

(5) Apply genetic operators to the k pairs selected at step 4, creating offspring classifiers.

(6) Replace the $2 \cdot k$ individuals selected at step 3 with the offspring created at step 5.

The only requirements the GA makes on the structure of a CS are that a population of rules exists, that the rules are in a syntax that allows for manipulation with genetic operators (the $\{1, 0, \#\}$ alphabet is such a syntax for binary alphabet, using $\#$ as a "don't care" symbol), and that a *fitness measure* for each rule is available as a basis for reproduction. Traditionally, a single measure, called a classifier *strength*, is used as a basis for resolving conflicts and as a fitness measure for the GA.

5.2.4 *Basic operations of a classifier system*

A CS basic execution cycle that combines the aforementioned systems consists of the following steps:

(1) Initially, a set of classifiers is created randomly or by some algorithm that takes into account the structure of the problem domain and they all have assigned the same strength.

(2) The input interface is allowed to code the current environmental output signals as messages.

(3) All messages from the input interface are added to the message list.

(4) The pattern-matching system determines the set of classifiers that are matched by the current messages of the message list.

(5) The CR system resolves conflicts between matched classifiers and determines the set of active classifiers.

(6) The message list is purged.

(7) The messages suggested by the active classifiers are placed on the message list.

(8) Any effectors that are matched by the current message list are allowed to submit their actions to the environment. In the case of inconsistent actions, the CR system is called.

(9) If a reward signal is present, it is distributed with the CA system.

(10) If the CA system has reached a steady state, the GA is applied over the classifier set.

(11) Return to step 1.

Complete descriptions on the basic definition and theory of CSs can be found in (Booker, Goldberg, and Holland, 1989; Goldberg, 1989; Forrest, 1991). Brown and Smith (1996) analysed the CS "renaissance" based on the inspirations from other fields. A state of the art of CSs can be found in

(Lanzi, Stolzmann, and Wilson, 2000). Moreover, a complete application of CS to robot shaping is described in (Dorigo and Colombetti, 1997).

5.2.5 *Classifier system extensions*

As we have mentioned, some aspects have been debated and are still subject of debate: alternatives for CS credit assignment (Grefenstette, 1988; Hewahi and Bharadwaj, 1996), bid-competition CR schemes (Smith and Goldberg, 1992), rules organised into a hierarchy (Dorigo and Sirtori, 1991), the implicit or "natural" niching in CS (incorporating some kind of niching mechanism) (Horn, Goldberg, and Deb, 1994; Horn and Goldberg, 1996).

Several CS extensions are actually under development. Two of them are briefly described in the following (XCS and ACS).

5.2.5.1 *ZCS and XCS*

ZCS and XCS, proposed by Wilson (1994, 1995, 1997), are an advance in learning CS research because (i) they have a sound and accurate generalisation mechanism, and (ii) their learning mechanism is based on Q-learning, a recognised learning technique. Many aspects of XCS are copied from ZCS, a "zeroth-level" CS intended to simplify Holland's canonical framework while retaining the essence of the CS idea. The differences between XCS and ZCS lie in the definition of classifier fitness, the GA mechanism and the more sophisticated action selection that accuracy-based fitness makes possible.

Shortly, XCS differs from Holland's framework in that (i) it has a very simple architecture, (ii) there is no message list, and (iii) the traditional *strength* is replaced by three different parameters.

The following paragraphs offer a brief review (Lanzi, 1999a) of the most recent version of XCS (Wilson, 1997). An algorithmic description of XCS can be found in (Butz and Wilson, 2000). In addition, Lanzi (1999b), and Lanzi and Perrucci (1999) proposed extensions of XCS with variable-length messy chromosomes and LISP S-expressions. Lanzi (1999a) introduced a better understanding of the generalisation mechanism of XCS, proposing a technique, based on Sutton's Dyna concept, through which wider exploration would occur naturally.

Classifier parameters

Classifiers in XCS have three main parameters: the prediction p_j, the prediction error ϵ_j and the fitness F_j. Prediction p_j gives an estimate of what is the reward that the classifier is expected to gain. Prediction error ϵ_j estimates how precise is the prediction p_j. The fitness parameter F_j evaluates the accuracy of the payoff prediction given by p_j and is a function of the prediction error ϵ_j.

Performance component

At each time step the system input is used to build a match set $[M]$ containing the classifiers in the population whose condition part matches the detectors. If the match set is empty, a new classifier that matches the input sensors is created through covering. For each possible action a_i, the *system prediction* $P(a_i)$ is computed as the fitness weighted average of the classifier predictions that advocate the action a_i in the match set $[M]$. The value $P(a_i)$ gives an evaluation of the expected reward if action a_i is performed. Action selection can be *deterministic*, the action with the highest system prediction is chosen, or *probabilistic*, the action is chosen randomly among the actions with a not null prediction.

The classifiers in $[M]$ that propose the selected action are put in the *action set* $[A]$. The selected action is performed and an immediate reward r_{inm} is returned to the system together with a new input configuration.

Reinforcement component

The reward received from the environment is used to update the parameters of the classifiers in the action set corresponding to the previous time step $[A]_1$. Classifier parameters are updated in the following order: first the prediction, then the prediction error, and finally the fitness.

First, the maximum system prediction is discounted by a fact γ ($0 \leq \gamma < 1$) and added to the reward returned in the previous time step. The resulting quantity, simply named P, is used to update the prediction p_j by the Widrow-Hoff delta rule with learning rate β ($0 < \beta \leq 1$) : $p_j \leftarrow p_j + \beta \cdot (P - p_j)$. Then, the prediction error ϵ_j is adjusted using the delta rule technique: $\epsilon_j \leftarrow \beta \cdot (|P - p_j|) - \epsilon_j$. Fitness update is slightly more complex. Initially, the prediction error is used to evaluate the classification accuracy k_j of each classifier as $k_j = exp(ln\ \alpha \cdot (\epsilon_j - \epsilon_0)/\epsilon_0)$ if $\epsilon_j > \epsilon_0$ or $k_j = 1$ otherwise. Subsequently, the relative accuracy k'_j of the classifier is computed from k_j and, finally, the fitness parameter is adjusted by the rule $F_j \leftarrow F_j + \beta \cdot (k'_j - F_j)$.

Covering

Covering acts when the match set $[M]$ is empty or the system is stuck in a loop. In both cases, a classifier, created with a condition that matches the system inputs and a random action, is inserted in the population while another one is deleted from it. The situation in which the system is stuck in a loop is detectable because the predictions of the classifiers involved start to diminish steadily. To detect this phenomenon when $[M]$ is created, the system checks whether the total prediction of $[M]$ is less than Ψ times the average prediction of the classifier in the population.

Genetic algorithm

The GA in XCS is applied to the action set. It selects two classifiers with probability proportional to their fitness, copies them, and with probability λ performs crossover on the copies while with probability μ mutates each allele.

Subsumption deletion

Subsumption deletion acts when classifiers created by the genetic component have to be inserted in the population. Offspring classifiers created by the GA are replaced with clones of their parents if: (i) they are a specialisation of the two parents, i.e., they are "subsumed" by their parents, and (ii) the parameters of their parents have been sufficiently updated. If both these conditions are satisfied, the offspring classifiers are discarded and copies of their parents are inserted in the population; otherwise, the offspring classifiers are inserted in the population.

Macro-classifiers

Whenever a new classifier has to be inserted in the population, it is compared to existing ones to check whether there already exists a classifier with the same condition/action pair. If such a classifier exists then the new classifier is not inserted but the weight of the existing classifier is incremented. If there is no classifier in the population with the same condition/action pair, then the new classifier is inserted in the population. Macro-classifiers are essentially a programming technique that speeds up the learning process reducing the number of real, macro, classifiers XCS has to deal with. The number of macro-classifiers is a useful statistic to measure the degree of generalisation obtained by the system. In fact, as XCS converges to a population of accurate and maximally general classifiers, the number of macro-classifiers decreases while the number of micro-classifiers is kept constant by the deletion/insertion procedures.

5.2.5.2 *Anticipatory Classifier System*

The Anticipatory Classifier System (ACS) is a new type of learning CS developed by Stolzmann (1997, 1998). In the following, a short description is introduced (Butz, Goldberg, and Stolzmann, 1999). An ACS is based on the psychological learning mechanism of anticipatory behavioural control. The knowledge of an ACS is represented by situation-reaction-effect units (SRE-units). By comparing the differences of its anticipations and the perceived consequences after acting on an environment, an ACS builds such SRE-units. It always starts with a most general knowledge and learns by continuously specifying the knowledge.

The ACS has proven to solve suitable problems quickly, accurately, and reliably. It is capable of latent learning —i.e., learning without getting any reward— in addition to the common reward learning. Recently, it has been applied in various experiments with different mazes. In these mazes, it was able to learn latently (Stolzmann, 1998) and to solve non-Markov decision problems (Stolzmann, 1999). Moreover, it has been applied on a hand-eye-coordination task of a robot arm interacting with a camera. In this environment, the action planning was successfully achieved, as shown by (Butz and Stolzmann, 1999).

A complete detail description of ACS can be found in (Stolzmann 1998, 1999; Butz and Stolzmann, 1999).

Recently, a large number of developments on the ACS algorithm have been presented (Butz, Goldberg, and Stolzmann, 1999, 2000a, 2000b, 2000c, 2000d; Stolzmann, Butz, Hoffmann, and Goldberg, 2000).

5.3 The Pittsburgh Approach

After the previous section introduced the main characteristics of the Michigan approach, we describe the Pittsburgh approach (Smith, 1980, 1983) in the following.

In some sense, the two ideas of learning have partially different fields of application. While the Michigan approach is mainly concerned with continuous (on-line) learning in non-inductive problems, the Pittsburgh one is particularly adapted for training in both inductive and non-inductive problems. The underlying concepts are quite different as well, while Michigan approach is closer to the idea of representing the knowledge of a single entity that learns through its interaction with the environment (and consequently its adaptation to it), Pittsburgh is closer to the idea of evolution through competition among individuals and adaptation to the environment.

In the Pittsburgh approach, each individual represents a complete entity of knowledge and, consequently, interaction among different individuals does not occur in the evaluation of that knowledge. That situation avoids the need of credit assignment and thus the definition of complex algorithms with that purpose. Instead, the performance measure is directly assigned to the sole individual responsible for the behaviour generating the evaluation.

To clarify the coincidences and the differences between both approaches, the Pittsburgh approach is described using a similar structure to that used for the Michigan one. Figure 5.3 uses a similar structure to that of Fig. 5.1, but in this case considering a Pittsburgh-type learning system. It is possible to consider quite different structures to describe this approach, but the one selected allows an easier comparison of both approaches.

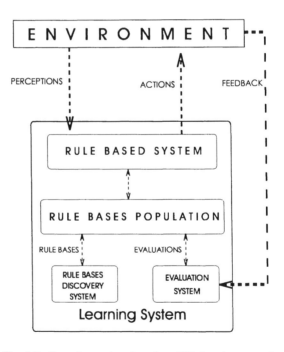

Fig. 5.3   Learning system based on Pittsburgh approach

The clearest difference, comparing the figures, is that in the Pittsburgh approach the population of candidate solutions exists as a separate, independent entity, whereas in the Michigan approach it is embedded in the performance system. This distinctive feature results from the fact that the Pittsburgh approach employs a population of RBs (individuals are RBs) which are evaluated within the rule-based system one at a time. In a CS, the whole population constitutes the RB and consequently the population is a part of the performance system as it is considered as a single entity.

A second question is the distribution of the feedback within the evaluation system that in the Pittsburgh case is directly assigned to the RB under evaluation (at each time there is only a RB, an individual of the population, at work).

The elements added to the rule-based system in order to obtain the learning system shown in Fig. 5.3 are:

- The population of RBs.
- The evaluation system.
- The RBs discovery system.

In the following subsections, we analyse these elements together with the representation of the RB. Then, three specific genetic learning processes based on the Pittsburgh approach (DeJong, Spears, and Gordon, 1993; Janikow, 1993; Corcoran and Sen, 1994) are briefly described.

5.3.1 *The population of rule bases*

The learning process works on a population of potential solutions to the problem. In this case, a potential solution is an RB. From this point of view, the learning process operates on a population of RBs.

Each RB in the population is evaluated by applying it to solve the problem, through the rule-based system, and considering the feedback from the environment. Each RB is evaluated independently and no interaction between individuals of the population occurs during evaluation. To start the learning process, an initial population of RBs is required. In some cases, it is obtained from available knowledge, while in other cases it is randomly generated.

5.3.2 *The evaluation system*

The evaluation of each RB is based on the effect of the interaction of the rule-based system, applying the corresponding RB, with the environment. As a result of this interaction, the environment generates a feedback that is used by the evaluation system to generate the evaluation of the RB. The evaluation is quite different depending on the application and the environment. In learning from examples applications (inductive learning), the evaluation is based on error measures. In control applications (non-inductive learning), the evaluation relies on a more complex performance index that captures the behaviour of the controlled system according to some specific design criteria.

Depending on the application, the evaluation system often becomes the most time consuming element of the process. Usually, the reduction in complexity (comparing with the Michigan approach) that produces the independent evaluation without interactions and conflicts between individuals, is accompanied and compensated by a larger computational effort as a result of the larger number of independent evaluations to be performed.

5.3.3 *The rule base discovery system*

Once a complete population, all the RBs composing it, has been evaluated, it is time to search for new RBs. This task is developed by the RB discovery system that generates a new population (a generation) of RBs, applying a set of genetic or evolutionary operators to the previous one.

The system presents some differences with the classifier discovery system, the first of which relates to the level of replacement. In the Michigan approach, the number of replaced individuals at each generation has to be low enough to preserve the system performance, as it is the result of the interaction between individuals. The system performance in the Pittsburgh approach can be considered as the performance of the best individual and consequently there is no degradation of performance as long as the best individual is maintained. Still, the replacement scheme has to balance between exploration of novel solutions and exploitation of the best solutions found so far in a way that prevents premature convergence.

A second question is related to the timing of evaluation and discovery. In a CS, discovery is applied with a lower frequency than credit assignment, usually between 1000 and 10000 times. This ratio is related to the idea of continuous learning and the need of reaching a steady-state situation before creating a new generation. On the other hand, the idea of training that is applied in Pitt approach is closer to that of a predefined training cycle that, in this case, has to be applied for each individual in the population. Consequently, the discovery phase takes place after a complete training cycle for each individual.

5.3.4 *Representation*

As analysed by different authors (DeJong, 1988; Michalewicz, 1996), one of the main concerns when applying the Pittsburgh approach to rule-based system learning is that of representation. A natural choice involves considering individual rules as genes and the RB as a string of these genes. With this representation, crossover provides new combinations of rules and mutation generates new rules. However, with this representation, genes can take on many values and premature convergence is a possible result of this choice. To solve this problem, a suitable selection of the genetic operators is required to allow crossover and mutation to generate new, potentially improved rules.

A second interesting question is that of using fixed or non-fixed length codes. The use of fixed-length codes reduces the complexity of crossover and allows using classical genetic operators, but implies working with RBs with a fixed number of rules, where each rule has a fixed length code. Obviously this choice is highly restrictive and several proposals using non-fixed length codes have been described in the literature. The counterpart when breaking the restriction of length is the need for new definitions of genetic operators adapted for such a code.

5.3.5 *Genetic learning processes based on the Pittsburgh approach*

Different kinds of rule-based systems have been coded and evolved within this approach. In this subsection, we briefly describe three of them:

- RBs with Boolean DNF rules for concept learning evolved by the GABIL system (DeJong, Spears, and Gordon, 1993),
- RBs with rules coded via a logical representation for learning concept descriptions in the VL_1 language evolved by the GIL system

(Janikow, 1993),

- RBs with interval rules for classification with continuous variables (Corcoran and Sen, 1994).

5.3.5.1 *GABIL learning system*

DeJong, Spears, and Gordon (1993) proposed a system called GABIL, a GA-based concept learner that continually learns and refines concept classification rules from its interaction with the environment.

A chromosome codes a set of DNF rules. The left-hand side of each rule consists of a conjunction of one or more tests involving feature values. The right-hand side of a rule indicates the concept (classification) to be assigned to the examples that are matched (covered) by the left-hand side of the rule. Each chromosome is a variable-length string representing an unordered set of fixed-length rules. The number of rules in a particular chromosome can be unrestricted or limited by a user-defined upper bound.

The fitness of each individual is computed by testing the encoded rule set on the current set of training examples. This provides a bias toward correctly classifying all the examples while providing a non-linear differential reward for imperfect rule sets.

Crossover and mutation are then applied probabilistically to the surviving rule sets to produce a new population. This cycle continues until a largely consistent and complete rule set has been found within the time/space constraint given.

5.3.5.2 *GIL learning system*

Janikow (1993) proposed a system called GIL (Genetic-based Inductive Learning), a learning system that evolves a population of rules coded in the VL_1 language (Michalski, Mozetic, Hong, and Lavrae, 1986). In this language, the variables (attributes) are the basic units having multi-valued domains. According to the relationship among different domain values, such domains may be of different types: nominal, linear with linearly ordered values; or structured with partially ordered values.

The chromosomes remain of fixed length (as a parameter of the system). Initially, the population must be filled with potential solutions. This process is either totally random, or it incorporates some task-specific knowledge. GIL allows for three different types of chromosomes to fill the population: the first type is a random initialisation; the second type is an initialisation with data; and the third type is an initialisation with prior hypotheses, provided such are available.

The genetic operators transform chromosomes to new states in the search space. Since the system operates in the problem space, the operators directly follow the inductive methodology. Accordingly to the three syntactic levels of the rule-based framework (conditions, rules, rule sets), GIL divides the operators into three corresponding groups. In addition, each operator is classified as having either generalising, specialising, or unspecified or independent behaviours.

The evaluation function reflects the learning criteria associated to supervised learning from examples. The criteria normally include completeness, consistency, and possibly complexity.

5.3.5.3 *Corcoran and Sen's learning system*

Corcoran and Sen (1994) proposed a GA-based learning system with real genes to evolve classification rule sets.

Each chromosome represents a set of classification rules. Each one of them is composed of a set of A attributes and a class value. Each attribute in the rule has two real variables which indicate the minimum and maximum in the range of valid values for that attribute. A *"don't care"* condition occurs when the maximum value is lesser than the minimum value. The class value can be either an integer or a real. This manner of representing don't cares allow us to eliminate the need for special don't care symbols, and also obviates the need of enforcing the validity of rules after applying operators modifying the rule sets. When each attribute value in an input instance is either contained in the range specified for the corresponding attribute of a rule or the rule attribute is a don't care, the rule matches the instance and the class value indicates the membership class of the instance. A fixed-length coding is used, where each chromosome is made up of a fixed number n of rules encoded as above. The length of a chromosome is thus $n \cdot (2 \cdot A + 1)$.

Since each chromosome in the representation comprises an entire classification RB, the fitness function must measure the collective effectiveness of the whole RB. A rule set is evaluated by its ability to correctly predict class memberships of the instances in a pre-classified training set. The fitness of a particular chromosome is simply the percentage of test instances correctly classified by the chromosome's rule set. Each rule in the RB is compared with each of the instances. If the rule matches the instance, the rule's class prediction is compared with the actual known class for the instance. Since there are several rules on the RB, it is possible that multiple rules match a particular instance and predict different classifications. When matching rules do not agree on classification, some form of conflict resolution must be used.

Conflict resolution is perhaps the most important issue to consider in the design of the objective function for this representation. Corcoran and Sen (1994) used a form of weighted voting. A history of the number of correct and incorrect matches is maintained for each rule. Each matching rule votes for its class by adding the number of correct and incorrect matches to the corresponding variables for its class. The winning class is determined by examining the total number of correct and incorrect matches assigned to each class. The first objective is to select the class with the least number of incorrect matches by rules predicting that class. If there is a tie between two classes on this metric, the second objective is to select the class with the largest number of correct matches. If there is still a tie between two or more classes, one of the two classes is arbitrarily selected. In either case, the winning class is compared with the actual known class. If they are identical, this contributes to the percentage of correct answers for this classifier set. Finally, the number of correct or incorrect matches is incremented appropriately for each of the matching rules.

Appropriate genetic operators are proposed to obtain new RBs in the search space.

5.4 The Iterative Rule Learning Approach

In the last few years, a third approach, the *iterative rule learning* (IRL) one, has been created trying to combine the best characteristics and to solve the problems associated to the two previous ones, the Michigan and Pittsburgh approaches. It was first proposed in *SIA* (Venturini, 1993) and it has been widely developed in the field of GFRBSs (Cordón, del Jesús, Herrera, and Lozano, 1999; Cordón, Herrera, González, and Pérez, 1998; González, Pérez, and Verdegay, 1993; González and Herrera, 1997).

In this last approach, as in the Michigan one, each chromosome in the population represents a single rule, but contrary to the latter, only the best individual is considered to form part of the final solution, i.e., of the RB ultimately generated, discarding the remaining chromosomes in the population. Therefore, in the iterative model, the GA provides a partial solution to the problem of learning and, contrary to the two previous approaches, it is repeated multiple times to obtain the complete set of rules. This operation mode substantially reduces the search space, because in each sequence of iterations, the learning method only searches for a single best rule.

In order to generate an RB, which constitutes a true solution to the learning problem, the GA is embedded into an iterative scheme similar to the following:

(1) Use a GA to obtain a rule for the system.
(2) Incorporate this rule into the final set of rules.
(3) Penalise this rule.
(4) If the set of rules generated till now is adequate to solve the problem, end up returning it as the solution. Otherwise, return to step 1.

When the genetic learning algorithm is guided by numerical information, a very easy and direct way to penalise the rules already obtained, and thus being able to learn new rules, involves eliminating from the data set all those examples that are covered by the set of rules previously obtained. Some learning algorithms not based on GAs, such us the *CN2* (Clark and Niblett, 1986) or those included in the *AQ* family (Michalski, 1983; Michalski, Mozetic, Hong, and Lavrae, 1986), use this way of penalising rules as well.

As it may be seen, the operation mode of the IRL approach causes the *niche* and *species* formation in the search space. These concepts were introduced in GAs in order to improve their behaviour when dealing with multi-modal functions with peaks of unequal value (Deb and Goldberg, 1989; Goldberg, 1989). As in nature, the formation of stable sub-populations of organisms surrounding separate niches by forcing similar individuals to share the available resources is induced. In practice, this is to say that the genetic population is divided into different sub-populations (species) according to the similarity of the individuals, and the members in each sub-population are forced to share the payoff given by the fitness function amongst them (niches) (see Sec. 2.2.3.4).

In fact, Beasly, Bull, and Martin (1993) proposed an iterative method for optimising multi-modal functions, allowing us to obtain the desired number of optima by using a niche GA, which worked in a very similar way to genetic learning processes based on the IRL approach. Of course, inducing the formation of niches in the search space seems to be a very adequate way of solving the concept learning problem, when this process is considered to be the learning of multi-modal concepts.

As regards the GA, the main difference with respect to the Michigan approach is that the fitness of each chromosome is computed individually, without taking into account any kind of cooperation with the remaining population members. Thus, since the rule generation is made without taking into account the global cooperation of the RB generated, i.e., each new rule is obtained without taking into account the previous ones already generated, an undesirable over-learning phenomenon may appear in the set of rules given as process output. With the aim of solving this problem, genetic learning processes following this approach usually include a post-processing process with the purpose of simplifying this rule set, removing the redundant rules and selecting the subset of the rules that demonstrate optimal cooperation. This is why these kinds of genetic learning processes usually have a two-level structure composed of the *generation* and *post-processing* processes.

On the other hand, another difference that exists between both approaches is their area of application. Contrary to the genetic learning approaches based on the Michigan approach that are usually applied to non inductive on-line learning problems, the IRL approach is considered to design genetic processes for off-line inductive learning problems due to the fact these are the kinds of problems that best fit its characteristics, as we are going to see in the following.

From the description shown above, we notice that in order to implement a genetic learning algorithm based on the IRL approach, we need, at least, the following components:

(1) a criterion for selecting the best rule in each iteration,
(2) a penalty criterion, and
(3) a criterion for determining when there are enough rules available to represent the existing knowledge about the problem.

The first criterion is normally related to one or several characteristics that are associated to a desirable rule. Usually, criteria about the rule strength (e.g., number of examples covered), the rule consistency and completeness, or the rule simplicity have been proposed. Some examples of these kinds of criteria are presented in Sec. 8.2.1.1.

As regards the penalty criterion, it is often associated with the elimination of the examples covered by the previous rules generated. Removing an example from the training set causes the elimination of the payoff associated to the region in which this example was located, thus translating the search focus to another, previously uncovered region in subsequent runs. This modification encourages an adequate space exploration and promotes the appearance of niches in the search space.

Finally, the third criterion is associated with the completeness of the whole set of rules (Cordón and Herrera, 1997c; Cordón, del Jesús, Herrera, and Lozano, 1999; González and Pérez, 1998a; Herrera, Lozano, and Verdegay, 1998a). It accomplishes that all the training examples in the data set are covered by the rule set and that the number of rules is sufficient to represent the underlying concept. This criterion is often associated, although it is not necessary, with the elimination of the examples related to a concept covered by the previous rules generated.

Chapter 6

Genetic Fuzzy Rule-Based Systems Based on the Michigan Approach

The general structure of a GFRBS based on the Michigan approach is shown in Fig. 6.1 which is an adapted version of Fig. 5.1.

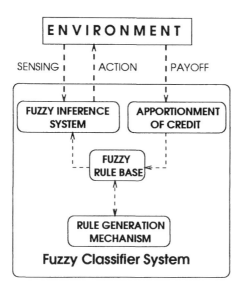

Fig. 6.1   FCS: Learning FRBSs based on Michigan approach

The Michigan learning algorithm for fuzzy rules combines the credit assignment mechanisms of CSs and FRBSs. Each fuzzy rule plays the role of a classifier. The Michigan approach gradually evolves a single RB constituted by the entire population of fuzzy rules. The fuzzy inference system replaces the production system in a classical CS. These types of learning processes therefore receive the name of fuzzy classifier systems (FCSs).

In the following subsection, we introduce the basic features of FCSs and discuss the two distinctive approaches: the learning of the RB coding a linguistic rule per chromosome and the learning of an approximate Mamdani-type FRB coding a fuzzy rule and its underlying membership functions per chromosome.

The Michigan approach does not allow it to simultaneously evolve the whole KB of a linguistic FRBS, i.e., the DB and the RB, since each chromosome only contains the code for a single fuzzy rule, but not the definition of the linguistic variables that provide the underlying global semantics.

6.1   Basic Features of Fuzzy Classifier Systems

The FCS provides a remedy to the limitations of the conventional CS that, due to their binary logic, are not particularly suitable for continuous variables. A binary classifier either triggers or not, whereas fuzzy classifiers become active to a certain degree, that reflects the similarity between the classifier condition part and the current input state.

As mentioned, a classifier can either code the labels of linguistic rules or the membership function parameters of approximate fuzzy rules. In order to illustrate the basic operation of an FCS, this section focuses on the coding scheme proposed by Valenzuela-Rendón (1991b) in which a classifier is a DNF rule (see Sec. 1.2.6.1) of the form:

$$IF\ X_1\ is\ \{C_{11}\ or\ C_{13}\}\ AND\ X_3\ is\ C_{32}\ THEN\ y_1\ is\ D_{14}$$

The classifiers are represented as binary strings that encode the linguistic terms involved in the condition and action parts of the fuzzy rule. Fuzzy classifiers are formed by a list of conditions, separated by commas and an action. The action and the condition parts are separated by a slash (/) and multiple conditions are assumed to be connected by an implicit "and".

The number of bits in a condition or an action part is identical to the number of linguistic terms defined over the given variable. An "1"-bit indicates the presence of the corresponding label in the condition or the action. Each condition or action segment is preceded by a binary code that indicates the variable to which the condition or action is referring to. The code segments for the variable identifier and the corresponding linguistic labels are syntactically separated by a colon.

Assume a DB for the three linguistic variables X_0, X_1 and Y, each one partitioned into three linguistic fuzzy terms $\{Low, Medium, High\}$.

The code segment

$$00 : 110, \ 01 : 101 \ / \ 11 : 001$$

represents the following DNF fuzzy rule

IF X_0 is $\{Low \ or \ Medium\}$ and X_1 is $\{Low \ or \ High\}$ THEN Y is High

in which the first two bits 00 identify the input variable X_0 and the following three bits indicate the presence of the fuzzy terms *Low* and *Medium* and the absence of the fuzzy term *High*.

In the following, we study the operation of FCSs and some specific features of this methodology.

Fuzzy messages

Fuzzy messages possess the same structure as the previously defined fuzzy classifiers, except that a fuzzy message only refers to a single variable rather than multiple ones. Therefore, a fuzzy message is composed of two sub-strings, of which the first identifies the variable to which the message refers and the second describes the linguistic term connected to that variable. Assuming the same DB structure as above, the message 01 : 010 corresponds to the expression "variable X_1 acquires the linguistic term *Medium*".

The overall length of a fuzzy message depends on the number of variables, namely the number of bits required to binary code their index, and the number of linguistic terms associated to the referred variable (one bit per linguistic term). Messages with exactly one linguistic term are called *minimal messages*.

Messages created by the input unit are called input messages. The rôle of the input unit is to generate messages according to the state of the input variables. For each linguistic term of each variable, one minimal input message is generated according to the above coding scheme. In the previous example, the three messages: (a) 01 : 100, (b) 01 : 010, and (c) 01 : 001 for the input variable X_1 are candidates for being added to the message list. For each input message, the input unit computes the membership value of the corresponding linguistic term for the current crisp value of the input variable. This degree of membership determines the activation level of the message. Messages with a zero membership degree are immediately discarded, whereas the other ones are added to the current message list.

Fuzzy matching of fuzzy classifiers and defuzzification

In addition to the externally created input messages, the FCS also adds messages to the list that originate from the firing of the fuzzy classifiers. The FCS matches the condition parts of the classifiers and the list of current messages. To compute the activity level of a classifier, the FCS considers the degree of activation of those messages that match the classifier condition part.

- First, the level of satisfaction of each individual condition in the classifier is computed as the maximum of the activity levels of messages that match the condition, i.e. denote the same variable and linguistic term.
- Second, the activity level of the entire classifier rule is computed as the minimum over the activation of the individual conditions.

Each time a classifier fires, its action part is translated into a set of minimal messages that are posted to the message list. The activation of each of those minimal messages is proportional to the strength and the activation of the classifier that generated it. Messages that are already on the list are not duplicated but rather contribute by adding their activity level to that of the existing message.

The output unit scans the message list for messages that refer to output variables and aggregates their outputs using max-min inference. The output fuzzy set to which the linguistic label in the message refers is clipped by the level of message activation. The crisp output is computed as the centre of gravity of the union (over each variable) of those fuzzy sets.

Credit assignment and fuzzy bucket brigade

Credit assignment is a mechanism to reward or penalise individual classifiers based on their contribution to the overall success or failure of the entire FCS. The bucket brigade algorithm (Sec. 5.2.2) adjusts the strength of individual classifiers by distributing the obtained reward across the sequence of active classifiers that directly or indirectly contributed to the past actions taken by the FCS.

Initially, each classifier rule starts with a fixed strength. Subsequent to the generation of a crisp output, the environment provides a feedback signal, by passing a reward to the output unit that indicates the desirability of the previously taken action. This reward is distributed to those classifiers that generated the output message in the first place. The reward is shared among all classifiers that contributed to the action, in a way that classifiers whose output are similar to the actual output generated by the output unit obtain a larger proportion of the reward than those whose output does not coincide with the actual one.

Classifiers not directly involved in generating the output, but which previously posted messages that in turn activated other classifiers, obtain their reward by means of the bucket brigade algorithm. Each message that fires bids a small portion of its strength (credit) to those classifiers that previously posted messages that triggered their own activation in the first place. In addition, a small living tax gradually reduces the strength of all classifiers, even those that do not become active.

Creation of new rules through a GA
The strength of each classifier is evaluated for a certain number of time steps before the FCS invokes a genetic process to generate new fuzzy classifiers that substitute the weaker one. The GA employs a steady-state generational replacement scheme (see Sec. 2.2.3.3) in which a certain fraction of the current population is replaced by offspring. The currently weakest classifier in the population is replaced by an offspring generated by means of crossover and mutation from two parent classifiers selected according to their strength.

Before a new classifier is added to the population, it is checked for its syntactical correctness. Recombination and mutation might generate classifiers that are invalid for one the following two reasons:

- classifiers whose code segment for the variable part of a condition or action refers to a non-existent variable, or
- classifiers whose action or conditions refer to an index of a non-existent linguistic term.

Invalid classifiers are discarded, and a new offspring is generated in their place.

Solving of the CCP (Sec. 3.4.6) is very difficult in these kinds of systems. As the evolutionary process is made at the individual rule level, it is difficult to obtain a good cooperation among the fuzzy rules that are competing among them. To do so, there is a need to have a fitness function able to measure both the goodness of an individual rule and the quality of the cooperation it presents with the remaining ones in the population to provide the best possible output. As said by Bonarini (1996a), it is not easy to obtain a fitness function of this kind.

The basic cycle of the FCS
The basic cycle of the FCS is composed of the following steps:

> *(1) The input unit receives input values, encodes them into fuzzy messages, and adds these messages to the message list.*
>
> *(2) The population of classifiers is scanned to find all classifiers whose conditions are at least partially satisfied by the current set of messages on the list.*
>
> *(3) All previously present messages are removed from the list.*
>
> *(4) All matched classifiers are fired producing minimal messages that are added to the message list.*
>
> *(5) The output unit identifies the output messages.*
>
> *(6) The output unit decomposes output messages into minimal messages.*
>
> *(7) Minimal messages are defuzzified and transformed into crisp output values.*
>
> *(8) External reward from the environment and internal taxes raised from currently active classifiers are distributed backwards to the classifiers active during the previous cycle.*

Geyer-Schulz (1995) investigated how the syntactic structure of the formal language of the production system influences the computational complexity of FCSs.

6.2   Fuzzy Classifier Systems for Learning Rule Bases

Different FCSs have been proposed for learning the RB. Each chromosome codes an individual linguistic rule, such that the complete RB is formed by the population (Valenzuela-Rendón, 1991b; Bonarini, 1993; Furuhashi, Nakaoka, Morikawa, and Uchikawa, 1994; Ishibuchi, Nakashima, and Murata, 1999).

Valenzuela-Rendón (1991a, 1991b) presented the first GFRBS based on the Michigan approach for learning RBs with DNF fuzzy rules. They later extended the original proposal, in order to enable true reinforcement learning (Valenzuela-Rendón, 1998).

Furuhashi, Nakaoka, Morikawa, and Uchikawa (1994) proposed a basic FCS based on the usual linguistic rule structure with a single label per variable. In (Furuhashi, Nakaoka, Morikawa, Maeda, and Uchikawa, 1995), it was shown that the feature of this FCS is its capability to find suitable rules. Their FCS was able to identify proper fuzzy rules for a collision avoidance controller of a ship based on the successes or failures of the manœuvre executed by the controller. The objective was to design an FLC that avoids collisions with other ships, while maintaining a commanded course as close as possible. The FCS receives a positive reward in the case of a successful avoidance manœuvre, and a negative penalty in case of a collision with another vessel. The authors describe a method of credit apportionment that solves the control problem based on these rewards. In accordance with this simple reward scheme, the FCS was able to automatically discover the correct set of fuzzy rules. In (Nakaoka, Furuhashi, and Uchikawa, 1994; Nakaoka, Furuhashi, Uchikawa, and Maeda, 1996; Furuhashi, Nakaoka, and Uchikawa, 1996), the original approach is extended by a novel credit assignment mechanism. The new method includes the following two steps:

(1) Each fuzzy rule receives reward proportional to the number of membership functions with a non-zero degree of activation;
(2) the rule receives reward depending on the degree of its contribution to the success or failure of the manœuvre.

With this method, fuzzy rules which describe knowledge in complex multi-input/output systems can be obtained.

Furuhashi, Nakaoka, and Uchikawa (1994b) presented a method for using multiple stimulus-response type FCSs (MFCSs) in which several FCSs are connected in series. The paper shows an example of an MFCS with three FCSs for steering the ship.

Bonarini (1993) developed ELF (Evolutionary Learning of Fuzzy Rules), which evolves a population of fuzzy rules in order to identify those fuzzy rules that are optimal according to a scalar reward function. ELF has proven to be robust, in the sense that it is able to deal with a reward function that is afflicted by noise and uncertainty. The method either starts from scratch, is initialised with a predefined RB, or generates the initial rules according to a set of constraints. ELF has the ability to generalise rules by the use of wild-card symbols. In ELF, the population is partitioned into *sub-populations* of rules that have different consequents but share the same antecedents, i.e., they fire in the same fuzzy state. Rules within the same sub-population compete to propose the best action for the state described by the antecedent. ELF has been tested to evolve robotic behaviours on a number of different platforms, both real and simulated. ELF is currently also being used to learn a fuzzy coordinator that blends the control actions of multiple basic fuzzy controller in order to accomplish a more complex, high-level behaviour (Bonarini and Trianni, 2001).

ELF demonstrates a number of characteristic properties making it particularly suitable for the design of control systems for autonomous agents:

- it is robust with respect to the imprecision and uncertainty of sensors, actuators and the environment, typical for the controller design of autonomous agents;
- it produces FLCs that show robustness and smoothness;
- it is flexible with respect to the learning needs;
- it is robust with respect to imperfect reinforcement programs;
- it can learn generalisations;
- it can accept a priori knowledge in terms of fuzzy rules and constraints on their shape;
- it has been applied in a wide range of learning tasks, including: delayed reinforcement learning (fuzzy Q-learning) (Bonarini, 1996a), constrained learning (Bonarini, 1993), learning to coordinate behaviours (Bonarini, 1994, 1996c, 1997; Bonarini and Basso, 1997), or anytime learning, comparative learning, and strategical learning (Bonarini, 1996c, 1997).

On the other hand, Ishibuchi, Nakashima, and Murata (1999) proposed an FCS for learning fuzzy linguistic rules for classification problems.

In the following we present two of these approaches:

- The FCS based on true reinforcement learning (Valenzuela-Rendón, 1998) (Sec. 6.2.1).
- The FCS for learning classification rules (Ishibuchi, Nakashima, and Murata, 1999) (Sec. 6.2.2).

6.2.1 *Valenzuela-Rendón's FCS: Introducing reinforcement learning*

In the first proposal by Valenzuela-Rendón (1991b, 1991a), the FCS employed a reward distribution scheme that required knowledge of the correct action, and thus, must be considered as a supervised learning algorithm. The new credit assignment mechanism capable of true reinforcement learning proposed by Valenzuela-Rendón (1998) distributes the reward as follows.

Activity levels of input messages
The activity level of an input message is computed as the membership degree of the crisp input for the corresponding input fuzzy set.

Activity level of classifier
The aggregated activity level of a classifier is the minimum of the satisfaction degree of its individual conditions. The satisfaction degree of a condition is the maximum of the activity levels of all the matching classifiers.

Classifiers' bid
Classifiers with an activity level greater than zero bid an amount of

$$B = k \cdot S \cdot \alpha \cdot (1 + N(\sigma^2_{noise}))$$

where k is a small constant, S is the classifier strength, α is the classifier activity level and $N(\sigma^2_{noise})$ is normally distributed noise with σ^2_{noise} variance and zero mean. This bid is equally divided among all messages with maximal activity levels that satisfied the classifiers conditions.

Posted messages
The activity level of messages posted by a classifier is equal to the classifier bid divided by the number of messages posted.

Reward distribution to output messages
Reward from the environment is distributed to output messages proportionally to their activity levels.

Payment from messages to classifiers
Messages divide their reward between the classifiers that posted them proportionally to the classifiers bids.

The reward issued to the FCS depends on a combination of the suggested actions of all firing classifiers. Depending on the membership functions, several classifiers might be active at the same time. In FCSs, conflict resolution is handled differently compared to conventional CSs, as all active fuzzy classifiers contribute to the output, rather than letting a single winner takes all classifier determine the next control action. In the new payoff distribution scheme, a noise term is added to each classifier bid, such that weak classifiers have a non zero probability to influence the output.

The activation and reward mechanism are best explained by the following example. Suppose that the classifier list contains two classifiers,

$$C_1 = 0:110 \;/\; 1:110 \;\; \text{and} \;\; C_2 = 0:001 \;/\; 1:001$$

with strengths S_1 and S_2, respectively. Suppose that the input variable X_0 assumes the current value x_0. The crisp value x_0 is fuzzified and three minimal messages are created and posted to the message list:

- $M_1 = 0:100$ with activity level of $\mu_{0,1}(x_0)$,
- $M_2 = 0:010$ with activity level of $\mu_{0,2}(x_0)$, and
- $M_3 = 0:001$ with activity level of $\mu_{0,3}(x_0)$.

Then, the classifier list is scanned for classifiers whose condition match any of the current messages. In this case, classifier C_1 matches the messages M_1 and M_2, and classifier C_2 matches the message M_3. The activity levels of the matched classifiers are computed as $\alpha_1 = \max(\mu_{0,1}(x_0), \mu_{0,2}(x_0))$ and $\alpha_2 = \mu_{0,3}(x_0)$, respectively. In the next step, the classifiers strengths are reduced by the amount of their bids:

- classifier C_1 bids $B_1 = k \cdot S_1 \cdot \alpha_1 \cdot (1 + N(\sigma_{noise}^2))$, and
- classifier C_2 bids $B_2 = k \cdot S_2 \cdot \alpha_2 \cdot (1 + N(\sigma_{noise}^2))$.

Classifiers are fired by posting the three messages indicated in the action part to the message list:

- classifier C_1 fires and post messages $M_4 = 1:100$ and $M_5 = 1:010$ with activity level of $\frac{B_1}{2}$ each, which corresponds to their share of the total classifier bid B_1;
- classifier C_2 fires and posts message M_6 with activity level of B_2.

Messages M_4, M_5, and M_6 are defuzzified and, in response to the crisp action, the FCS receives a payoff P from the environment. Messages M_4

and M_5 each obtain a share $\frac{P \cdot B_1}{2 \cdot (B_1 + B_2)}$ of the total payoff P; message M_6 obtains a proportional reward of $\frac{P \cdot B_2}{B_1 + B_2}$. Finally, the two classifiers C_1 and C_2 receive, through the mediation of their messages, payoffs of $\frac{P \cdot B_1}{B_1 + B_2}$ and $\frac{P \cdot B_2}{B_1 + B_2}$, respectively.

6.2.2 *Fuzzy classifier systems for learning fuzzy classification rules*

Ishibuchi, Nakashima, and Murata (1999) proposed an FCS for designing FRBCSs (see Sec. 1.5.3), in which the population consists of a predefined number of fuzzy classification rules. Thus, the search for a compact FRBCS with high classification accuracy becomes equivalent to the evolution of an FCS. Genetic operations such as selection, crossover and mutation are used to generate new combinations of antecedent linguistic terms for each fuzzy rule. The output class and certainty degree of a rule are determined by the heuristic procedure described by Ishibuchi, Nozaki, and Tanaka (1992); Ishibuchi, Nozaki, Yamamoto, and Tanaka (1995).

The outline of the FCS is as follows:

(1) Generate an initial population of linguistic classification rules;
(2) Evaluate each linguistic classification rule in the current population;
(3) Generate new linguistic classification rules by means of genetic operations;
(4) Replace a part of the current population with the newly generated rules;
(5) Terminate the algorithm if a stopping condition is satisfied, otherwise return to Step 2.

The following subsections explain the individual parts of the FCS, along with a method for coding the rules.

6.2.2.1 *Coding the linguistic classification rules and initial population*

Since the output class and the degree of certainty of a fuzzy rule are already determined by the heuristic procedure (Ishibuchi, Nozaki, and Tanaka, 1992; Ishibuchi, Nozaki, Yamamoto, and Tanaka, 1995), only antecedent linguistic terms are altered by the genetic operations in the FCS. Table 6.1 shows six symbols representing a linguistic term set with five labels.

Table 6.1   Coding example of five terms

Labels	Symbols
S: small	1
MS: medium small	2
M: medium	3
ML: medium large	4
L: large	5
DC: don't care	#

Each fuzzy classification rule is denoted by a string composed of these six symbols. For example, a string "1#3#" denotes the following fuzzy classification rule for a four-dimensional pattern classification problem:

IF X_1 is Small and X_2 is don't care and X_3 is Medium
and X_4 is don't care THEN Class C_j with $CF = CF_j$

Since the fuzzy clauses containing a "don't care" term can be omitted, this rule is rewritten in a more compact notation as:

IF X_1 is Small and X_3 is Medium THEN Class C_j with $CF = CF_j$

The number of fuzzy rules in the population of the FCS is denoted by N_{pop}. To construct an initial population, N_{pop} fuzzy rules are generated by randomly selecting their antecedent labels from the set of possible symbols containing the five linguistic values and the additional "don't care" term with equal probability. The consequent class C_j and the degree of certainty CF_j of each fuzzy rule are determined by the heuristic procedure.

6.2.2.2   *Evaluation of each rule*

Let us denote the set of N_{pop} fuzzy rules in the current population by S. The set of training examples is classified by the FRBCS with the rule set S using the classical fuzzy reasoning method where an example is classified according to the class suggested by the single winner rule that best matches the example (see Sec. 1.5.3). This feature greatly simplifies the credit assignment mechanism, as there is always a unique winner rule to be rewarded or penalised depending on whether it correctly predicted the class of the training example. The winner rule obtains a reward of '1' if it correctly classifies the training example and a reward of '0' otherwise.

The fitness value of each fuzzy rule R_j is determined by the total reward assigned to that rule as follows:

$$fitness(R_j) = NCP(r_j)$$

where $NCP(R_j)$ is the number of training patterns that are correctly classified by R_j. The fitness value of each fuzzy rule is updated at each generation of the FCS.

An alternative credit assignment scheme that explicitly penalises misclassification will be discussed in the subsection devoted to the *extensions* of this model.

In case more than one fuzzy rule with the same consequent class have the same activation level, the tie is broken by assuming that the fuzzy rule with the smaller index j is the winner. For example, suppose that R_3 and R_9 in the rule set S are fuzzy rules with the same antecedent and the same consequent (i.e., R_3 and R_9 are the same fuzzy classification rule). In this case, R_3 and R_9 always have the same degree of activation, and R_3 is always chosen as the winner if no other rule has a larger activation. This means that the reward is assigned only to R_3. Therefore, only the unique rule with the smaller index j survives the generation update in the FCS as the other duplicate fuzzy rules necessarily obtain a total reward of zero. Thus, the FCS implicitly prevents that homogeneous, identical fuzzy rules dominate the population.

6.2.2.3 *Genetic operations for generating new rules*

In order to generate new linguistic classification rules, a pair of parent fuzzy rules is selected from the current population. Each fuzzy rule in the current population is selected according to the following selection probability based on the roulette wheel mechanism with linear scaling:

$$P(R_j) = \frac{fitness(R_j) - fitness_{min}(S)}{\sum_{R_i \in S}\{fitness(R_i) - fitness_{min}(S)\}}$$

where $fitness_{min}(S)$ is the minimum fitness value of the fuzzy classification rules in the current population S.

From the selected pair of fuzzy classification rules, two new rules are generated by means of uniform crossover applied to their antecedent linguistic terms. Notice that only the antecedent linguistic terms of the selected pair of rules are mated.

According to a small predefined mutation probability, each antecedent label of the fuzzy rules generated by the crossover operation might be replaced with a new random antecedent label. This mutation operator randomly changes a rule position. As in the crossover operation, the mutation operation is only applied to the antecedent labels. The consequent class and the certainty degree of each of the newly generated fuzzy rules are determined by the heuristic procedure after crossover and mutation were applied.

These genetic operations (i.e., selection, crossover, and mutation) are repeatedly applied until as many new fuzzy rules are generated as there are individuals to be replaced in the current population.

6.2.2.4 *Rule replacement and termination test*

At each generation, the N_{rep} worst fuzzy rules in the population, are replaced by the newly generated offspring rules.

In order to generate N_{rep} rules, the genetic operations are iterated $N_{rep}/2$ times. That is, $N_{rep}/2$ pairs of rules are selected from the current population by the selection operation, and two new linguistic rules are generated from each pair by the crossover and mutation operators.

The algorithm terminates after a maximum number of generations elapsed. In the Pittsburgh approach with an elite strategy, the final population necessarily always includes the best rule set. However, since evolution in the Michigan approach operates on individual rules rather than the entire RB, the final population does not necessarily represent the overall best solution. Therefore, the RB with the best classification rate over the training patterns that emerged in the course of all generations constitutes the final solution of the FCS. This problem of how to identify the optimal RB will be addressed in the subsection that analyses the extensions of the basic approach that includes a penalty term for misclassified training examples.

6.2.2.5 *Algorithm*

The FCS is composed of the following steps:

> *(1) Generate an initial population of N_{pop} fuzzy rules by randomly specifying the antecedent linguistic term of each rule. The consequent class and the degree of certainty are determined by the heuristic procedure.*

(2) Classify all the given training patterns by the fuzzy rules in the current population, then calculate the fitness value of each rule.

(3) Generate N_{rep} fuzzy rules from the current population by means of the selection, crossover and mutation operator. The consequent class and the degree of certainty of each fuzzy rule are again determined by the heuristic procedure.

(4) Replace the worst N_{rep} fuzzy rules with the smallest fitness values in the current population with the newly generated rules.

(5) Terminate if the maximum number of generations elapsed, otherwise return to Step 2.

6.2.2.6 *Fuzzy classifier system for learning classification rules: extensions*

The previous subsections explained the basic version of the FCS proposed by Ishibuchi, Nakashima, and Murata (1999). They also suggested some directions to extend the FCS, that are briefly introduced in the following.

Misclassification penalty

Since each fuzzy rule receives a reward of '1' when it correctly classifies a training pattern, it is possible that a rule achieves a high fitness value even though it misclassified a large number training patterns. In the basic FCS, such fuzzy rules survive the generation update because of their high fitness value.

The idea of a negative reward for misclassification is to penalise those fuzzy rules in the selection and replacement step. Let us denote the penalty for the misclassification of a single training pattern by w_{error}. Using the misclassification penalty, the fitness value of each rule is modified as follows:

$$fitness(R_j) = NCP(R_j) - w_{error} \cdot NMP(R_j)$$

where $NCP(R_j)$ is the number of correctly classified training patterns by the rule R_j, and $NMP(R_j)$ is the number of misclassified patterns.

Learning of the degree of certainty

It has been demonstrated that the performance of fuzzy *"IF-THEN"* rules is improved by adapting the degree of certainty of each rule (Nozaki, Ishibuchi, and Tanaka, 1996). The learning procedure introduced in that paper employs the following simple reward-and-penalty scheme:

- When a training pattern is correctly classified by a fuzzy rule (say, R_j), the degree of certainty is increased as a reward for the correct classification by

$$CF_j^{new} = CF_j^{old} + \mu_1 \cdot (1 - CF_j^{old})$$

where μ_1 is the learning rate for increasing the degree of certainty.

- On the other hand, if the training pattern is misclassified by the fuzzy rule R_j, its degree of certainty is reduced as a penalty for the misclassification by

$$CF_j^{new} = CF_j^{old} - \mu_2 \cdot CF_j^{old}$$

where μ_2 is the learning rate for decreasing the degree of certainty.

This mechanism can be combined with the FCS to design a hybrid algorithm. In the hybrid algorithm, the adaptation of the rule certainty is carried out prior to the computation of its fitness value.

Alternative coding

It is possible to introduce an alternative coding scheme in the actual FCS such as the one proposed by Valenzuela-Rendón (1991b), which allows more than one label per variable and is therefore able to represent DNF fuzzy classification rules.

Heuristic procedure for generating an initial population

As was pointed out by Nakaoka, Furuhashi, and Uchikawa (1994), the performance of a randomly generated initial population is usually very poor. Therefore, it is possible to use an initialisation mechanism to seed the first generation with rules that better capture the underlying training examples: i) to generate m rules from the given m patters, ii) to replace with a pre-specified probability antecedent fuzzy sets with "don't care" conditions, and iii) to select randomly N_{pop} rules from the m generated.

Variable population size

Since the number of rules in each population is constant, the basic FCS can not simultaneously address the conflicting objectives of rule selection, namely to maximise the classification rate and to minimise the number of rules. The extension to variable population size is another problem to be tackled.

6.3   Fuzzy Classifier Systems for Learning Fuzzy Rule Bases

A number proposals are concerned with FCSs that learn the fuzzy rule set of an approximate Mamdani-type FRBS (Parodi and Bonelli, 1993; Velasco and Magdalena, 1995; Velasco, 1998). Similar to the linguistic FCS, each chromosome represents an individual fuzzy rule, and the entire population forms the approximate FRB. The main difference between the approximate and the linguistic FCS is that the rule code encodes the membership function parameters of the underlying fuzzy sets rather than just the linguistic labels that refer to the globally defined fuzzy sets.

Parodi and Bonelli (1993) introduced the first proposal of classifiers that code a fuzzy rule with parameters. Velasco and Magdalena (1995) and Velasco (1998) described an FCS particularly suited for on-line learning in control applications. Chapter 11 describes an application of this approach to the supervisory control of a power plant.

In the following we describe these two approaches.

6.3.1   *Parodi and Bonelli's fuzzy classifier system for approximate fuzzy rule bases*

Parodi and Bonelli (1993) presented an FCS that learns an approximate Mamdani-type FRB in which the relevance of a rule is determined by its weight. The message list contains only input messages as their classifier rules only refer to the output variables rather than posting messages themselves. Therefore, the FCS realises a direct mapping from inputs to outputs, rather than a decision process based on a chain of sequentially activated classifiers. The dynamics of the resulting FCS are less complex and therefore avoid the need for a sophisticated credit assignment scheme such as the bucket brigade mechanism.

6.3.1.1   *Basic model*

The two main features of the system are the rule syntax and semantics, and the computation of output weights.

Rule syntax and semantic
The GA adapts the parameters of the membership functions of the approximate fuzzy rules R_k defined over the input X_i and output variables Y_j. The triangular symmetric membership function of the i-th input in the

k-th fuzzy rule is specified by its centre point x_{cik} and its width x_{wik}.

In the implementation by Parodi and Bonelli (1993), the GA only adapts the centre positions using a real-valued code, whereas the widths of the membership functions remain constant and are defined in advance. The rule R_k is therefore described by the parameters:

$$R_k : ((x_{c1k}, x_{w1k}); (x_{c2k}, x_{c2k}); ...; (x_{cnk}, x_{cnk})) \rightarrow (y_{ck}, y_{wk})$$

Computing the output weights

Most practical applications of FRBSs assume that all fuzzy rules are equally important and ignore the issue of separate rule weights. However, this approach adapts the weight s_k of a rule R_k based on its strength (fitness) acquired during evaluation.

The classifier population contains a fixed number of rules. The rule strength s_k has a dual purpose: first it forms the fitness criteria for selection and replacement in the GA, and second it modifies the inference mechanism in favour of stronger rules as the contribution of a rule to the overall output is weighted by its strength:

$$B' = max_k\{s_k \cdot B'_k\}, \qquad \text{or} \qquad B' = \sum_k s_k \cdot B'_k$$

with B'_k being the output fuzzy set of rule R_k resulting from the normal max-min inference step.

6.3.1.2 *Description of the algorithm*

The rule R_k is expressed as:

$$R_k = A_{1k}, A_{2k}, ..., A_{nk} \rightarrow B_k$$

with

$$\forall i \in \{1, 2, ..., n\}, A_i k = (x_{cik}, x_{wik}) \text{ and } B_k = (y_{ck}, y_{wk})$$

In the following we describe the three components of the learning algorithm.

Performance component

The antecedents of the K fuzzy rules in the population are compared with the input vector $x = (x_1, ..., x_n)$ and the matching value h_k is computed as:

$$h_k = min_i\{\mu_{A_{ik}}(x_i)\}$$

The output fuzzy set of the rule is multiplied by its matching degree

$$B'_k = h_k \cdot B_k$$

The output fuzzy sets B'_k are weighted by their respective rule strengths s_k and aggregated into the final output fuzzy set:

$$B' = \sum_k s_k \cdot B'_k$$

Finally, B' is defuzzified in order to obtain the crisp output y.

Reinforcement component

As the FCS learns in a supervised fashion, the reward is computed based on the discrepancy between the desired crisp output y' and the FCS output y as:

(1) The overall reward is computed as:

$$Total\ Payoff = R \cdot \frac{c}{c + distance(y, y')}$$

where R (maximal reward) and c (payoff constant) are predefined parameters.

(2) For every individual rule R_k, calculate its measure of goodness (MG), defined by:

$$MG(R_k) = \begin{cases} 1 - |\mu_{B_k}(y') - h_k| & \text{if } h_k > 0 \\ 0 & \text{if } h_k = 0 \end{cases}$$

(3) Deduct a tax t from all matched rules R_k (i.e., for which $h_k > 0$).

(4) Distribute to each rule R_k its share of the total payoff:

$$Total\ Payoff \cdot \frac{MG(R_k)}{\sum_k MG(R_k)}$$

The *MG* function not only determines the quality of the rule output fuzzy set, but also the accuracy h_k with which the antecedent matches the current crisp input.

Discovery component

The GA has the basic following features:

- In crossover, randomly selected fuzzy sets (input and output) are exchanged between pairs of rules.

- Mutation adds a small, random value to the continuous input and output fuzzy set parameters. As said, in the implementation considered in this paper, width values x_{wik} and y_{wk} are not altered by the GA.

6.3.2 *Fuzzy classifier system for on-line learning of approximate fuzzy control rules*

An FCS for on-line learning of FRBs in control applications is analysed by Velasco and Magdalena (1995); Velasco (1998). In this approach —based on working with approximate Mamdani-type FRBs— new rules generated by the GA are placed in a so-called *Limbo*, an intermediary repository where rules are evaluated off-line before being applied to the real process. Only when a rule demonstrated its usefulness in the off-line evaluation, it is accepted and inserted into the on-line FRB. The evaluation algorithm assigns a strength to every classifier with a non-zero degree of activation, regardless of whether the rule is currently under evaluation in the Limbo or used by the actual FLC. In fact, the algorithm evaluates the quality of a rule based on the similarity between its own proposed action and the actual action applied by the controller. This means that the strength of the rules in the Limbo is updated even though they do not contribute to the decision taken by the on-line FLC that governs the process. The GA generates new rules which are tested in the Limbo first, before they might eventually be incorporated into the FRB of the controller.

6.3.2.1 *The proposed fuzzy classifier system*

The rules
The FLC operates with approximate rules based on trapezoidal fuzzy sets, such as the one:

$$IF \ V_1 \in [0.1, 0.2, 0.4, 0.5] \ and \ V_3 \in [0.3, 0.3, 0.4, 0.6]$$
$$THEN \ V_7 \in [0.2, 0.3, 0.3, 0.4]$$

The antecedents (conditions) are formed by a variable number of clauses (from one to the number of input variables) arranged in arbitrary order*. The rule consequents (actions) are unique, i.e., even in the case of multiple-

---

*The genetic operators operate individually on the term of a particular variable, and are not affected by the position of that variable in the code.

output systems, each rule refers to a single output variable only. This restriction facilitates the credit assignment step and allows it to evaluate the strength of a rule solely based on the unique action it proposes. Each rule is associated to a set of parameters: the rule strength (credibility of the rule), the average rule strength over the course of evaluation, life span, etc. These performance parameters affect the learning algorithm as well as the fuzzy inference mechanism itself.

The rules are coded using a variable-length chromosome, containing a list of pairs: variable label (an integer), and parameters of the corresponding fuzzy set (four reals to represent the characteristic points). The last pair in the rule always refers to an output variable. It is important to notice that some variables might simultaneously assume the role of input and output variables[†]. If such a variable is located in the last position, it is automatically considered as an output variable, whereas at any other location within the rule, it is assumed to be an input variable.

Genetic operators

The genetic operators are particularly adapted to the specific coding scheme:

- *Selection:* Selection is based on classical roulette wheel.
- *Crossover:* The chromosome is reordered such that the locations of identical input variables in the antecedents of the two rules become aligned before crossover is applied. The chromosomes are expanded such that input variables that appear in the partner rule but are missing in the own rule code are represented by empty fuzzy clauses depicted by the empty rectangles in Fig. 6.2. Subsequent to this rule code unification, uniform crossover is applied to the parent rules (see Sec. 2.2.3.5). As the rule output always refers to a single variable, reordering and expansion for the consequent are obsolete.

- *Soft and Hard Mutation:* Soft mutation adjusts the fuzzy membership function whereas hard mutation changes the variable of the fuzzy clause (see Fig. 6.3). Soft mutation modifies one of the characteristic points (a, b, c, d) of the trapezoidal fuzzy set. The parameter range of the soft mutation is limited to the intervals

---

[†]If it is interpreted as an input, it denotes the current value of the variable at time $(x(t))$. If it is interpreted as an output, its value defines the set-point (desired value) of the variable in the next time step $(x(t + 1))$.

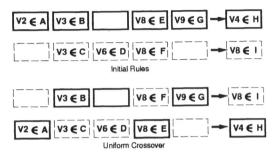

Fig. 6.2   Uniform crossover

$([0, a + \frac{b-a}{2}], [a + \frac{b-a}{2}, b + \frac{c-b}{2}], [b + \frac{c-b}{2}, c + \frac{d-c}{2}], [c + \frac{d-c}{2}, 1])$. Hence, this operator induces a soft constrained learning process (see Sec. 1.4.2). On the other hand, hard mutation assigns a new random fuzzy set to either an input or output variable.

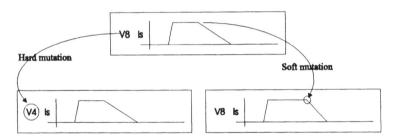

Fig. 6.3   Hard and soft mutation

Fitness function

The objective of the system is to minimise a certain cost function characterising the behaviour of the system. In the case described in Sec. 11.3.3, the objectives were to minimise the fuel consumption and the emissions caused by the plant operation.

Minimisation of a dynamical cost function such as fuel consumption is not equivalent to the case of fuzzy modelling (see Sec. 11.2) which has the objective to minimise the error between the model output and the desired output over the static set of training examples. The main difference is related to the fact that the modelling error has a known lower limit (zero), while in case of an accumulated cost function we only know that a lower cost (e.g. fuel consumption) is preferable to a higher one. It is usually impossible to estimate the optimal fuel consumption in advance, as it depends on the particular circumstances of the plant operation.

To resolve the problem of an unknown target cost, the system performance is evaluated by means of a performance indicator in the range [-1,1], that describes how good (1) or bad (-1) the current value of the cost function (C) is, according to its range of values (Velasco and Magdalena, 1995). Reducing the immediate cost alone does not necessarily result in the overall optimal solution, as the control action depends on the current operating conditions and effects the future states of the process which in turn determine the future costs. The proposed system minimises the cost function over a time horizon of the last N control steps: in other words, the performance measure SP compares the current cost C_t with respect to the minimal and maximal costs C_{min}, C_{max} that occurred within the past N steps (Magdalena, Velasco, Fernández, and Monasterio, 1993):

$$SP = \begin{cases} 1 & \text{if } C_t \leq C_{min} \\ \frac{C_t - C_{t-1}}{C_{min} - C_{t-1}} & \text{if } C_t > C_{min} \wedge C_t \leq C_{t-1} \\ \frac{C_{t-1} - C}{C_{max} - C_{t-1}} & \text{if } C_t < C_{max} \wedge C_t > C_{t-1} \\ -1 & \text{if } C_t > C_{max} \end{cases}$$

The performance measure SP evaluates the overall behaviour of the system. However, the Michigan approach requires it to estimate the impact of the individual rule actions on the current process state. To evaluate that influence, the fuzzy set (A) in the consequent of rule R (remember that each rule refers to a single output variable) suggested for the variable Y is compared with the current crisp output y of Y at times t and $t-1$. The rule influence is computed as:

$$RI_R = \begin{cases} \mu_A(y_t) & \text{if } \mu_A(y_t) > 0 \\ 0 & \text{if } \mu_A(y_t) = 0 \wedge |A - y_t| \leq |A - y_{t-1}| \\ \frac{y_t - y_{t-1}}{y_{t-1} - Y_l} & \text{if } \mu_A(y_t) = 0 \wedge |A - y_t| > |A - y_{t-1}| \end{cases}$$

where $|A - y|$ is the distance between the crisp action y and the fuzzy set A (i.e., the closest distance between y and any point in the support of A), and Y_l is either the upper (if $y_t > y_{t-1}$) or lower (if $y_t < y_{t-1}$) limit of the universe of discourse of variable Y. The index RI_R assigns positive influence to a rule R, in case the actual applied control action lies within the support of the output fuzzy set, a zero influence in case it lies outside of the support but is moving towards it (i.e., the current control action is closer to the support than the previous control action), and negative influence in case the control action lies not only outside the support but, in addition, is moving away from it.

The rule evaluation E_R is the product between the estimated rule influence and the current performance measure:

$$E_R = SP \cdot RI_R \qquad (6.1)$$

In case of a positive performance (the process is working well), rule evaluation depends on the rule influence: if the rule contributed to the actual control action and thereby the desirable evolution of the system state, the final evaluation becomes positive, in the opposite case, it becomes negative. Dual considerations apply to the case of negative performance. Those rules that proposed control actions different from the actually applied actions that caused the undesirable behaviour of the plant, obtain a positive reward, as the poor performance suggests that the controller should evaluate some other, untested action in that particular state. The rule evaluation is weighted by its overall truth value, to obtain the final value[†].

For modelling problems and for regulation problems in control, a different evaluation function is proposed (Velasco, 1998). In this case, the evaluation of the rules depends on its power to predict the correct output associated to the current input. The evaluation of a rule measures the degree to which its predicted output matches the actual observed output. For a rule consequent $< Y$ is $A >$, the evaluation function is defined as:

$$E_R = \begin{cases} \mu_A(y) & \text{if } \mu_A(y) > 0 \\ \frac{y-a}{D_{max}} & \text{if } y < a \\ \frac{d-y}{D_{max}} & \text{if } y > d \end{cases} \qquad (6.2)$$

in which y is the observed target value, a and d are the left and right points of the support of A, and D_{max} is the maximum distance between the outer points of the support to the limits that define the universe of discourse of Y (see Fig. 6.4).

[†]The most significant aspect of this evaluation scheme is that it implicitly evaluates the performance of every rule with a positive truth degree, regardless of whether it has been actually applied to the plant or not.

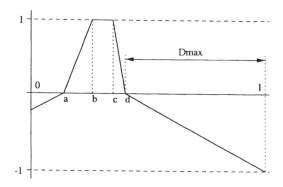

Fig. 6.4   Evaluation function

The fitness of a rule is adapted according to its rule evaluation (E_R) obtained by either of the performance measures given in Eqs. 6.1 and 6.2. Rules increase their strength if their evaluation is positive, and decrease it otherwise. The rule strength is modified according to:

$$S_{R,t} = \begin{cases} S_{R,t-1} + K \cdot T_R \cdot E_R \cdot (1 - S_{R,t-1}) & E \geq 0 \\ S_{R,t-1} + K \cdot T_R \cdot E_R \cdot S_{R,t-1} & E < 0 \end{cases} \tag{6.3}$$

in which the constant K determines the adaptation rate —for small values of K the rule strength changes slowly over time—, T_R is the truth value of rule R, and E_R is its evaluation.

6.3.2.2   *On-line learning of fuzzy rules using the Limbo*

The GA generates new classifier rules which are first stored in the *Limbo*, an intermediary repository with the purpose to evaluate these candidate fuzzy rules off-line before they are finally integrated into the on-line controller that regulates the real process. The evaluation algorithm adapts the strength (from its initial value of 0.5) of rules in the Limbo, even though they momentarily do not contribute to the decision taken by the on-line FLC that regulates the process. The performance of a candidate rule is estimated, based on the similarity between its own proposed action and the actual action applied by the on-line controller in the same situation. The decision to incorporate a rule from the Limbo into the actual FLC depends on the following parameters:

Rule parameters

- Rule Age ($R.A.$): Overall number of control steps since the rule was placed into the Limbo.

- Rule Activations ($R.Ac.$): Number of control steps in which the rule would have fired in the on-line controller, i.e. the rule antecedent matches the current input state to a non-zero degree.

- Equivalent Rule Evaluation ($E.R.E.$): Constant evaluation that the rule should have obtained to reach its present strength value after $R.Ac.$ activations. It is computed as the value E_r that if placed in Eq. 6.3, starting from an initial strength $S_{R,0} = 0.5$, evolves to the current strength $S_{R,t}$ within $R.Ac.$ time steps for $T_R = 1$.

Limbo parameters

- Limit Age ($L.A.$): Defines the maximum evaluation period. After $L.A.$ control steps, the rule is removed from the Limbo, being either promoted to the FRB or discarded.

- Minimum Rule Activations ($M.R.Ac.$): Minimum number of activations that are required for a rule to be considered for promotion from the Limbo to the FRB.

- Minimum Equivalent Evaluation ($M.E.E.$): Minimum value of the $E.R.E.$ that is required for a rule to be considered for promotion from the Limbo to the FRB.

Entering the FRB

Considering these parameters, whenever the GA generates new rules that are added to the Limbo, the quality of the current rules in the Limbo is estimated. The worst performing rules are discarded, whereas the best performing rules are promoted to the on-line FLC according to the following performance criteria:

- Rules whose evaluation period expired ($R.A. \geq L.A.$), and that were either mostly inactive ($R.Ac. < M.R.Ac.$) or obtained a poor evaluation ($E.R.E. < M.E.E.$) are removed from the Limbo and discarded.

- Sufficiently tested rules that underwent a minimum number of activations ($R.Ac. \geq M.R.Ac.$) and obtained a good evaluation ($E.R.E. \geq M.E.E.$) are promoted to the on-line controller.

- Rules that are not promoted and whose evaluation period did not expire yet are kept in the Limbo for further evaluation.

Chapter 7

Genetic Fuzzy Rule-Based Systems Based on the Pittsburgh Approach

The general structure of a GFRBS based on the Pittsburgh approach is shown in Fig. 7.1 which is an adapted version of Fig. 5.3. The major difference between the two figures is that the classical rule-based system has been replaced by an FRBS and that it includes additional components required for learning. The KB discovery system, implemented by an EA, maintains

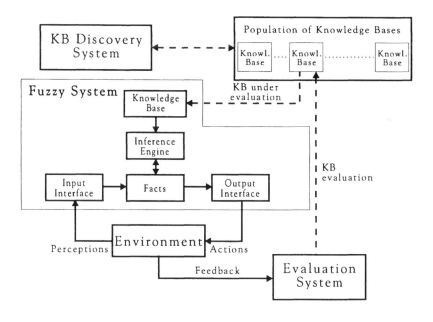

Fig. 7.1   Fuzzy system with learning based on Pittsburgh approach

and updates a population of KBs based on their individual performance observed by the evaluation scheme. The FS can be either an approximate or descriptive Mamdani-type FRBS (Fig. 1.1) or a TSK FRBS (Fig. 1.9).

As described in Sec. 5.3, the Pitt learning process operates on a population of candidate solutions, in this case FRBSs. As the inference engine itself is not subject to adaptation, all FRBSs employ the identical fuzzy reasoning mechanism and the individuals in the population only encode the fuzzy rule sets themselves rather than the entire FS.

GFRBS based on the Pitt approach focus the learning process on the RB (Sec. 7.1), but might in addition incorporate a part of the DB (scaling or membership functions) to that process. In the second case, the chromosomes contain heterogeneous information about the RB and the corresponding membership functions, which requires the use of multi-chromosome codes (Sec. 7.2). Therefore, the representational issue becomes more complex than for the case in which the chromosome encodes the RB alone as described in Sec. 5.3. Finally, Sec. 7.3 describes in more detail some particular applications of the approaches described in Secs. 7.1 and 7.2.

As GFRBSs based on the Pittsburgh approach employ a population of KBs, they properly solve the CCP (described in Sec. 3.4.6). The fitness function explicitly measures the cooperation of fuzzy rules that belong to an individual as the evaluation scheme observes the performance of the RBs or KBs as a whole, rather than that of isolated rules. Therefore, the GFRBS evolves the RBs or KBs in a way that automatically achieves an optimal cooperation among the fuzzy rules.

However, this approach faces the drawback that the dimension of the search space increases significantly, making it substantially more difficult to find good solutions. The genetic representation used in GFRBSs based on the Pittsburgh approach includes a large number of necessary parameters in order to define the KB. In fact, the FRBS parameters considered in the learning process condition the size and properties of the search space, which in turn determines the complexity and feasibility of the optimisation process.

7.1   Coding Rule Bases as Chromosomes

All those questions described in Section 5.3.4 concerning the representation aspects of Pittsburgh approach when working with rule-based systems apply now. The two main issues, which are tightly interrelated, are the use of fixed or non-fixed length codes, and the use of positional or non-positional representation schemes.

Fixed length codes are applicable to RB representations that possess a static structure such as a decision table (see Sec. 1.2.1) or a relational matrix. The obtained code is positional, that is, the functionality of a gene is determined by its position, and its value defines the corresponding attribute. A specific location in a positional chromosome refers to a certain entry in the decision table or the relational matrix, and the gene value defines the output fuzzy set associated to that entry. The number of genes in the chromosome is identical to the number of elements in the decision table or the relational matrix. It is also possible to rely on a fixed length code when working with a list of rules, if the number of rules is known in advance and remains constant. The code is non-positional since it represents a list of rules, and the order of rules in an FS is immaterial to their interpretation.

When working with RBs represented as a list of rules with variable number of rules, the code length becomes variable as well. Again, the code of an individual rule can be either represented with a fixed or non-fixed length code. In the first case, the sub-code of each rule is positional, considering one particular gene per variable or per fuzzy set. In the second case, genes employ a composed code that includes both the variable and the associated fuzzy set. This code explicitly indicates to which part of the fuzzy rule it refers to, in contrast to a positional coding scheme in which this information is implicit in the gene position.

7.1.1 *Positional semantics*

This situation refers to FSs that operate with a complete RB represented by means of a decision table or a relational matrix. A tabular representation guarantees completeness of the knowledge of a FRBS in the sense that the coverage of the input space (the Cartesian product of universes of the input variables) is only related to the level of coverage of each input variable (the corresponding fuzzy partitions), and not to a proper combination of rules.

7.1.1.1 *Decision tables*

Many GFRBSs employ the decision table as the common, classical representation for the RB of an FRBS. A fuzzy decision table represents a special case of a crisp relation (the ordinary type of relations we are familiar with) defined over the collections of fuzzy sets corresponding to the input and output variables. Fig. 7.2 shows a fuzzy decision table for an FRBS with two inputs (X_1, X_2) and one output (Y) variable, with three fuzzy sets $(A_{11}, A_{12}, A_{13}; A_{21}, A_{22}, A_{23})$ related to each input variable and four fuzzy sets (B_1, B_2, B_3, B_4) related to the output variable.

	A_{21}	A_{22}	A_{23}
A_{11}		B_1	B_2
A_{12}	B_1	B_2	B_3
A_{13}	B_1	B_3	B_4

Fig. 7.2   A fuzzy decision table

Representation

A chromosome is obtained from the decision table by going row-wise and coding each output fuzzy set as an integer or any other kind of label. It is possible to include the "no output" definition in a certain position, using a "null" label (Ng and Li, 1994; Thrift, 1991). Applying this code to the fuzzy decision table represented in Fig. 7.2, one obtains the string (0,1,2,1,2,3,1,3,4).

Operators

Standard crossover operators such as one-point, two-point, multi-point and uniform crossover described in Sec. 2.2.3.5 operate on these strings in the usual manner, although the code is not binary ($\{0,1\}$) but takes values from a finite and ordered set (usually $\{0, \ldots, N\}$). Figure 7.4 shows the effect of two-point crossover applied to the decision tables in Fig. 7.3 to generate those in Fig. 7.5.

	A_{21}	A_{22}	A_{23}		A_{21}	A_{22}	A_{23}
A_{11}		B_1	B_2	A_{11}		B_2	B_2
A_{12}	B_1	B_2	B_3	A_{12}	B_3	B_1	B_3
A_{13}	B_1	B_3	B_4	A_{13}		B_3	B_4

Fig. 7.3   Parents (decision tables)

Fig. 7.4   Two-point crossover

	A_{21}	A_{22}	A_{23}			A_{21}	A_{22}	A_{23}
A_{11}		B_2	B_2		A_{11}		B_1	B_2
A_{12}	B_3	B_2	B_3		A_{12}	B_1	B_1	B_3
A_{13}	B_1	B_3	B_4		A_{13}		B_3	B_4

Fig. 7.5   Children (decision tables)

The mutation operator for this coding scheme usually changes the code of the fuzzy label encoded in the mutated gene either one level up or one level down, or to zero. In case the code is already zero, a non-zero code is chosen at random.

7.1.1.2   *Relational matrices*

A fuzzy relationship over the collections of fuzzy sets $A_1, \ldots, A_m, B$, is a fuzzy subset over their Cartesian product, $A_1 \times \ldots \times A_m \times B$. In a fuzzy relationship, each element of the Cartesian product $(A_{1i}, \ldots, A_{mj}, B_k)$ has a degree of membership to the relation. This degree of membership is represented by $\mu_R(A_{1i}, \ldots, A_{mj}, B_k) \in [0, 1]$. Fig. 7.6 shows the tabular representation of a fuzzy relation for an FRBS with one input (X) and one output (Y) variable, with three fuzzy sets (A_{11}, A_{12}, A_{13}) related to the input variable and four fuzzy sets (B_1, B_2, B_3, B_4) related to the output variable.

R	B_1	B_2	B_3	B_4
A_{11}	0.5	0.8	0.2	0.0
A_{12}	0.0	0.3	1.0	0.1
A_{13}	0.0	0.0	0.3	1.0

Fig. 7.6   A fuzzy relation R

Representation

Occasionally, GAs are used to modify the fuzzy relational matrix R of an FS with one input and one output. The chromosome is obtained by concatenating the $M \cdot N$ elements of R, where M and N are the number of fuzzy sets associated with the input and output variables respectively. The elements of R that will make up the genes may be represented by binary codes (Pham and Karaboga, 1991) or real numbers (Park, Kandel, and Langholz, 1994), e.g., (0.5,0.8,0.2,0.0,0.0,0.3,1.0,0.1,0.0,0.0,0.3,1.0) for the relation defined by Fig. 7.6.

Operators

In the work by Park, Kandel, and Langholz (1994), one-point crossover is applied while mutation is performed by replacing a gene with a new value being 90% or 110% of the previous one (fifty-fifty chance for each case). Considering that the system can be classified as an RCGA, any of the operators defined in Sec. 2.2.4 is applicable. Pham and Karaboga (1991) applied one-point crossover working with a binary code.

7.1.1.3 *TSK-type rules*

FSs working with TSK-type rules can be coded by a positional or a non-positional code in the same way as Mamdani-type rules.

Considering that the antecedents of Mamdani and TSK FRBSs share an identical structure, a TSK system can be represented as a sort of decision table. That way, the mapping from antecedents to cells in the decision table is exactly the same as in a positional code. The only difference lies in the contents of each cell, the label to a linguistic term of the output variable (Mamdani rules) or the set of parameters of a linear equation (TSK rules).

After these considerations, it is possible to define approaches to learn the RB of a TSK system using positional coding schemes.

Representation

Assume a TSK system has n input variables $X_1, \dots, X_n$, with $N = \{N_1, \dots, N_n\}$ linguistic terms per variable and a single crisp output variable Y. The RB of the system is represented as a decision table having $L_r = \prod_{i=1}^{n} N_i$ cells or positions. Each of these positions contains $n + 1$ real numbers representing the parameters of the equation $Y = w_0 + \sum_{i=1}^{n} w_i \cdot X_i$ that constitute the output of the system. As a result, a TSK RB, assuming a predefined DB, can be described as a list of $L_r \cdot (n+1)$ real numbers.

Lee and Takagi (1993c) applied this idea by coding the $L_r \cdot (n+1)$ numbers with a binary representation using eight bits per real. This code maps a binary string of length 2400 bits to the hundred rules of a two input TSK FRBS with ten linguistic terms for each variable.

Operators

Depending on the code, real or binary crossover and mutation operators are used. In the cited reference, two-point binary crossover is applied.

7.1.2 *Non-positional semantics (list of rules)*

Neither the relational nor the tabular representations are applicable to systems with more than two or three input variables because the number of cells in a complete RB grows rapidly with the dimension of the input. This drawback stimulates the idea of working with sets of rules, rather than decision tables or relation matrices. In a *set of rules* representation, the absence of applicable rules for a certain input that was perfectly covered by the fuzzy partitions of individual input variables is possible. This loss of completeness is compensated by allowing the *compression* of several rules with identical outputs into a single rule. The consequence is the reduction of the number of rules that becomes an important issue as the dimension of the system grows. As a counterpart, the use of non-positional semantics induces some problems in crossover process. The fact is that traditional crossover operators are positional and consequently, they exchange genes in chromosomes according to their position and not according to their functionality. This problem has to be considered when defining non-positional crossover operators.

When working with a multiple input system, decision tables become multidimensional. A system with three input variables produces three-dimensional decision tables. The number of "cells" (L_r) coded in an n-dimensional decision table is

$$L_r = \prod_{i=1}^{n} N_i$$

where the i-th component of vector N ($N = \{N_1, \ldots, N_n\}$) is the number of linguistic terms associated with the i-th input variable. Each cell of the decision table describes a fuzzy rule. We will refer to these fuzzy rules as elemental fuzzy rules.

The rule described by Eq. 7.1 is not an elemental fuzzy rule, but is a DNF rule (Sec. 1.2.6.1), a kind of rule applied by different GFRBSs.

$$IF\ X_i\ is\ \{C_{io}\ or\ C_{ip}\}\ THEN\ Y_j\ is\ \{D_{jq}\ or\ D_{jr}\} \qquad (7.1)$$

An elemental fuzzy rule must contain fuzzy inputs for all input variables, without using 'or' operators. A rule containing an 'or' operator replaces two elemental fuzzy rules (the rule is equivalent to the aggregation of the elemental rules when working with the connective *also* as the union operator). A rule that has no fuzzy input for a certain input variable replaces as many elemental rules as the number of fuzzy sets defined for the input variable. The rule includes the aggregation of the elemental rules. The number of elemental rules replaced by a certain rule is obtained by multiplying the number of rules replaced for each variable. Considering an FRBS with three inputs and $N = \{5, 3, 5\}$ ($L_r = 75$), the fuzzy rule

$$\begin{aligned} IF\ X_1\ is\ &\{C_{13}\ or\ C_{14}\}\ and\ X_3\ is\ \{C_{31}\ or\ C_{32}\} \\ THEN\ Y\ is\ &\{D_{14}\ or\ D_{15}\} \end{aligned} \qquad (7.2)$$

replaces $2 \cdot 3 \cdot 2 = 12$ elemental fuzzy rules.

When the FS works with a few input variables, the fuzzy relational matrix or the fuzzy decision table allow an appropriate representation to be used by the GA. When the number of input variables increases, the size of the decision table and thereby the number of elemental fuzzy rules grows exponentially. Keeping in mind that the rules acquired from expert knowledge usually only contain a subset of the input variables, it is possible to argue that the number of elemental rules replaced by such a more comprehensive rule increases exponentially as well. As a consequence, the number of rules grows at a considerably slower rate when working with a set of rules rather than complete decision tables. As a counterpart of this important advantage, the set of rules representation has to deal with the issues of incompleteness or inconsistency. These problems may either be avoided through an appropriate design of the FRBS, or by explicitly incorporating some consistency and completeness criteria (see Sec. 1.4.6) into the learning strategy (usually into the fitness function).

Before starting on describing specific models, a common point must be stated: the natural way to code a list of rules is by concatenating the individual code of each rule. Consequently, given a system with k rules, the overall code of the RB will be $(r_1|r_2|\ldots|r_k)$, where r_i represents the code of the i-th rule, and $|$ defines a concatenation operator. In order to mark the start of a *new rule*, the code might contain a separate symbol for the concatenation operator.

7.1.2.1 *Rules of fixed length*

Representation

There are many different methods to code the RB in this kind of evolutionary system. The code of the RB is usually obtained by concatenating individual rule codes. No *new rule* symbol is needed since in a positional code of fixed length, the rule code is fixed and consequently the locations on the chromosome indicate the start and end of a particular rule code. The next step is, then, to define the code of a single rule.

The overall number of fuzzy sets in the DB is L:

$$L = L_a + L_c$$

$$L_a = \sum_{i=1}^{n} N_i$$

$$L_c = \sum_{j=1}^{m} M_j$$

where n and m are the number of input and output variables, N_i represents the number of linguistic terms associated to input variable X_i, and M_j the number of linguistic terms associated to output variable Y_j. Consequently, the DB of an FS with parameters $n = 3$, $m = 1$, $N = \{5,3,5\}$, $M = 7$ contains $L = 20$, $L_a = 13$, $L_c = 7$ fuzzy sets and a decision table with $L_r = 75$ entries. The number of entries provides an upper limit to the number of rules codes considering the rule matrix representation shown in Sec. 1.2.1.

A first approach is to represent a rule with a code of fixed length and position-dependent meaning. The code contains as many elements $n+m$ as there are fuzzy variables in the system. In a straightforward representation, each element carries a label j that points to a particular fuzzy set C_{ij} in the fuzzy partition $\{C_{i1}, C_{i2}, \ldots, C_{in_i}\}$ of variable X_i. The code (3 1 4 5) refers to the fuzzy sets C_{13}, C_{21}, C_{34}, D_{15} that constitute the following rule

IF X_1 is C_{13} and X_2 is C_{21} and X_3 is C_{34} THEN Y is D_{15}

It is also possible to use a binary coding representation of these labels (Satyadas and KrishnaKumar, 1996) generating, for the previous example, a code of $3 + 2 + 3 + 3$ bits.

This standard code suffers from the drawback that it can only form elemental rules. As each label (in integer or binary representation) refers to a singular fuzzy set in the partition, it can not encode DNF rules as in Eq. 7.1 with 'or' operators. We postpone the discussion of rule codes with labels to the next section, which introduces a non-positional coding scheme that overcomes the aforementioned limitation.

As an alternative to the above coding scheme, González, Pérez, and Verdegay (1993) and Magdalena and Monasterio (1995) proposed a bit-string code that is able to represent DNF rules of the form shown in Eq. 7.1. For each input variable X_i, the code contains a binary sub-string of length N_i that refers to the fuzzy partition $\{C_{i1}, C_{i2}, \ldots, C_{in_i}\}$. Each bit s_k denotes the absence or presence of the linguistic term C_{ij} in the rule as shown in Fig. 7.7. The output linguistic labels D_{ij} are coded in a similar fashion. In our example, the code of a single fuzzy rule is composed of two binary strings, one string S^a of length $L_a = 5 + 3 + 5$ for the antecedent part and another string S^c of length $L_c = 7$ for the conclusion part.

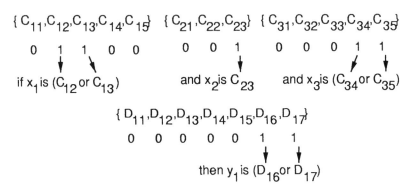

Fig. 7.7   Binary code for DNF rules of fixed length

In the example, bit $s_4 = 0$ in string S^a indicates the absence of the linguistic term C_{14}, whereas bit $s_8 = 1$ indicates the presence of term C_{23} in the rule. The coding scheme employs a similar mapping from the string S^b to the conclusion of the fuzzy rule. Considering all bits, the string

$(01100 : 001 : 00011 :: 0000011)$ encodes the fuzzy rule

> *IF X_1 is $\{C_{12}$ or $C_{13}\}$ and X_2 is $\{C_{23}\}$ and X_3 is $\{C_{34}$ or $C_{35}\}$*
> *THEN Y is $\{D_{16}$ or $D_{17}\}$*

A fuzzy (input or output) variable is omitted in the rule if all its bits in the code are zero. For example, since bits s_6, s_7, s_8 are all zero in the string $(00110 : 000 : 11000 :: 0110000)$, the variable X_2 does not occur in the fuzzy rule

$$
\begin{aligned}
&IF\ x_1\ is\ \{C_{13}\ or\ C_{14}\}\ and\ x_3\ is\ \{C_{31}\ or\ C_{32}\}\\
&THEN\ y_1\ is\ \{D_{12}\ or\ D_{13}\}
\end{aligned}
\tag{7.3}
$$

The RB contains a variable number of rules, with an upper limit given by the number of entries in the decision table $L_r = 75$ in order to avoid redundant rules. Therefore, the entire RB code is composed of up to 75 substrings of length $L = 20$ each. In principle, the entire input space can be covered with fewer than 75 rules, as a single DNF rule can replace multiple elemental fuzzy rules. Figure 7.8 shows how the comprehensive DNF rule shown in Eq. 7.3 replaces twelve elemental fuzzy rules. Nevertheless, the coding scheme does not guarantee the completeness of the fuzzy rule set as there might exist input states for which none of the DNF rules triggers. Therefore, the fitness function should incorporate a term that penalises incomplete KBs.

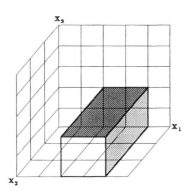

Fig. 7.8   The rule covers multiple *cells*

Classical operators

- *Crossover:* The usual way of merging lists of rules is to apply the binary crossover operators taking rules rather than bits as the atomic entity of information.

  Starting from two chromosomes representing two RBs:

$$r_i = \{r_{i1}, \ldots, r_{ik}\}$$
$$r_j = \{r_{j1}, \ldots, r_{jl}\}$$

two cutting points* are randomly defined. Working with cutting points α and β, we obtain:

$$r_i = \{r_{i1}, \ldots \ldots, r_{i\alpha} | r_{i\alpha+1}, \cdots, r_{ik}\} \qquad (7.4)$$
$$r_j = \{r_{j1}, \ldots, r_{j\beta} | r_{j\beta+1}, \ldots \ldots, r_{jl}\}$$

and the result after crossover is:

$$r_u = \{r_{i1}, \ldots \ldots, r_{i\alpha} | r_{j\beta+1}, \ldots \ldots, r_{jl}\} \qquad (7.5)$$
$$r_v = \{r_{j1}, \ldots, r_{j\beta} | r_{i\alpha+1}, \cdots, r_{ik}\}$$

This kind of crossover operator has an important drawback, since the recombination mechanism is solely based on the position rather than a similarity measure between rule codes. For lists of rules, the functionality of a gene is not related to its locus, and consequently, a positional recombination operates in completely blind and random manner. The effect is similar to that of independently selecting arbitrary genes from the two parents without considering their mutual role, such that the resulting offspring may end up with duplicate genes for the eye colour, whereas none that defines the hair colour.

- *Mutation:* The mutation operator applied to an RB works at the level of bits or integers that compose a rule.

- *Reordering:* In an FS characterised by a set of fuzzy rules, the order of the rules is immaterial to its behaviour since the applied connective *also* has properties of commutativity and associativity. Nevertheless, the order of the rules in the chromosome imposes a bias on the crossover operation. Therefore, a special operator that reorders rules is added to the system. It allows rules in subsequent crossover operations to be grouped in different ways. Such a reorder

---

*It is important to notice that we work with two independent cutting points, one per RB, because of the different length that each RB could have.

operator provides at least a partial remedy to the said problem in positional crossover.

To reorder an RB (r_i), a cutting point (γ) is randomly selected and a new RB (r_j) is created to replace (r_i).

$$
\begin{aligned}
r_i &= \{r_{i1}, \ldots, r_{i\gamma} | r_{i\gamma+1}, \ldots, r_{ik}\} \\
r_j &= \{r_{i\gamma+1}, \ldots, r_{ik} | r_{i1}, \ldots, r_{i\gamma}\}
\end{aligned}
$$

- *Alignment:* A reordering operator works on a single RB by changing the order of the list of rules. The effect is an increase in diversity but still without obtaining a true context-dependent crossover (the operator is still position-dependent). A variety of alignment procedures have been proposed in order to realise context-dependent crossover operators. Alignment takes two parent chromosomes and searches for similarities among their RBs. The objective is to reorder the two RBs in a way that rules with closely matching antecedent occupy similar positions on the corresponding chromosomes. This alignment prevents a pair of similar rules originating from both parents from being inherited by the same child. This method tries to assign similar rules to nearby loci.

Non-positional crossover operators

Reordering and alignment are only intermediate steps between a position-driven crossover and a functional-driven crossover. Reordering does not constitute a full, exhaustive solution, and in addition alignment in itself is a computationally complex operation. Magdalena (1998) proposed a functional-driven crossover operator for list of rules, by arranging the rules in the order they would assume in an equivalent decision table.

Considering a two-input one-output system with five linguistic terms per variable, the set of rules:

R_1 :   IF X_1 is $\{C_{11}$ or C_{12} or $C_{13}\}$ and X_2 is $\{C_{21}$ or C_{22} or $C_{23}\}$
          THEN y is D_5

R_2 :   IF X_1 is $\{C_{11}$ or C_{12} or $C_{13}\}$ and X_2 is $\{C_{23}$ or C_{24} or $C_{25}\}$
          THEN y is D_4

R_3 :   IF X_1 is $\{C_{13}$ or C_{14} or $C_{15}\}$ and X_2 is $\{C_{21}$ or C_{22} or $C_{23}\}$
          THEN y is D_2

R_4 :   IF X_1 is $\{C_{13}$ or C_{14} or $C_{15}\}$ and X_2 is $\{C_{23}$ or C_{24} or $C_{25}\}$
          THEN y is D_1

$$(7.6)$$

generates the distribution of rules and the decision table shown in Figs. 7.9
and 7.10.

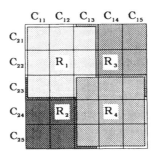

Fig. 7.9   Distribution of rules in the decision table

	C_{11}	C_{12}	C_{13}	C_{14}	C_{15}
C_{21}	D_5	D_5	$D_5 \cup D_2$	D_2	D_2
C_{22}	D_5	D_5	$D_5 \cup D_2$	D_2	D_2
C_{23}	$D_5 \cup D_4$	$D_5 \cup D_4$	$D_5 \cup D_4 \cup D_2 \cup D_1$	$D_2 \cup D_1$	$D_2 \cup D_1$
C_{24}	D_4	D_4	$D_4 \cup D_1$	D_1	D_1
C_{25}	D_4	D_4	$D_4 \cup D_1$	D_1	D_1

Fig. 7.10   Generated decision table

Once the set of rules is mapped onto the decision table, it is possible to
apply to this particular representation, a crossover operator that splits the

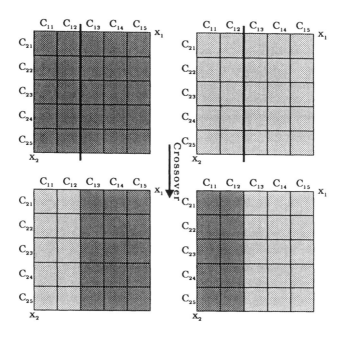

Fig. 7.11   Crossover of decision tables using a cutting surface

cells along a cutting surface rather than a cutting point (Fig. 7.11). This mechanism can be considered as assigning a virtual locus to each rule, and then cutting the RB according to these loci. There are two main problems to solve before this kind of operator can be applied:

(1) How to code the rules in a way that easy crossover is applicable without requiring the intermediate step of first transferring the rules into the equivalent decision table.
(2) How to resolve those conflicts that emerge when a rule is divided by the cutting surface.

• **Coding the rules and the cutting surface**

The code is partitioned into two disjoint fragments according to a pair of complementary *cutting rules* of the form: *IF the premise of the rule contains a certain fuzzy proposition THEN the rule goes to a specific fragment of code.*

Using the example of Fig. 7.11, those two cutting rules are:

$CutR_1$: IF the premise contains any of the propositions
                          $(X_1$ is $C_{11})$ or $(X_1$ is $C_{12})$
              $THEN$ the rule belongs to the first fragment of code
$CutR_2$: IF the premise contains any of the propositions
                          $(X_1$ is $C_{13})$ or $(X_1$ is $C_{14})$ or $(X_1$ is $C_{15})$
              $THEN$ the rule belongs to the second fragment of code

$$(7.7)$$

According to $CutR_1$ and $CutR_2$, the first fragment of the first parent will go to the first child (and the second to the second) and the first fragment of the second parent will go to the second child (and the second to the first).

With this interpretation in mind, the cutting process can be reduced to a matching operation among the codes of both the fuzzy and cutting rules.

With the coding scheme represented in Fig. 7.7, the rules described by Eq. 7.6 have the following code:

$$11100\ 11100\ 00001$$
$$11100\ 00111\ 00010$$
$$00111\ 11100\ 01000$$
$$00111\ 00111\ 10000$$

composed of four strings (one per rule) of $5 + 5 + 5$ bits.

If a similar code is applied to the antecedents of the cutting rules defined by Eq. 7.7, the result is

$$11000\ 00000$$
$$00111\ 00000$$

To enable the use of cutting rules, a binary *and* operator is applied to the pair of strings coding the antecedent of the fuzzy rule and the cutting rule.

$$
\left.
\begin{array}{l}
R_1 \text{ and } CutR_1 = 11000\ 00000 \\
R_1 \text{ and } CutR_2 = 00100\ 00000
\end{array}
\right\} \rightarrow \text{fragment ?}
$$
$$
\left.
\begin{array}{l}
R_2 \text{ and } CutR_1 = 11000\ 00000 \\
R_2 \text{ and } CutR_2 = 00100\ 00000
\end{array}
\right\} \rightarrow \text{fragment ?}
$$
$$
\left.
\begin{array}{l}
R_3 \text{ and } CutR_1 = 00000\ 00000 \\
R_3 \text{ and } CutR_2 = 00111\ 00000
\end{array}
\right\} \rightarrow \text{fragment 2}
$$
$$
\left.
\begin{array}{l}
R_4 \text{ and } CutR_1 = 00000\ 00000 \\
R_4 \text{ and } CutR_2 = 00111\ 00000
\end{array}
\right\} \rightarrow \text{fragment 2}
$$

$$(7.8)$$

Rules R_3 and R_4 are assigned to the second fragment of code but the classification of R_1 and R_2 remains ambiguous because the cutting surface cuts the rules in a way that parts of them belong to both sides of the surface. Whenever a rule is split by the surface, it is classified according to an external decision criterion.

- **Rules divided by the cutting surface**

  In the situation of rules that are divided by the cutting surface, it is possible to apply different criteria but there are three main possibilities: split the rule, make a stochastic decision or employ a deterministic criterion.

  - *Splitting the rule.* This method possesses the drawback of increasing fragmentation, such that after a number of generations merely elementary rules remain and ultimately the advantage of the list of rules gets lost. Still, this method can be used occasionally as some sort of mutation rather than a crossover operator.
  - *Stochastic decision.* A second option is to employ a probabilistic criterion based on the volume of the rule and of each resulting part. Using the results shown in Eq. 7.8, rules R_1 and R_2 produce a ratio of 2 to 1 active bits when evaluating to what portion of the rule belongs to which fragment of the code. This ratio resembles the distribution of volume that the cutting surface produces and is consequently interpreted as the likelihood for the rule to be assigned to a particular fragment. In this case, the probability is $p_1 = 2/3$ for the first fragment and $p_2 = 1/3$ for the second fragment.
  - *Deterministic decision.* Based on the same concept of distribution of volumes, it is possible to use a deterministic criterion that assigns the rule to the fragment that contains a larger number of 1's. For the above situation, rules R_1 and R_2 end up in the first fragment of code since the ratio is 2 to 1. If both fragments carry the same number of 1's, allocating the rule to the first fragment breaks the tie. Therefore, a rule that is cut in half and stems from the first parent is inherited to the first child, whereas the same rule in the second parent would go to the second child. That way crossover of identical parents produces children that themselves are identical copies of their parents. A stochastic decision criterion does not obey this paradigm, as both children although functionally identical to their parents contain a larger number of rules.

• Cutting in several variables

The cutting structure defined in the previous example produces a cut along a single variable, in this case X_1. It is possible to include conditions related to several variables in the definition of the cutting rules. That is the case of Fig. 7.12, representing a cut defined by means of two variables.

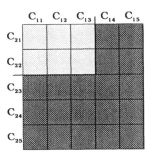

Fig. 7.12   Cut in two variables

The cutting rules corresponding to the situation depicted in Fig. 7.12 are described by the following expression:

$CutR_1$:   *IF*   the premise contains any of the propositions
                        $(x_1$ *is* $C_{11})$ *or* $(x_1$ *is* $C_{12})$ *or* $(x_1$ *is* $C_{13})$
                        and any of $(x_2$ *is* $C_{21})$ *or* $(x_2$ *is* $C_{22})$
          THEN   the rule goes to the first fragment of code

$CutR_2$:   *IF*   the premise contains any of the propositions
                        $(x_1$ *is* $C_{14})$ *or* $(x_1$ *is* $C_{15})$
          THEN   the rule goes to the second fragment of code

$CutR_3$:   *IF*   the premise contains any of the propositions
                        $(x_2$ *is* $C_{23})$ *or* $(x_2$ *is* $C_{24})$ *or* $(x_2$ *is* $C_{25})$
          THEN   the rule goes to the second fragment of code

By using these cutting rules, it is possible to obtain the proportion of volume contained in the two regions defined by the cutting surface. It is not necessary to use the whole set of cutting rules, the same result is obtained by using only the first cutting rule, obtaining the proportion of volume contained in the first region and comparing it with the global volume of the fuzzy rule. To obtain the volume of the intersection we take the cutting rule and the antecedent of the fuzzy rule, and compare the content of the corresponding positions for each variable. The comparison of 11100 11000 (cutting rule) and 11100 11100 (antecedent of the first fuzzy rule), produces, for the first five bits (first input variable, x_1), three corresponding positions that simultaneously contain the value '1', among the three values '1' appearing in that fragment of the fuzzy rule. That means that three out of three parts of the volume of the fuzzy rule (related to the first variable) are contained in the fragment of code generated by the first cutting rule. For the second five bits (corresponding to input variable x_2), two out of three parts of the volume are contained. The result is that the fuzzy rule and the cutting rule match $3 \cdot 2 = 6$ of $3 \cdot 3 = 9$ positions (six out of nine parts of the volume are contained), and consequently mismatch three out of nine. The two fragments of the RB are illustrated in Fig. 7.13 and the process to obtain them in Table 7.1.

Table 7.1   Cutting an RB

Fuzzy rule	Cutting rule	var. x_1	var. x_2	Match	Destination
11100 11100	11100 11000	3 of 3	2 of 3	6 of 9	1st fragment
11100 00111	11100 11000	3 of 3	0 of 3	0 of 9	2nd fragment
00111 11100	11100 11000	1 of 3	2 of 3	2 of 9	2nd fragment
00111 00111	11100 11000	1 of 3	0 of 3	0 of 9	2nd fragment

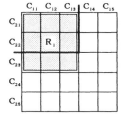 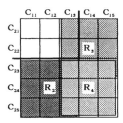

Fig. 7.13   The two fragments of the RB

7.1.2.2 *Rules of variable length*

Representation

Codes of variable length imitate the type of representation employed by messy GAs described in Sec. 2.2.5. In a position-dependent coding scheme, the functionality of a gene is determined by its locus on the chromosome. In the bit-string code of the previous section for example (see Fig. 7.7), the fourth bit always refers to the linguistic term C_{14}. A messy coding scheme relaxes this assumption, as it allows genes to be located arbitrarily across the chromosome. Hoffmann and Pfister (1996) use codes with position independent meaning. The elements in the rule code are pairs {*variable, label*}, that point to a particular fuzzy set in the fuzzy partition of the variable.

We restrict the discussion to the case of a descriptive RB with predefined fuzzy sets and DNF rules of the form shown in Eq. 7.2. Notice however that the non-positional coding scheme easily extends to rules of approximate nature as will be in next section.

First of all, fuzzy variables and linguistic terms are enumerated in a consecutive fashion by integers. The n input variables X_i are labelled from $1, \ldots, n$, the m output variables Y_i from $n + 1, \ldots, n + m$. The example depicted in Fig. 7.14 employs two input variables X_1, X_2 and a single output variable Y_1. They are labelled by the integers $1, 2, 3$ denoted by boxes with rounded corners. Each variable comprises five linguistic terms $C_{11}, \ldots, C_{15}, C_{21}, \ldots, C_{25}$ and $D_{11}, \ldots, D_{15}$ which are enumerated by integers $1, \ldots, 5$, marked by the rectangular boxes.

The basic element of the non-positional rule code is a pair {*variable, label*} that refers to a fuzzy statement of the form X_i is C_{ij}. The first integer value specifies the fuzzy variable; the second value points to the associated linguistic term in the distribution of that particular variable.

In the example depicted in Fig. 7.14, the pair $(1, 3)$ encodes the fuzzy statement X_1 is C_{13}, in which the first entry 1 refers to X_1 as the variable, and the label 3 points to the third linguistic label C_{13} in the partition $C_{11}, C_{12}, C_{13}, C_{14}, C_{15}$ of x_1.

A sequence of pairs constitutes the code of an elemental fuzzy rule as shown in Fig. 7.15. In the first example, the pairs $(2, 1), (3, 1), (1, 3)$ form the fuzzy rule

$$IF\ X_1\ is\ C_{13}\ and\ X_2\ is\ C_{21}\ THEN\ Y_1\ is\ D_{11}$$

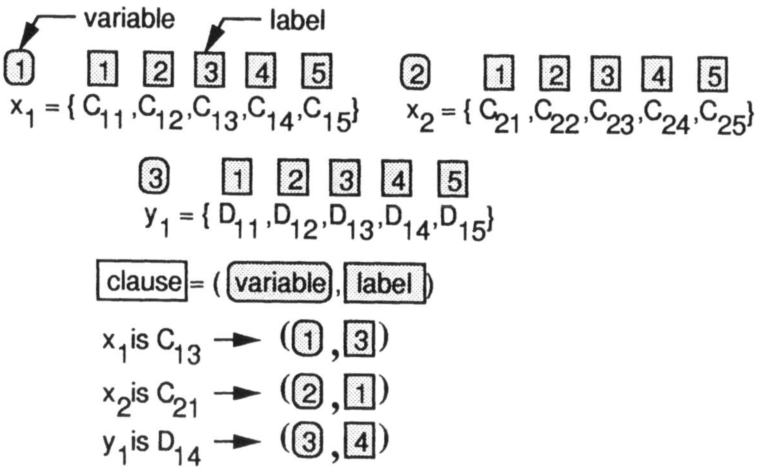

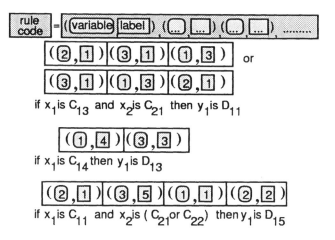

Fig. 7.14 Enumeration of input and output variables and their fuzzy sets (top), rules codes (bottom)

Fig. 7.15 Genetic representation of fuzzy rules

This rule can also be coded by the sequence $(3,1)(1,3)(2,1)$ with the same set of pairs but arranged in a different order.

As discussed in Sec. 2.2.5, position-independent coding and variable-size genotypes are a distinctive feature of messy GAs. The rule code might have multiple elements of the same variable that refer to different linguistic labels. Sometimes, a rule code contains no reference to some variable resulting in an incomplete fuzzy rule. In both cases, the coding scheme has to resolve these conflicts in order to enforce or preserve a meaningful mapping between the code and the fuzzy rule itself.

The *incomplete* rule code $(3,3)(1,4)$ contains no pair of the type $(2,*)$ pointing to the second input variable X_2. In this case of under-specification, the non-positional coding scheme considers X_2 as a "don't care" or wild-card variable that is omitted in the fuzzy rule

$$IF \; x_1 \; is \; C_{14} \; THEN \; y_1 \; is \; D_{13}$$

As stated earlier, such a more general rule of the type given in Eq. 7.2 simultaneously covers multiple elemental fuzzy rules.

An under-specified rule code such as $(2,4)(1,2)$ that does not comprise a reference $(3,*)$ to the output variable, can obviously not be transcribed into a valid fuzzy rule. In order to avoid these *meaningless* fuzzy rules, the genetic operators for initialisation and recombination of rule genes are augmented in a way that protects output genes from distinction.

In a non-positional coding scheme there is a dual problem to under-specification, over-specification, which arises whenever multiple genes with a conflicting fuzzy set label refer to the same variable. In the lower example in Fig. 7.15, the rule code $(2,1)(3,5)(1,1)(2,2)$ includes two conflicting genes $(2,1)$, $(2,2)$ for the second input variable X_2. The conclusion part of the corresponding rule aggregates the two complementary fuzzy clauses $X_2 \; is \; C_{21}$ and $X_2 \; is \; C_{22}$ with a fuzzy *or* as in Eq. 7.2.

Similar to the notion of a *list of rules* described earlier in this section, the complete RB is obtained by concatenating individual rule codes as depicted in Fig. 7.16. The rule list can be converted into a fuzzy decision table as indicated by the arrows. Notice that in the variable-length coding scheme, over-specified rules with disjunctive terms in the antecedent cover multiple entries in the decision table as for the two leftmost cells in the third row. Under-specified rules with wild-cards correspond to entire rows, columns or blocks in the decision table as shown in the fourth column. That way, non-positional semantics are able to reduce the size and complexity of the RB and thereby to overcome the course of dimensionality that usually occurs in positional coding schemes.

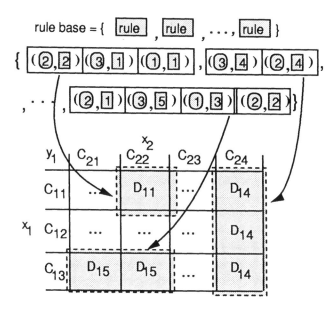

Fig. 7.16 Genetic representation of the RB and equivalent fuzzy decision table

Operators

In addition to a modified coding scheme, messy GAs also employ a *cut and splice* operator for recombination as introduced in Sec. 2.2.5. Cut and splice operations occur on the level of rule codes, as well as on the level of the rule list.

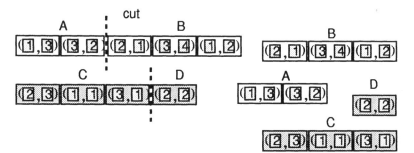

Fig. 7.17 The cut operator applied to the parent rule codes generates four fragments A, B, C, D

The parent rule codes are cut at a random, not necessarily identical location, resulting in four rule fragments as shown in Figure 7.17. The

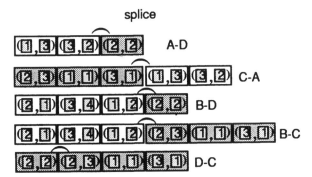

Fig. 7.18   The splice operator concatenates the four fragments A, B, C, D to form the possible offspring $A - D, C - A, B - D, B - C, D - C$

splice operator concatenates two randomly selected fragments to form a new offspring. Fig. 7.18 shows five possible ways to splice the four fragments. In the first case for which fragments A and D are spliced to form the offspring $(1, 3)(3, 2)(2, 2)$, the following rule is obtained

$$IF\ x_1\ is\ C_{13}\ and\ x_2\ is\ C_{22}\ THEN\ y_1\ is\ D_{12} \qquad (7.9)$$

Since the code is position-independent, the order of fragments in the offspring can be reversed as it is the case for the splices $B - C$ and $D - C$. Notice that in the later case both fragments stem from the same parent, so that cut and splice lead to an inversion of the original gene order.

Cut and splice operations on the rule code level modify the rule antecedent by adding or deleting pairs inherited from the parents. A rule might incorporate additional clauses as in the case $B - C$ with offspring $(2, 1)(3, 4)(1, 2)(2, 3)(1, 1)(3, 1)$ and rule

$$IF\ x_1\ is\ \{C_{11}\ or\ C_{12}\}\ and\ x_2\ is\ \{C_{21}\ or\ C_{23}\}$$
$$THEN\ y_1\ is\ \{D_{11}\ or\ D_{14}\}$$

or might become more compact as in the above case $A - D$ in Eq. 7.9. That way, cut and splice adjust the level of granulation since the number of clauses in the code determine whether the rule antecedent covers a small or large region of the input space.

7.1.2.3 *Rules of approximate type*

Up to this point, this chapter only discussed GFRBSs of the descriptive (linguistic) type. Although the majority of methods based on the Pittsburgh approach use linguistic rules, it is also possible to integrate GFRBSs with approximate rules into this framework (Carse, Fogarty, and Munro, 1996; Kang, Woo, Hwang, and Woo, 2000).

Approximate fuzzy rules are coded by structures of either fixed or variable length. In both cases, clauses of the form "x_i *is* C_j" refer to a fuzzy set C_j whose parameters are an integral part of the chromosome segment that encodes the clause, rather than having a label pointing to a fuzzy set that is defined outside the rule code as in the case of descriptive rules. For rules of fixed length, the code contains a segment to represent the fuzzy set associated to each variable (input or output). Each fuzzy set is encoded by a predefined number of parameters that specify the characteristic points of the fuzzy membership function. An entire approximate rule is formed by the concatenation of the fuzzy set code segments. Section 8.1.2 describes this coding scheme in more detail.

In the case of variable-length codes, a clause is composed of a pair of elements: a label (or code) defining the variable, and a set of parameters defining the fuzzy set. A possible method of coding these kinds of rules is described in Sec. 6.3.2. Despite the fact that the method proposed in (Velasco and Magdalena, 1995) employs the Michigan approach, the coding scheme itself remains applicable for approximate rules in the Pittsburgh approach. The code describes trapezoidal fuzzy sets by their four characteristic points, resulting in approximate rules of the type

$$IF \ V_1 \ \in \ [0.1, 0.2, 0.4, 0.5] \ and \ V_3 \ \in \ [0.3, 0.3, 0.4, 0.6]$$
$$THEN \ V_7 \ \in \ [0.2, 0.3, 0.3, 0.4]$$

Quite related to the previous ways of coding rules is the method proposed by Kang, Woo, Hwang, and Woo (2000) that operates on approximate TSK FRBSs. The fixed-length rule code employs four parameters to represent each trapezoidal fuzzy set in the premise part and a single parameter to encode the singleton fuzzy sets, in the consequent part. The list of rules is coded by concatenating the codes of individual rules. In addition, the chromosome contains the code of a two-dimensional matrix, whose rows correspond to fuzzy rules, and whose columns refer to the input variables. The element $M_{ij} \in [0, 1]$ of the so-called connection matrix M specifies the relative importance of the variable j-th, in the i-th rule. Consequently, a zero entry in this matrix means that the value of the corresponding input variable in that particular rule is ignored when computing the output of the system.

7.2   Multi-chromosome Genomes (Coding Knowledge Bases)

The KB of an FRBS as described in Sec. 1.2.1 is composed of two parts, the DB and the RB. That structure is particularly applicable to linguistic FRBSs in which the DB and the RB have independent representations. As described in section 3.4.2, the architecture of the KB opens a wide range of approaches to the tuning/learning process. To this point, the methods of adapting the DB or the rules of either a descriptive RB or an approximate FRB have been analysed. All previous approaches focus on learning a single structure of the KB. This section is devoted to the description of learning processes that simultaneously consider the two components RB and DB of a descriptive KB.

The GFRBSs discussed in the previous sections learn the RB that is coded in a single chromosome genome. In contrast, a learning process that in addition incorporates the DB usually employs a multi-chromosome genome, in which one of the chromosomes codes the RB, and the other the scaling factors or membership functions. Yet another approach is to apply a multi-stage design process, which first adapts the RB and afterwards optimises the DB.

At this point, the question of a proper representation that is always present in GFRBSs becomes of particular importance. As in any learning system based on a search process, two opposing factors have to be considered:

- A faster, simpler learning process is obtained when the dimension of the search space is reduced.
- Due to constraints in the representation of solutions, the search space is a subspace of all possible KBs (possible solutions). When this subspace is reduced, some of the potential solutions become *unreachable* for the learning system and it might turn out that the optimal solution is part of those unreachable regions.

The idea of obtaining an "unlimited" accuracy in FSs leads to the definition of an *approximate* FRBS. These systems prefer accuracy to description ability (from the view point of interpretability) and to simplicity of the search process (from the view point of learning). Linguistic systems where the RB and the DB are simultaneously learnt belong to an intermediate category with lower accuracy and complexity (at different levels depending on the approach) but nevertheless complete description ability.

7.2.1 *Representation*

Systems that simultaneously adapt DB and RB (from a learning point of view) usually emerge from the amalgamation of methods that in itself operate on one of the components alone.

In this case, the KB is not regarded as a homogeneous piece of information but the union of qualitatively different elements, namely the RB and DB. These pieces of information are coded by applying a coding scheme that generates a non-homogeneous code that better fits the structure of the KB. This structured code contains different elements representing the various kinds of information contained in the KB (rules, fuzzy sets and normalisation functions). Consequently, the representation is merely the *concatenation* of two structures that may follow any of the different coding schemes widely described along the book. Nevertheless, the relation between the applied code and the search space induces the use of codes that reduce the dimension of the search space.

The method to reduce the search space is to include only some of the components of the KB in the code, but trying to maintain the highest possible level of *accuracy*. The learning process does not necessarily have to consider all three components of the KB, namely scaling functions, membership functions and fuzzy rules, but rather aims to achieve a maximal reduction of complexity for a minimal loss of accuracy.

The first consideration when reducing the complexity of the search is related to the tight interaction between the two components of the DB. Simultaneous learning of scaling and membership functions is not considered because of the tightly coupled information represented by them, and the fact that under certain assumptions learning membership or scaling functions is equivalent.

As described in Sec. 4.1.1.2, the effects of modifying contexts (scaling functions) and semantics (membership functions) of an FRBS are tightly coupled. This is reflected in the idea of considering context and semantics information as a single block for the learning purpose. As a consequence, two basic possibilities are open when focusing on simultaneous learning of RBs and DBs: to learn rules and membership functions (Park, Kandel, and Langholz, 1994), or rules and scaling functions (Magdalena and Monasterio, 1997).

When coding these two pieces of information, multiple kinds of code have been considered: hybrid codes where binary and real representations are mixed (Magdalena and Monasterio, 1997), pure real codes (Park, Kandel, and Langholz, 1994), pure binary codes (Lee and Takagi, 1993c), integer codes, etc.

7.2.2 *Operators*

There are three different levels of integration of the sub-chromosomes from the point of view of the genetic operators. A first approach is that applied by Lee and Takagi (1993c), who consider the whole chromosome as a single, homogeneous entity with complete integration. As a result, crossover operates on the whole code, disregarding the substructures for rules and membership functions.

A different point of view applied by other authors is to consider that, when working simultaneously with RBs (in the form of decision table, relational matrix or list of rules) and DBs, the operators must take into account that the code contains different substructures. As described by Park, Kandel, and Langholz (1994) (working with relational matrices and membership functions):

> Since each chromosome consists of two different sub-chromosomes (one for the fuzzy relational matrix and the other for the fuzzy membership functions), we treat the two sub-chromosomes as independent entities as far as crossover and mutation are concerned, ...

But, at the same time, it is important to notice that these two substructures are components of a higher level entity, the KB, and consequently (continuing the previous quote):

> ... but treat both as a single entity as far as the reproduction operator is concerned.

The result of this compound (multi-chromosome) structure is the use of two pairs of crossover and two mutation operators (one for the RB and the other for the DB). Each operator is adapted to the kind of code used for the respective chromosome (sub-chromosome) of the compound structure.

In some cases, crossover for the RB and the DB is not a parallel process but a sequential one, in which, the result of the crossover of one of the chromosomes (sub-chromosomes) affects the crossover of the other. This is the method proposed by Magdalena and Monasterio (1997), considering KBs (k_i) coded through a RB and a DB ($k_i = (r_i, d_i)$).

In that case, after crossing RBs (coded as lists of rules) using the crossover operator described in Eqs. 7.4 and 7.5, the effect of parent RBs on the child RBs is considered to modify the process of crossing DBs (the ranges played the role of a linear scaling function):

> After RBs are crossed, the process of crossing DBs will con-
> sider what rules from r_i and r_j go to r_u or r_v. The rules
> we use contain fuzzy inputs and fuzzy outputs for only a
> subset of the input and output variables. Then, normali-
> sation limits for the remaining variables have no influence
> on the meaning of the rule. A larger influence of a certain
> variable on rules that, proceeding from r_i, go to r_u, pro-
> duces a higher probability for this variable to reproduce its
> corresponding range from d_i, in d_u. The influence is evalu-
> ated by simply counting the number of rules containing the
> variable, that are reproduced from r_i to r_u. The process of
> selection is independent for each variable and for each de-
> scendent, so it is possible for both descendents to reproduce
> the range of a certain variable from the same parent.

Then, it is possible to define three different kinds of crossover processes:

(1) Single crossover (as in (Lee and Takagi, 1993c)).
(2) Parallel compound crossover (as in (Park, Kandel, and Langholz, 1994)).
(3) Sequential compound crossover (as in (Magdalena and Monasterio, 1997)).

7.3   Examples

In this section, some of the previously mentioned references are described to exemplify how to apply the Pitt approach to different FRBSs. The examples cover methods for learning:

(1) decision tables, proposed by Thrift (1991),

(2) relational matrices, proposed by Pham and Karaboga (1991),

(3) TSK-type rules and membership functions, proposed by Lee and Takagi (1993c),

(4) DNF Mamdani-type rules of fixed length and membership functions, proposed by Magdalena and Monasterio (1995),

(5) DNF Mamdani-type rules of variable length, proposed by Hoffmann and Pfister (1996), and

(6) approximate Mamdani-type fuzzy rules, proposed by Carse, Fogarty, and Munro (1996).

7.3.1   *A method to learn decision tables*

This method, proposed by Thrift (1991), works on the basis of a phenotype constituted by a complete decision table. These decision tables are encoded in the chromosomes using a positional code and establishing a mapping between the label set associated to the system output variable and an increasing integer set taking 0 as first element, representing the allele set. The first n elements of this set ($\{0, \ldots, N-1\}$) represent the N elements of the term set of a certain input variable, while the last element plays the role of the "null" symbol that indicates the absence of a value for the output variable Let $\{NB, NS, ZR, PS, PB\}$ be the term set associated to the output variable, and using the null symbol "$-$" to indicate the absence of a term for the output variable, the mapping from the decision table to the chromosome transforms $\{NB, NS, ZR, PS, PB, -\}$ into $\{0, 1, 2, 3, 4, 5\}$. Hence, the label NB is associated with the value 0, NS with 1, $\ldots$, PB with 4 and the blank symbol ($-$) with 5.

Therefore, the GA employs an integer coding. Each of the chromosomes is constituted by concatenating the partial coding associated to each one of the linguistic labels contained in the decision table cells. A gene presenting the allele "$-$" denotes the absence of the fuzzy rule for that particular cell.

The GA employs an elitist selection scheme and uses genetic operators of different nature. While the crossover operator is the standard two-point crossover, the mutation operator is specifically designed for the process. When it is applied over an allele different from the blank symbol, changes it either up or down one level or to the blank code. When the previous gene value is the blank symbol, it selects a new value at random.

Finally, the fitness function is based on an application specific measure. A measure of convergence is considered for this task. The performance of the FLC being designed is evaluated in a closed loop system[†] using the RB coded in its genotype. The system is started from several different initial states and the fitness function computes how fast the output converges to the desired set point.

7.3.2 *A method to learn relational matrices*

In this approach proposed by Pham and Karaboga (1991), the GA is used to modify the fuzzy relational matrix (R) of an one-input, one-output FLC. The chromosome is obtained by concatenating the $M \cdot N$ elements of R, where M and N are the number of linguistic terms associated with the input and output variable, respectively. The elements of R are real numbers in the interval $[0,1]$, which are represented by alleles using a binary code (eight bit representation). With this code, a system that has seven terms for the input variable, and eleven terms for the output, produces a bit string of length $l = 616$.

The approach uses the standard genetic operators for bit strings: one-point crossover and binary mutation. It employs the same elite strategy used by Thrift (1991).

The authors introduce an improved GA by dividing the process into two stages. In the first one, k GAs are executed in parallel, starting with k random initial populations. After a fixed number of generations, a new *crossbred* population is created from the k final populations of the GAs in the first stage (i.e. the fittest $1/k$ of each population). In the second stage, the *crossbred* population is used as the initial population for a new GA executed a fixed number of generations.

[†]The same idea shown in Fig. 7.1 if considering that the actions are control actions, the environment is the controlled system, and the perceptions are the status of the controlled system.

7.3.3   *A method to learn TSK-type knowledge bases*

Lee and Takagi (1993c) introduced a GFRBS that automatically learns the complete KB. The method works on an RB composed of TSK rules and simultaneously adapts the membership functions describing the elements of the term set and the coefficients of the linear expression defining the output of each rule.

For the linguistic terms in the rule antecedents, the authors consider triangular-shaped membership functions although they note that in principle the method works with any kind of parameterised membership functions, such as Gaussian, bell, trapezoidal or sigmoidal ones. Each triangular membership function associated to one of the primary fuzzy sets (linguistic terms) is represented by three parameters. The first one is the centre of the triangle, i.e., its modal point. Only the centre of the triangle associated to the first primary fuzzy set is given as an absolute position, while the parameters associated to the other ones represent the distance from the centre point of the previous triangle to the current one. The other two parameters represent the left and right points of the triangle base respectively at which the membership function goes to zero. A given initial fuzzy partition of the input space is assumed, with the cardinality of the term sets being a parameter to the learning system which is known in advance. As we shall see in the following, this value is an initial maximum threshold, since the system is able to automatically learn the optimal number of terms in each input variable fuzzy partition (and, consequently, the number of fuzzy rules in the RB). In the experiments shown in the reference, the maximum number of fuzzy sets per input variable was set to ten.

The consequents of a TSK fuzzy system are linear combinations of the input variables. A system with n input variables $(X_1, \ldots, X_n)$ has $n+1$ parameters per consequent $(C_j = w_{n+1,j} + \sum_{i=1}^{n} w_{ij} \cdot X_i)$. That way, the rule consequents are derived by learning the weights w_{ij}, where i ranges from 1 to the number of input variables plus one, and j ranges from 1 to the maximum number of rules in the system (equal to the product of the number of linguistic terms per variable).

The GA works on a binary coding. First, the membership functions are encoded by taking the binary values of the three associated parameters and joining them into a binary sub-string. Eight bits are used to encode each parameter value. Then, the complete DB is encoded by concatenating the partial codes of the membership functions associated to each one of the input variable primary fuzzy sets one after the other. The second part of the chromosome is built by coding the parameters w_{ij} associated to each rule into a new binary sub-string. Eight bits are used again to encode the values of these parameters. That way, each chromosome represents a complete KB. The number of rules L_r forming the KB depends on the number M_i of primary fuzzy sets associated to each one of the input variables and is equal to their product $L_r = \prod_{i=1}^{n} M_i$ (the RB is complete).

Once the complete binary string composed of the code for DBs and RBs has been generated, binary genetic operators are used. In the examples described by Lee and Takagi (1996), the applied operator (two-point crossover) works on the binary string without taking into account the internal structure of the code. Consequently, any point in the string is a potential cutting point, and not just rather points that separate two membership functions or two rule consequents.

As said, the code is able to adjust the number of fuzzy rules forming the RB in the following way. Membership functions whose centre point lies outside a predefined interval are discarded and do not contribute with a rule to the RB. Only rules for which the antecedent is a combination of valid linguistic terms are incorporated into the RB.

The FLC in the reference tries to stabilise the cart-pole system (the problem described in Sec. 11.3.1). The cart and pole are started from different initial states. The fitness function tries to optimise multiple criteria. The controller obtains a score that is proportional to the time it prevents the pole from falling over. An additional term in the fitness function tries to minimise the steady-state error. Finally, the fitness function penalises the size of the RB in order to obtain a compact FRBS with a small number of rules.

The authors propose two different ways of incorporating previous knowledge into the GFRBS in order to improve the system behaviour. They mention that this previous knowledge results in a significant speedup of the GA if initial solutions are approximately correct. On the one hand, it is possible to incorporate knowledge via the initial settings of the KB parameters. This knowledge is used to seed the initial population. Thus, not all the individuals forming the first generation are generated at random but several of them are obtained by partitioning the input spaces with equidistant membership functions. This knowledge can also be used to find appropriate initial values for the parameters w_i in the rule consequents. On the other hand, the previous knowledge can be incorporated via the structural representation of the KBs. For a control problem that is symmetric in some of the input variables as the inverted pendulum, the FLC is constrained by imposing a symmetric partition of the input space around the origin. This reduces the dimensions of the search space since the number of membership functions is cut in half.

7.3.4   A method to learn DNF Mamdani-type knowledge bases (with rules of fixed length)

Magdalena and Monasterio (1995, 1997) proposed a bit-string code to represent DNF fuzzy rules. As seen in Sec. 7.1.2.1, the antecedent of a rule is coded by concatenating, for each input variable X_i, a binary sub-string of length N_i (number of linguistic terms for that variable) denoting the absence or presence of a fuzzy set C_{ij} ($j \in \{1, \ldots, N_i\}$) in the rule[‡]. A similar process is applied to the output fuzzy sets D_{ij} in the consequent. This coding scheme was shown in Fig. 7.7.

The code of the RB is obtained by concatenating the binary codes of the individual rules. That produces a code of variable length composed of a list of rules (variable number of rules) where each rule has a fixed length code.

In addition, the DB (through the scaling functions) is also adapted in the genetic process. For this purpose, the scaling function related to each system variable is parameterised using two (linear normalisation), three (non-linear and symmetric normalisation) or four (non-linear normalisation) parameters per function. The parameters for the different scaling functions are concatenated into a string of real numbers. The four parameters are $\{V_{min}, V_{max}, S \text{ and } a\}$, where V_{min} and V_{max} are real numbers defining the upper and lower limits of the universe of discourse, a is a real, greater than zero, that produces the non-linearity and S is a parameter in $\{-1,0,1\}$ to distinguish between non-linearities with symmetric shape (lower granularity for middle or for extreme values) and asymmetric shape (lower granularity for lowest or for highest values).

The scaling or normalisation process, as described in Sec. 4.1.1.2, includes three steps:

[‡]We should remind that a code of all zeros for a certain input variable means that the variable is not included in the rule (the output of the rule is independent on the value of that variable), that is equivalent to a don't care code.

(1) The first step based on the parameters $\{V_{min}, V_{max}$ and $S\}$ produces a linear mapping from $[V_{min}, V_{max}]$ interval to $[-1,0]$ (when $S = -1$), to $[-1,1]$ (when $S = 0$) or to $[0,1]$ (when $S = 1$).

(2) The second step introduces the non-linearity (using the parameter a) through the expression

$$f(x) = sign(x) \cdot |x|^a$$

an odd function that maintains the extremes of the interval unchanged in any case ($[-1,0]$, $[-1,1]$ or $[0,1]$).

(3) Finally, a second linear mapping transforms the resulting interval ($[-1,0]$, $[-1,1]$ or $[0,1]$) into $[-1,1]$, or any other normalisation interval where the fuzzy partitions are defined.

In case of two parameters (V_{min}, V_{max}), only the first step is needed, assuming $S = 0$. If a third parameter is added (a), the second step is included (assuming again $S = 0$). The overall result (when including the fourth parameter, S) is a non-linear mapping from the interval $[V_{min}, V_{max}]$ to the interval $[-1,1]$.

The complete code of a KB has the structure

$$Sc_1, \ldots, Sc_n, Sc_{n+1}, \ldots, Sc_{n+m}; r_1, \ldots, r_k$$

where Sc describes the set of two, three or four parameters of a scaling function, n and m are the number of input and output variables, respectively, and r_j is the $j - th$ rule of the list composing the RB.

A one-point crossover operator (Magdalena and Monasterio, 1995) including reordering, or the non-positional crossover operator introduced in Sec. 7.1.2.1 (Magdalena, 1998), is applied to the list of rules. The crossover operator only splits entire rules and never divides the bit string of a rule itself.

After RBs are crossed, crossover of DBs takes place in a way that takes into account the influence of information coming from first and second parent to first or second child. Each rule in the RB only refers to a subset of the input and output variables (not to those generating an *all zero* sub-string), therefore scaling functions for the remaining variables have no influence on the rule. A rule inherited by a child from a particular parent increases the probability that a scaling function which effects this particular rule is also inherited from the very same parent. The influence of a scaling function is evaluated by simply counting the number of rules containing the variable, which are reproduced from each parent to each child. The process of selection is independent for each variable and for each offspring. Therefore, it is possible for both offspring to inherit an identical scaling function from the same parent.

The mutation operator has different components:

- The rule mutation works at the level of bits that compose a rule, being each one of them a candidate to be muted by a classical mutation operator.
- The mutation of gain (over the values of V_{min} and V_{max}) applies the expressions:

$$
\begin{aligned}
V_{min}(t+1) &= V_{min}(t) + \frac{K \cdot P_1 \cdot S_1}{2}(V_{max}(t) - V_{min}(t)) \\
V_{max}(t+1) &= V_{max}(t) + \frac{K \cdot P_2 \cdot S_2}{2}(V_{max}(t) - V_{min}(t))
\end{aligned}
\tag{7.10}
$$

where $K \in [0,1]$ is a parameter of the learning system that defines the maximum variation (shift, expansion or shrinkage). P_1 and P_2 are random values uniformly distributed on $[0,1]$, and S_1 and S_2 take values -1 or 1 by a 50% chance. For symmetric variables, it is possible to maintain the symmetry by imposing: $P_2 = P_1$ and $S_2 = -S_1$.

- The mutation of sensitivity applies the following expression:

$$
a(t+1) = a(t) \cdot (1 + P \cdot (\alpha - 1))^S
\tag{7.11}
$$

where $\alpha \in [1,10]$ is a parameter of the learning system that defines the maximum variation (increase or decrease) of the sensitivity parameter, and P and S are equivalents to those in gain mutation.

7.3.5   *A method to learn DNF Mamdani-type rule bases (with rules of variable length)*

As seen in Sec. 7.1.2.2, Hoffmann and Pfister (1996, 1997) introduced an RB coding scheme in which the number of rules as well as the complexity of an individual rule varies. It offers the largest amount of flexibility for GFRBS design as it imposes no constraints on the kind of rules and RBs that can be represented by the code.

In the GFRBS proposed in those references, the fuzzy variables and their partitions in the DB are defined in advance, so that only the RB itself becomes subject to optimisation. The rule code employs a list of integer pairs {*variable, label*} that refer to a particular fuzzy set in the fuzzy distribution of the variable. Each pair P_j codes a primitive fuzzy clause of the type "*variable* is *label*" which are then combined to form rules.

The code for an entire rule $R_i = \{P_{i1}, \ldots, P_{in_i}\}$ is a list of variable length n_i of primitive clauses P_{ij}. The sequence $(1,2)(2,4)(3,3)$ corresponds to the fuzzy rule

$$IF\ X_1\ is\ C_{12}\ and\ X_2\ is\ C_{24}\ THEN\ Y_1\ is\ D_{13}$$

A rule code that contains multiple labels for the same input variable is translated into a disjunction of the fuzzy sets in the antecedent. Multiple instances of output variables are handled by a dominance scheme such that only the leftmost pair is expressed in the rule consequence whereas the remaining ones are ignored. An input variable that does not occur in the rule code is considered a don't care and is therefore discarded in the antecedent. The sequence $(1,1)(3,4)(1,3)(3,2)$ is interpreted as the DNF fuzzy rule

$$IF\ X_1\ is\ \{C_{11}\ or\ C_{13}\}\ THEN\ Y_1\ is\ D_{14}$$

in which C_{11} and C_{13} are combined using the fuzzy or-operator, the input variable X_2 is a wild-card, and the leftmost output fuzzy set D_{14} dominates D_{12} which is not expressed in the conclusion.

As previously discussed in Sec. 7.1.2.2, the scheme enables the formation of more or less specific fuzzy rules, depending on the number n_i of primitive clauses P_{ij} in the rule code $R_i = \{P_{i1}, \ldots, P_{in_i}\}$. This property is helpful to reduce the overall number of rules in the RB, as coarse rules cover a larger region of the input space. Nevertheless, very specific rules can be generated for those situations that require a finer level of granulation. That way, a coding scheme of variable length constitutes a promising approach to overcome the course of dimensionality inherent to the fixed length rule representations of fuzzy decision tables and fuzzy relation matrices.

The complete RB is composed of a list

$$\begin{aligned} RB\ &=\ \{R_1, R_2, \ldots, R_k\} \\ &=\ \{\{P_{11}, \ldots, P_{1n_1}\}, \{P_{21}, \ldots, P_{2n_2}\}, \ldots, \{P_{k1}, \ldots, P_{kn_k}\}\} \end{aligned}$$

of variable length k, whose elements are the rule codes $R_i = \{P_{i1}, \ldots, P_{in_i}\}$. Non-positional semantics on the RB level are very natural as the order of rules R_i in the rule list RB has no influence on the fuzzy inference process.

Since the number of rules k as well as the composition of individual rules R_i varies, the completeness of the resulting RB can not be guaranteed. One can either incorporate a penalty term in the fitness function for incomplete RBs or define a default output fuzzy set D_0 that is only activated when no other rule triggers.

At this stage it also becomes necessary to resolve conflicts among rules that trigger for the same input but which suggest contradicting output sets D_{ij}. The inference scheme is modified in a way that prioritises more specific rules with many clauses P_{ij} over coarser ones with fewer clauses. A rule of higher resolution such as

$$IF \ X_1 \ is \ C_{13} \ and \ X_2 \ is \ C_{24} \ THEN \ Y_1 \ is \ D_{11}$$

obtains priority over a less specific rule such as

$$IF \ X_1 \ is \ C_{13} \ THEN \ Y_1 \ is \ D_2$$

This hierarchical inference scheme ignores the conclusion of a coarse rule whenever it triggers at the same time as a more specific one. This type of reasoning is very natural to humans, as they tend to ignore a default behaviour such as *"drive in the right lane"* in rare exceptional situations such as *"if right lane is blocked and left lane is empty then drive in the left lane"*.

The method uses the cut and splice operators introduced in Sec. 2.2.5. Cut and splice operators are applied on the level of rule codes as discussed in Sec. 7.1.2.2 as well as on the list of rules. The size of the RB grows or shrinks as cut and splice operations recombine RB fragments taken from the parent rule lists. Assume two parent RBs $\{R_1^A, \ldots, R_k^A\}$ and $\{R_1^B, \ldots, R_l^B\}$ with a respective number of rules k and l. Cut produces the four fragments $\{R_1^A, \ldots, R_a^A\}$, $\{R_{a+1}^A, \ldots, R_k^A\}$, $\{R_1^B, \ldots, R_b^B\}$ and $\{R_{b+1}^B, \ldots, R_l^B\}$. The following splice operation might select the third and first fragment and concatenate them to the offspring RB $\{R_1^B, \ldots, R_b^B, R_1^A, \ldots, R_a^A\}$ which contains $a + b$ rules.

A positional coding scheme restricts crossover to segmentations of the decision table along a certain cutting point or surface. This problem does not occur in the non-positional coding which contains no bias for a particular order of rules. A non-positional coding scheme possesses the capability to optimise not only the chromosome itself but, by means of cut and splice, its structure as well. That way, the GA is able to form building blocks, namely subsets of advantageous rule codes that are highly interdependent and occupy adjacent loci on the chromosome. Assume two parents already contain building blocks of proper fuzzy rules for complementary regions of input space. By means of cut and splice, they merge their adapted rules to form a potentially superior offspring RB.

Rather than training all elemental rules of the decision table as in a positional scheme, the non-positional GA starts with a few general rules, which in the course of evolution are gradually refined and supplemented by more specific rules. As a result of cut and splice operations, rule lists and rule codes might grow or shrink in size which enables the GA to adapt the number of rules and the complexity of rules to the problem structure. That way, the genetic search can exploit regularities in the training data or control problem that facilitate the FRBS design task.

7.3.6 A method to learn approximate Mamdani-type fuzzy rule bases

The P-FCS1 (Pittsburgh-style Fuzzy Classifier System #1), proposed by Carse, Fogarty, and Munro (1996), is a novel GFRBS based on the Pittsburgh approach to generate approximate fuzzy rules. It is characterised by a unique coding scheme and crossover operator. Its main features are described as follows.

The P-FCS1 is based on a generational RCGA. Each triangular-shaped membership function is encoded by two real-valued parameters (x_{ik}^C, x_{ik}^W) that represent its centre and width. Each individual rule is represented by a fixed-length vector of these kinds of membership functions. The entire chromosome is a variable-length string concatenated of an arbitrary number of rule code sub-strings. That is, r approximate fuzzy rules with n input variables and m output variables are encoded as

$$(x_{1,1}^C, x_{1,1}^W) \ldots (x_{1,n}^C, x_{1,n}^W)(x_{1,n+1}^C, x_{1,n+1}^W) \ldots (x_{1,n+m}^C, x_{1,n+m}^W)$$
$$\vdots$$
$$(x_{r,1}^C, x_{r,1}^W) \ldots (x_{r,n}^C, x_{r,n}^W)(x_{r,n+1}^C, x_{r,n+1}^W) \ldots (x_{r,n+m}^C, x_{r,n+m}^W)$$

which involves $r \cdot (n + m)$ fuzzy sets, i.e., pairs of centre and width parameters.

The crossover operator employed is similar to the classical two-point crossover. Instead of cutting out a linear segment in the genotype, the two

$n - dimensional$ cut points define two opposite corners of a hyper-cube in the n-dimensional fuzzy input space. The offspring inherits rules whose antecedents lie inside the hyper-cube from one parent and the complementary rules with antecedents outside the hyper-cube from the other parent. First, the rules are sorted according to the centres of the input membership functions. Then, two cut points C^1 and C^2 located within the range of each input variable are randomly generated as follows:

$$C_i^1 = MIN_i + (MAX_i - MIN_i) \cdot (r_1)^{\frac{1}{n}}$$
$$C_i^2 = C_i^1 + (MAX_i - MIN_i) \cdot (r_2)^{\frac{1}{n}}$$

with $[MIN_i, MAX_i]$ being the domain of the variable and with r_1 and r_2 being two random numbers drawn from the interval $[0, 1]$ with uniform probability.

The first offspring contains all rules of the first parent that satisfy

$$\forall i, \left(\left(x_{ik}^C > C_i^1 \right) \; AND \; \left(x_{ik}^C < C_i^2 \right) \right) \; OR \; \left(\left(x_{ik}^C + MAX_i - MIN_i \right) < C_i^2 \right)$$

and in addition all complementary rules of the second parent that on contrast do not satisfy this condition, i.e.,

$$\exists i, \left(\left(x_{ik}^C \leq C_i^1 \right) \; OR \; \left(x_{ik}^C \geq C_i^2 \right) \right) \; AND \; \left(\left(x_{ik}^C + MAX_i - MIN_i \right) \geq C_i^2 \right)$$

The second offspring contains the rest of the rules from both parents not already inherited to first offspring.

The mutation operator adds a small random real number to the fuzzy set membership function centres and widths. Therefore, mutation is used for fine tuning rather than for introducing radically different individuals into the population. The authors also propose additional operators for the creation and deletion of rules, and a coverage operator to ensure that all the examples in the training data set are covered by at least one rule.

Chapter 8

Genetic Fuzzy Rule-Based Systems Based on the Iterative Rule Learning Approach

Genetic learning processes based on the iterative rule learning (IRL) approach are characterised by tackling the learning problem in multiple steps (see Sec. 5.4). Therefore, they are composed of, at least, two stages:

- a *generation process*, that derives a preliminary set of fuzzy rules representing the knowledge existing in the data set, and
- a *post-processing process*, with the function of refining the previous rule set in order to remove the redundant rules that emerged during the generation stage and to select those fuzzy rules that cooperate in an optimal way. This second stage is necessary as the generation stage merely ignores the cooperation aspect.

Due to this structure, GFRBSs based on the IRL approach are usually called *multi-stage GFRBSs* (González and Herrera, 1997).

As may be noticed, this multi-stage structure is a direct consequence of the way in which GFRBSs based on the IRL approach solve the CCP. These kinds of systems try to solve the CCP in a way that combines the advantages of the Pitt and Michigan approach. The objective of the IRL approach is to reduce the dimension of the search space by encoding individual rules in the chromosome like in the Michigan approach, but the evaluation scheme take the cooperation of rules into account like in the Pitt approach. Hence, the CCP is solved by partitioning the learning process into two steps that act on different levels:

- *the generation stage forces competition between fuzzy rules*, as in genetic learning processes based on the Michigan approach, *to obtain a fuzzy rule set composed of the best possible fuzzy rules*. To do so, a fuzzy rule generating method is run several times (in each run, only the best fuzzy rule as regards the current state of the example set is obtained as process output) by an iterative covering method that wraps it and analyses the covering that the consecutively rules learnt cause in the training data set. Hence, the co-operation among the fuzzy rules generated in the different runs is only briefly addressed by means of a rule penalty criterion.

- *the post-processing stage forces cooperation between the fuzzy rules generated in the previous stage* by refining or eliminating the previously generated redundant or excessive fuzzy rules *in order to obtain a final fuzzy rule set that demonstrates the best overall performance.*

The iterative operation mode followed by the generation stage in multi-stage GFRBSs based on the IRL induces the formation of niches and substantially reduces the size of the search space. The post-processing stage deals with a simple search space as well because it focuses on the fuzzy rule set obtained in the previous step.

The structure of this chapter is somewhat different from the two previous ones due to the fact that there are only two different examples of application of the IRL approach to the design of GFRBSs (González and Herrera, 1997; Cordón, Herrera, González, and Pérez, 1998): MOGUL (Cordón, del Jesús, Herrera, and Lozano, 1999) and SLAVE (González and Pérez, 1998a, 1999a).

On the one hand, as mentioned in Chapter 4, MOGUL (Methodology to Obtain GFRBSs Under the IRL approach) is a methodology composed of some design guidelines that allow different users to customise their own GFRBSs based on the IRL approach, in order to design FRBSs able to cope with their specific problems. To do so, the user only has to define his evolutionary process in each of the GFRBS learning stages, ensuring that they verify MOGUL assumptions. Working in this way, the authors of MOGUL have created different GFRBSs based on the IRL approach to design different types of FRBSs: descriptive Mamdani-type (Cordón and Herrera, 1997c; Cordón, del Jesús, and Herrera, 1998), approximate Mamdani-type (Cordón and Herrera, 1997c, 2001; Herrera, Lozano, and Verdegay, 1998a) and TSK FRBSs (Cordón and Herrera, 1999b).

On the other hand, SLAVE (Structural Learning Algorithm in Vague Environment) is a genetic learning process based on the IRL approach to design DNF Mamdani-type FRBSs that was proposed by González, Pérez, and Verdegay (1993) and has been extended by González and Pérez (1997b, 1998a, 1999a, 1998c) and Castillo, González, and Pérez (2001). Initially, it was designed for fuzzy classification purposes but it was also extended to deal with fuzzy control and fuzzy modelling problems (González and Pérez, 1996b) obtaining good results.

In view of this, this chapter first analyses the aspects common to both models, MOGUL and SLAVE, in the design of GFRBSs based on the IRL paradigm, before it studies the particular aspects of each individual method. The different ways to code fuzzy rules in the generation stage are introduced first. Then, the existing alternatives to design the two stages composing a GFRBS based on the IRL approach, generation and post-processing, are analysed. Some possible extensions based on inducing cooperation in the generation stage to solve better the CCP are proposed. Finally, the chapter discusses the particular aspects of MOGUL and SLAVE.

8.1 Coding the Fuzzy Rules

As in GFRBSs based on the Michigan approach, each individual in the population, evolved in the fuzzy rule generating process encodes a single fuzzy rule. Hence, the coding scheme is similar in both approaches. The type of FRBS determines how fuzzy rules are encoded in the chromosome, e.g., as linguistic labels in the case of descriptive Mamdani-type FRBSs, fuzzy membership functions in the case of approximate Mamdani-type FRBSs and linguistic labels and real-valued consequent parameters in TSK FRBSs.

In the following subsections, we summarise the different coding schemes that have been proposed for different types of fuzzy rules in the literature.

8.1.1 *Coding linguistic rules*

In this first case, the only information that has to be evolved by the GA in order to obtain the rule composition is the linguistic label associated with each linguistic variable. Thus, as seen in Sec. 7.1.2.1, it is usually considered an integer coding as follows (Cordón and Herrera, 1997c). The primary fuzzy sets belonging to each one of the variable fuzzy partitions considered are enumerated from 1 to the number of labels in each term set. A linguistic variable X_i taking values in a primary term set

$T(X_i) = \{L_1(X_i), \ldots, L_{n_i}(X_i)\}$ is associated with the ordered set $T'(x_i) = \{1, \ldots, n_i\}$.

Hence, the following fuzzy rule

$$IF\ X_1\ is\ L_{i_1}(X_1)\ and\ \ldots\ and\ X_n\ is\ L_{i_n}(X_n)$$
$$THEN\ Y\ is\ L_{i_{n+1}}(Y)$$

is encoded into a chromosome C of the form:

$$C = (i_1, \ldots, i_n, i_{n+1})$$

On the other hand, González, Pérez, and Verdegay (1993) proposed a binary coding scheme to encode DNF Mamdani-type rules (see Sec. 1.2.6.1) that has been used in *SLAVE*. Since in this model the class in the rule consequent is already determined by the iterative covering method, the chromosome only encodes the antecedent part of the DNF rule using a similar coding as the one described in Fig. 7.7. Having three input variables X_1, X_2 and X_3, such that the linguistic term set associated to each one is

$$D_1 = \{A_{11}, A_{12}, A_{13}\} \qquad D_2 = \{A_{21}, A_{22}, A_{23}, A_{24}, A_{25}\}$$
$$D_3 = \{A_{31}, A_{32}\}$$

the following DNF rule antecedent

$$X_1\ is\ \{A_{11}\ or\ A_{13}\}\ and\ X_2\ is\ \{A_{23}\ or\ A_{25}\}\ and\ X_3\ is\ \{A_{31}\ or\ A_{32}\}$$

is encoded in the chromosome

$$C = (1010010111)$$

As seen in Sec. 7.1.2.1, each variable is associated with a bit-string whose length is equal to the number of labels in its linguistic term set. A bit is set to 1 if the corresponding term is present in the rule antecedent, and is set to 0 otherwise.

Since X_3 takes all the elements in the domain D_3, the said antecedent is equivalent to the rule

$$X_1\ is\ \{A_{11}\ or\ A_{13}\}\ and\ X_2\ is\ \{A_{23}\ or\ A_{25}\}$$

which does not explicitly include the variable X_3.

Therefore, this coding scheme has the capability of selecting the relevant linguistic variables for each single rule.

Moreover, this capability is extended in the coding scheme considered in the GA of SLAVE-2, an extension to the basic SLAVE process proposed by González and Pérez (1997a, 2001) that is briefly described in Sec. 8.5.2. In this new coding, each element of the population is represented by two binary chromosomes: the variable chromosome (VAR) is a binary string of length n (the number of input variables) with an 1 indicating that the variable is active and a 0 that it is not active. The value chromosome (VAL) uses the binary coding previously shown.

The following example clarifies this novel coding. Assuming the three input variables, X_1, X_2 and X_3, considered in the previous example, the coding:

$$C = (\text{VAR } 110 \text{ VAL } 1111100010)$$

represents the following antecedent:

$$X_1 \text{ is } \{A_{11} \text{ or } A_{12} \text{ or } A_{13}\} \text{ and } X_2 \text{ is } \{A_{21} \text{ or } A_{22}\}$$

Since X_1 refers to every one of the elements in its domain D_1, the said antecedent is equivalent to:

$$X_2 \text{ is } \{A_{21} \text{ or } A_{22}\}$$

Notice that the X_3 variable is eliminated at the variable level, but the X_1 one is eliminated at the value level.

8.1.2 *Coding approximate Mamdani-type fuzzy rules*

When coding individual rules, approximate fuzzy rules require a different kind of representation scheme than linguistic rules, which encode the fuzzy set membership functions as part of the rules (see Sec. 7.1.2.3). For that task, a real-valued coding scheme is most appropriate to represent the characteristic points of the parameterised membership functions. The different GFRBSs based on the IRL approach proposed in the literature to learn approximate Mamdani-type FRBs (Cordón and Herrera, 1997c, 2001; Herrera, Lozano, and Verdegay, 1998a), work with linear piecewise membership functions, either normalised triangular or trapezoidal fuzzy sets, although in principle other shapes may be considered.

These GFRBSs take into account the computational way to characterise triangular and trapezoidal membership functions by using a parametric representation achieved by means of a 3-tuple (a, b, c) or a 4-tuple (a, b, c, d), respectively. Hence, an approximate Mamdani-type fuzzy rule of the form

$$IF \ X_1 \ is \ A_1 \ and \ \ldots \ and \ X_n \ is \ A_n \ THEN \ Y \ is \ B,$$

where the A_i and B are directly triangular-shaped fuzzy sets, is encoded into the following chromosome

$$C = (a_1, b_1, c_1, \ldots, a_n, b_n, c_n, a, b, c)$$

and, when normalised trapezoidal-shaped ones are considered, its representation is the one shown below

$$C = (a_1, b_1, c_1, d_1, \ldots, a_n, b_n, c_n, d_n, a, b, c, d)$$

However, the coding scheme considered by Cordón and Herrera (2001) constitutes an extension to the one shown above. The GFRBS proposed in that paper belongs to the *constrained learning* family (see Sec. 1.4.4) and, most specifically, to the group of *hard constrained learning methods*, due to the fact that each fuzzy set definition point has an interval associated specifying its possible location during the learning stage. In this case, these definition intervals are obtained from the knowledge existing in some descriptive initial domain fuzzy partitions associated to each variable. Notice that this constitutes another means to incorporate expert information in the design of an approximate FRBS as discussed in Sec. 1.4.4.

Thus, in order to design a constrained GFRBS, additional information that defines the intervals in which each parameter of the fuzzy set may vary has to be included in the chromosome. Hence, a chromosome C encoding a candidate rule is composed of two different parts, C_1 and C_2. The second part C_2 encodes the composition of the approximate fuzzy rule itself, i.e., the specific membership functions associated with each variable, whilst the first part C_1 encodes the primary terms specifying the intervals for the membership function parameters.

The first part is built by means of the integer coding previously introduced to encode linguistic rules. The second part adopts the real-valued representation introduced above as well. Therefore, an approximate rule

$$IF \ X_1 \ is \ A_1 \ and \ \ldots \ and \ X_n \ is \ A_n \ THEN \ Y \ is \ B$$

comprised by triangular fuzzy sets A_i, B generated by means of a constrained GFRBS from the intervals defined by the following combination of primary linguistic terms

$$\{L_{i_1}(X_1), \ldots, L_{i_n}(X_n), L_{i_{n+1}}(Y)\}$$

is encoded into a chromosome C of the form:

$$C_1 = (i_1, \ldots, i_n, i_{n+1})$$
$$C_2 = (a_1, b_1, c_1, \ldots, a_n, b_n, c_n, a, b, c)$$
$$C = C_1 C_2$$

8.1.3 *Coding TSK fuzzy rules*

TSK fuzzy rules are defined by the linguistic terms that the linguistic variables in the rule antecedent has associated and the values of the gain factors p_i in the rule consequent. Thus, a hybrid integer-real coding, similar to the one previously proposed for approximate rules, represents the TSK rule

$$IF\ X_1\ is\ L_{i_1}(X_1)\ and\ \ldots\ and\ X_n\ is\ L_{i_n}(X_n)$$
$$THEN\ Y = p_1 \cdot X_1 + \cdots + p_n \cdot X_n + p_0$$

in a chromosome C, composed of two parts C_1 and C_2 as follows

$$C = C_1 C_2 \ ; \ C_1 = (i_1, \ldots, i_n) \ ; \ C_2 = (p_0, p_1, \ldots, p_n)$$

The parameters p_i are encoded either using a binary, real or angular coding. We refer to the discussion in Sec. 4.4.1 that the binary and real codes require the GFRBS designer to specify proper parameter intervals in advance. As previously mentioned, this problem can be addressed by either choosing sufficiently large upper and lower bounds or by simply using the angular coding.

8.2 Learning Fuzzy Rules under Competition

The first learning stage in the IRL approach obtains a preliminary fuzzy rule set that captures the relationship between input and output data in the underlying training set. It consists of two components, a *generating method* that identifies individual desirable fuzzy rules from examples and an *iterative covering method* that progressively builds up a set of rules in a way that all the examples are captured by at least one fuzzy rule.

(1) The *fuzzy rule generating method* finds the best individual rule in every run over the set of examples according to the features included in a fitness function. It is based on any kind of EA encoding a single rule in each chromosome and its composition depends on the type of fuzzy rule being generated.

(2) The *iterative covering method* obtains a set of fuzzy rules that completely covers the set of examples. It puts into effect the iterative operation mode of GFRBSs following the IRL approach. In each iteration, it runs the generating method to obtain the best fuzzy rule according to the current examples in the training set. It penalises this rule according to the relative degree of coverage it causes on the current training examples. Then, it incorporates the generated rule into the preliminary fuzzy rule set and removes those examples from the training set that are covered by this set to a sufficient degree.

The following subsections describe both methods in detail.

8.2.1   *The fuzzy rule generating method*

Before this section introduces the principles and operation of the fuzzy rule generating method, it first reviews the criteria that are relevant for the design of the fitness function.

8.2.1.1   *Different criteria for the fitness function of the generating method*

Different aspects play a role when designing the multi-objective fitness function for the fuzzy rule generating method. Frequency criteria consider several indices about the examples that a rule covers. Additional criteria are concerned with properties of the entire fuzzy rule set such as completeness or consistency (see Sec. 1.4.6). Rule simplicity and local measures of error also enter the fitness function.

In the following, we introduce several of these criteria considering a training set E_p composed of p examples $e_l = (ex_1^l, \ldots, ex_n^l, ey^l)$.

- *High frequency value* (Cordón, del Jesús, Herrera, and Lozano, 1999; Herrera, Lozano, and Verdegay, 1998a).

The frequency of a descriptive or approximate fuzzy rule,

$$R_i : \ IF \ X_1 \ is \ A_{i1} \ and \ ... \ and \ X_n \ is \ A_{in} \ THEN \ Y \ is \ B_i,$$

through the set of examples, E_p, is defined as:

$$\Psi_{E_p}(R_i) = \frac{\sum_{l=1}^{p} R_i(e_l)}{p}$$

with $R_i(e_l)$ being the *compatibility degree* between the rule R_i and the example e_l:

$$A_i(ex^l) = T(A_{i1}(ex_1^l), ..., A_{in}(ex_n^l))$$
$$R_i(e_l) = T(A_i(ex^l), B_i(ey^l))$$

where T is a t-norm.

- *High number of positive examples* (Cordón, del Jesús, Herrera, and Lozano, 1999; González and Pérez, 1998a).

  As mentioned in Sec. 1.4.6, an example is considered positive for a fuzzy rule R_i when it matches its antecedent as well as its consequent. Hence, the set of positive examples to R_i is defined as:

$$E^+(R_i) = \{e_l \in E_p / R_i(e_l) > 0\}$$

  with $n^+(R_i) = |E^+(R_i)|$, or when working with a fuzzy set of positive examples (González and Pérez, 1998a):

$$\hat{E}^+(R_i) = \{(e_l, R_i(e_l))/e_l \in E_p\}$$

  with $|\hat{E}^+(R_i)| = \sum_{e_l \in E_p} R_i(e_l)$.

  In both cases, the objective becomes to maximise the number of positive examples.

- *High degree of completeness* (González and Pérez, 1999a).

  The degree of completeness of R_i is defined as:

$$\Lambda_{E_p}(R_i) = \frac{n^+(R_i)}{n_i}$$

  with n_i being the number of examples in E_p whose output matches the consequent of R_i.

- *High average covering degree over the most representative positive examples* (Cordón, del Jesús, Herrera, and Lozano, 1999).

The set of positive examples to R_i with a compatibility degree greater than or equal to $\omega \in (0, 1]$ is defined as:

$$E_\omega^+(R_i) = \{e_l \in E_p / R_i(e_l) \geq \omega\}$$

with $n_\omega^+(R_i)$ being equal to $|E_\omega^+(R_i)|$. The *average covering degree* on $E_\omega^+(R_i)$ can be defined as:

$$G_\omega(R_i) = \sum_{e_l \in E_\omega^+(R_i)} R_i(e_l)/n_\omega^+(R_i)$$

Again, the goal is to maximise this criterion.

- *Penalty associated to the non-satisfaction of the k-consistency property* (Cordón, del Jesús, Herrera, and Lozano, 1999).

  An example is considered negative for a fuzzy rule R_i when it matches its antecedent but not its consequent (see Sec. 1.4.6). Hence, the set of the negative examples for R_i is defined as:

$$E^-(R_i) = \{e_l \in E_p / R_i(e_l) = 0 \text{ and } A_i(ex^l) > 0\}$$

  This criterion penalises fuzzy rules with a high ratio of negative examples to positive examples with a compatibility degree larger or equal to ω. In this way, it penalises the non-satisfaction of the k-consistency property (González and Pérez, 1998a) (see Sec. 1.4.6). The *penalty function on the negative example set of the rule R_i* is given by:

$$g_n^-(R_i) = \begin{cases} 1 & \text{if } n_{R_i}^- \leq k \cdot n_\omega^+(R_i) \\ & (R_i \text{ is } k\text{-consistent}) \\ \frac{1}{n_{R_i}^- - kn_\omega^+(R_i) + exp(1)} & \text{otherwise} \end{cases}$$

  Notice that the negative example set is always computed over the entire, original training data set E_p and not just over the current set of remaining examples.

- *High degree of satisfaction of the soft consistency property* (González and Pérez, 1999a).

  It extends the consistency property making use of two noisy thresholds, a lower bound k_1 and an upper bound k_2, thereby admitting some noise in the rules (González and Pérez, 1999a). The other difference is that the negative example set considered by the authors is also a fuzzy set, defined in a similar way to the fuzzy set of positive examples to a fuzzy rule R_i shown above, and thus its cardinality is computed in a fuzzy way.

The *degree to which a rule satisfies the soft consistency condition* is

$$\Gamma_{k_1 k_2}(R_i) = \begin{cases} 1 & \text{if } n^-_{R_i} \leq k_1 \cdot n^+(R_i) \\ \frac{k_2 \cdot n^+_E(R) - n^-_E(R)}{n^+_E(R)(k_2 - k_1)} & \text{if } k_1 \cdot n^+(R_i) \leq n^-_{R_i} \leq k_2 \cdot n^+(R_i) \\ 0, & \text{otherwise} \end{cases}$$

with $k_1, k_2 \in [0, 1]$ and $k_1 < k_2$.

- *High degree of rule variable simplicity* (González and Pérez, 1998c). A variable X_j in a DNF fuzzy rule is considered to be irrelevant if it encompasses its entire linguistic term set, D_j (see Sec. 8.1.1 for an example). The number of irrelevant variables of a rule R_i is denoted as $i(R_i)$ (González and Pérez, 1998c; Castillo, González, and Pérez, 2001). Hence, the *simplicity degree in the variables of* R_i is defined as

$$svar(R_i) = \frac{i(R_i)}{n}$$

with n being the number of possible antecedent variables.

- *High degree of rule value simplicity* (González and Pérez, 1998c). A rule variable X_j is said to be *stable* in a DNF fuzzy rule if it takes as a value an *unique sequence of adjacent terms in the term set associated*. On the other hand, $e(R_i)$ is defined as the number of input variables X_j of the rule R_i such that D_j is an ordered term set and $\widetilde{A_j}$ or $\overline{\widetilde{A_j}}$ (i.e., the complement of $\widetilde{A_j}$) are stable values (Castillo, González, and Pérez, 2001; González and Pérez, 1998c). Finally, the *simplicity degree in the values of* R_i is

$$sval(R_i) = \frac{1 + e(R_i)}{1 + p}$$

with $p \leq n$ being the number of variables with an ordered term set associated.

- *Low local error of the rule* (Cordón and Herrera, 1999b). The local error of a rule is defined by:

$$LE(R_i) = \sum_{e_l \in E} h_l \cdot (ey^l - S(ex^l))^2 \tag{8.1}$$

with $E \subset E_p$ being the set of examples $e_l \in E_p$ located in the fuzzy input subspace defined by the rule antecedent, $h_l = T(A_1(ex_1^l), \ldots, A_{iv}(ex_n^l))$ being the matching degree between the antecedent part of the rule and the input part of the current data pair ex^l, and $S(ex^l)$ being the output generated by the fuzzy rule upon receiving ex^l as input.

This criterion is usually considered to generate TSK fuzzy rule consequents.

The fitness function is supposed to assign higher fitness values to rules that might contribute to an accurate fuzzy rule base, that ultimately satisfies the completeness and consistency properties. It is designed by aggregating some of the previous criteria into a multi-objective fitness function. To do so, an aggregation function is considered that monotonically increases in each of the individual performance indices, with the product being the usual choice (Cordón and Herrera, 1997c, 1997b, 2001; Cordón, del Jesús, Herrera, and Lozano, 1999; González and Pérez, 1999a; Herrera, Lozano, and Verdegay, 1998a). For example, it is possible to consider the following fitness function (Cordón and Herrera, 1997c) to generate linguistic rules:

$$F(R_i) = \Psi_{E_p}(R_i) \cdot G_\omega(R_i) \cdot g_n^-(R_i) \tag{8.2}$$

Another alternative is to order the multiple criteria according to their priority and to demand the satisfaction of the most important ones in order to give a a non-zero fitness score to the fuzzy rule. If all the required criteria are satisfied, the fitness value associated to the rule will depend on the remaining least important criteria. González and Pérez (1998a) proposed a fitness function to generate DNF fuzzy classification rules that gives the consistency criterion priority over the positive examples criterion:

$$F(R_i) = \begin{cases} n^+(R_i), & \text{if } R_i \text{ is } k\text{-consistent} \\ 0, & \text{otherwise} \end{cases} \tag{8.3}$$

Regardless of the particular type of fitness aggregation, the performance indices of individual criteria are normally designed such that higher values correspond to better solutions. Therefore, finding the best fuzzy rule is equivalent to maximising the overall fitness.

8.2.1.2 *Some examples of fuzzy rule generating methods*

Many different alternatives for the design of this component have been proposed in the specialised literature. In the following subsections, three fuzzy rule generating methods for fuzzy rules of different kinds are introduced.

The fuzzy rule generating method for DNF linguistic rules considered by SLAVE

This generating method is based on a binary coded GA that allows SLAVE to learn the antecedent part of DNF fuzzy rules for a specific concept (the consequent part is specified by the iterative covering method). It was first proposed by González, Pérez, and Verdegay (1993) and was later refined and described in papers such us those by González and Pérez (1996b, 1998a, 1999a). A final extension was introduced by González and Pérez (1997a, 2001) and will be described in Sec. 8.5.2.

- *Coding scheme*: As mentioned, SLAVE deals with DNF Mamdani-type fuzzy rules and hence employs the coding scheme described in Sec. 8.1.1.
- *Genetic operators*: Two classical genetic operators are used: the multi-point crossover, performed at two points, and the uniform mutation. SLAVE uses the standard proportional selection mechanism, linear ranking and an elitist scheme.
- *Generation of the initial population*: The initial population is generated by randomly selecting an example from the subset of the training data that matches the rule consequent value afforded by the iterative covering method and then taking the most specific rule antecedent that best describes the example, i.e., each input variable assumes a single fuzzy term, namely the one with the maximal membership degree for the corresponding example component.

  In order to clarify the previous procedure, let us consider an FS with three input variables, X_1, X_2, and X_3, with the corresponding fuzzy partitions shown in Fig. 8.1. Let $(r1, r2, r3)$ be the crisp input vector of the example randomly drawn from the training data set. The most specific rule antecedent that best matches the example is:

$$X_1 \text{ is } A_{13} \text{ and } X_2 \text{ is } A_{23} \text{ and } X_3 \text{ is } A_{31}$$

which is encoded into a chromosome with the following representa-

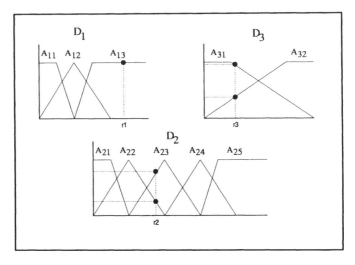

Fig. 8.1   Obtaining the most specific rule covering an example in the generating method of SLAVE

tion:

$$C = (0010010010)$$

- *Fitness function*: González and Pérez (1996b, 1998a) applied the fitness function described in Eq. 8.3. The generating process thus returns a rule that is k-consistent and covers the largest number of positive examples.
- *Termination criteria*: Two different situations are distinguished in order to stop the generating method run depending on the composition of the fuzzy rule set derived by the iterative covering method up to this point: the case in which at least one rule has been generated for the concept currently being learnt (i.e., for the current consequent value), and the case in which no such rule has been obtained yet. The generating process is allowed to pursue a more exhaustive search when trying to generate the first rule for a previously uncovered concept. The intensity of the search for subsequent rules is relaxed once the search process finds the first rule for a novel concept.

  Hence, if the current concept is already covered by at least one rule, the generating process terminates if the fitness of the best rule in the population does not increase its value within a fixed number of iterations. However, when no rule has been obtained for this concept yet, the generating process continues, until the current best rule eliminates at least one example from the training set, independent of the number of iterations. In both cases, the GA terminates anyway after a maximum number of generations.

González and Pérez (1999a) proposed an extension of the generating method which uses the three following novel genetic operators in order to improve the genetic population diversity and to increase the interpretability of the generated DNF fuzzy rules:

- *Rotation operator.* The role of the rotation operator, which is a modified version of the traditional inversion operator, is to increase the genetic diversity in the population. It randomly selects a cut-off point in the chromosome and swaps the position of the two segments as shown as follows:

$$1000 \mid 000110000011 \quad \rightarrow \quad 000110000011 \mid 1000$$

- *OR operator.* This is a special crossover operator that aggregates the fuzzy terms in the parent rules into a conjoint DNF fuzzy rule. To do so, two crossover-points are determined and a binary OR operation is performed on the two parent segments located between them. The outer segments are directly passed to the offspring without alteration and the result of the operation is copied in the inner segment of both offspring:

$$
\begin{array}{ccccccc}
1000 & \mid 000110 & \mid 000011 & \rightarrow & 1000 & \mid 001110 & \mid 000011 \\
0111 & \mid 001100 & \mid 001100 & \rightarrow & 0111 & \mid 001110 & \mid 001100
\end{array}
$$

- *Generalisation operator.* The generalisation operator improves the clarity and comprehensibility of the DNF fuzzy rules. It tries to maximise the number of stable variables in the rule, in the sense that a variable is considered stable if the terms associated to it form a unique, consecutive sequence without gaps (see Sec. 8.2.1.1). Stable variables offer the advantage that rules become more comprehensible. For example, a fuzzy clause such as X_1 *is* $\{A_{13}$ *or* A_{14} *or* $A_{15}\}$, can be interpreted as X_1 *is higher than or equal to* A_{13} (of course, when A_{15} is the last linguistic label in the term set).

SLAVE is biased towards more general rules as in case of identical fitness values, it always prefers the rule that has the larger number of fuzzy terms.

The role of the generalisation operator is to transform unstable variables into stable ones by eliminating their unstable regions. Given a binary representation of the value of a variable in a DNF rule, each consecutive sequence of '0'-bits is considered to be an unstable zone. Hence, the generalisation operator works as follows:

1. *A variable of a DNF rule antecedent encoded in the individual is selected at random.*
2. *If this variable is unstable, do the following:*
    2.1 *Detect its unstable regions and randomly select one of them.*
    2.2 *Replace all the '0'-bits in that region with a consecutive sequence of '1'-bits.*
    2.3 *If the fitness value of the original chromosome is smaller than or equal to that of the new chromosome, substitute the former by the latter*

Figure 8.2 shows an example of the behaviour of the generalisation operator, with three different options.

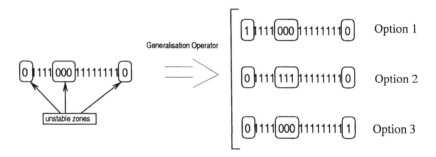

Fig. 8.2   Behaviour of the generalisation operator

The fuzzy rule generating methods for approximate Mamdani-type fuzzy rules designed from MOGUL

As shown in Sec. 1.4.4, there are unconstrained and constrained approximate fuzzy rule learning methods, with the latter ones restricting the fuzzy set parameters to a predefined interval during learning. One further distinguishes among hard constraints imposed on the individual membership functions parameters and soft constraints imposed on the collection of parameters that belong to one fuzzy set.

The learning processes for approximate Mamdani-type FRBSs based on MOGUL fall, accordingly, into three different categories: unconstrained, soft constrained and hard constrained. Cordón and Herrera (1997b) and Herrera, Lozano, and Verdegay (1998a) presented an unconstrained approximate fuzzy rule generating method using a RCGA. The soft constrained learning process applied by Cordón and Herrera (1996a, 1997c) employs an (1+1)-ES. Finally, the hard constrained generating method operates based on a *genetic local search* algorithm (Jog, Suh, and Gutch, 1989; Suh and Gutch, 1987), which combines a GA and an (1+1)-ES (Cordón and Herrera, 1996b, 2001). The remainder of this section focuses on the soft constrained generating method.

The *soft constrained fuzzy rule generating method* (Cordón and Herrera, 1996a; Cordón and Herrera, 1997c) starts by obtaining the linguistic rule that best matches the current training set, according to the rule selection function shown in Eq. 8.2. The parameters of the membership function associated to each label in this linguistic rule are used for two different tasks:

- to initialise the approximate fuzzy rule that is to be adapted by the $(1+1) - ES$, and
- to define the global adjustment intervals for the fuzzy sets composing the approximate fuzzy rule where their definition parameters may vary.

Let (a_i, b_i, c_i) be the three parameters defining the triangular membership function of the fuzzy set associated to the i-th variable in the linguistic fuzzy rule. The global adjustment interval $[D_i^{min}, D_i^{max}]$ for the fuzzy set in the approximate fuzzy rule is computed as follows:

$$[D_{ij}^{min}, D_{ij}^{max}] = [a_i - \frac{b_i - a_i}{2}, c_i + \frac{c_i - b_i}{2}]$$

The (1+1)-ES allows the three characteristic points (a_i, b_i, c_i) of the triangular membership function to vary freely in $[D_i^{min}, D_i^{max}]$ subject to the additional ordering restriction of a meaningful fuzzy set —i.e., that $a_i \leq b_i \leq c_i$—. Thus, the momentary, possible range for a particular parameter depends on the current value of the other parameters. Notice that the authors employ the same approach as in the evolutionary tuning process for TSK KBs introduced in Sec. 4.4.2.

Hence, the operation mode of the fuzzy rule generating method is summarised in the following algorithm:

(1) Start with an empty candidate fuzzy rule set B^c.

(2) For every example $e_l \in E_p$, generate the linguistic rule R_c whose fuzzy terms best match the input values of e_l. If the rule is not part of the fuzzy rule set yet, $R_c \notin B^c$, add it to B^c.

(3) Evaluate all the linguistic rules in B^c and select the prototype rule R_r that is optimal with respect to the rule selection function in Eq. 8.2.

(4) Use the prototype R_r to define the global adjustment interval for the characteristic points of the each triangular fuzzy set. Convert the linguistic prototype rule R_r into the equivalent approximate rule and locally tune its membership functions by means of an (1+1)-ES.

The $(1 + 1) - ES$ has the following components:

- *Coding scheme:* The initial linguistic rule is encoded in an individual C by means of the 3-tuple real coding scheme introduced in Sec. 8.1.2.

- *Mutation process:* The usual ES mutation scheme presented in Sec. 2.3.1 is modified in order to adapt fuzzy membership functions, whose parameters are interrelated —as said, the range of a particular parameter depends on the current values of the remaining parameters— and belong to different domains —each fuzzy set is defined in a different universe of discourse—. The mutation process is identical to the one shown in Sec. 4.4.2.3 for the evolutionary tuning of TSK KBs.

- *Fitness function:* The fitness function employs the same criteria previously used in the linguistic rule selection function. In addition, it incorporates a novel criterion that induces cooperation between rules during the generation stage. Cordón and Herrera (1996a, 1997c) initially proposed this so-called *low niche interaction rate.* It is based on the concept of *niches* in GAs (Deb and Goldberg, 1989; Goldberg, 1989) (see Sec. 2.2.3.4) and penalises the excessive interaction among the fuzzy rules being successively generated during the fuzzy rule generation process, in order to facilitate cooperation among them. This criterion is introduced in Sec. 8.4.2.

Thus, the final multi-objective fitness function becomes:

$$F(R_i) = \Psi_{E_p}(R_i) \cdot G_\omega(R_i) \cdot g_n^-(R_i) \cdot LNIR(R_i)$$

with $LNIR(\cdot)$ as the low niche interaction rate criterion.

The fuzzy rule generating method for TSK rules considered in MOGUL

This generating method for TSK fuzzy rules (Cordón and Herrera, 1997a, 1999b), following the design guidelines of MOGUL, uses a (μ, λ)-ES (see Sec. 2.3.1). The ES only adapts the parameters of the TSK rule consequent, whereas the antecedent parts are provided by the iterative covering method —notice that SLAVE works in the opposite way, i.e., the iterative covering method provides the consequent and the fuzzy rule generating method learns the antecedents—. The covering method determines the rule antecedents based on the presence or absence of data points in the different fuzzy input subspaces, i.e., every fuzzy input subspace containing at least one example is considered to generate a rule for the KB. To do so, the *TSK rule consequent learning method* is applied to determine the existing partial linear input-output relation, based on the data located in this input subspace, a subset of the global data set.

The (μ, λ)-ES that adapts the TSK rule consequent parameters has the following components:

- *Coding scheme:* The method employs the coding scheme presented in Sec. 8.1.3. The real-valued chromosome $\vec{x}$ encodes the weights p_i in the TSK rule consequent based on the angular coding introduced in Sec. 4.4.1.
- Recombination operators and mutation scheme: The usual recombination operators and mutation scheme of (μ, λ)-ESs are considered without modifications. In the experiments described by Cordón and Herrera (1999b), the ES employs the discrete and local intermediary recombination operators for the components of vectors $\vec{x}$ and $\vec{\sigma}$, respectively (refer to (Bäck, 1996) or Sec. 2.3.1, for more information). The $\vec{\alpha}$ vector is only subject to mutation, no recombination is applied.

- *Generation of the initial population*: The $\vec{x}$ part of the individuals in the first generation are initialised using the training data set. This data-driven initialisation computes the average y_{med}, minimum y_{min}, maximum y_{max} of the output y, the matching degree between the example best covered in the fuzzy input subspace and the rule antecedent h_{max}, and the subset E_θ of the most representative —best covered— examples in that subspace (Cordón and Herrera, 1997a, 1999b):

$$y_{med} = \frac{\sum_{e_l \in E} ey^l}{|E|} \quad ; \quad y_{min} = \min_{e_l \in E}\{ey^l\} \quad ; \quad y_{max} = \max_{e_l \in E}\{ey^l\}$$

$$h_{max} = \max_{e_l \in E}\{h_l\} \quad ; \quad E_\theta = \{e_l \in E / h_l \geq \theta \cdot h_{max}\}$$

The initial population of the ES is generated in three steps:

(1) Generate one individual setting parameters $p_i = 0$, with $i = 1, \ldots, n$, and parameter p_0 to the angular coding of the average output value y_{med}.

(2) Generate γ individuals, with $\gamma \in \{0, \ldots, \mu - 1\}$ defined by the GFRBS designer, initiating parameters p_i, $i = 1, \ldots, n$, to zero, and p_0 to a value computed at random in the interval $[y_{min}, y_{max}]$.

(3) Generate the remaining $\mu - (\gamma + 1)$ individual setting the parameters p_i, $i = 1, \ldots, n$, to values randomly drawn from the interval $(-\frac{\pi}{2}, \frac{\pi}{2})$ (angular coding), and p_0 to a value computed from a randomly selected element e in E_θ (with E_θ being the set of positive examples for the rule antecedent (see Sec. 8.2.1.1) and with $\theta \in [0.5, 1]$ being provided by the GFRBS designer as well) in such a way that e belongs to the hyper-plane defined by the TSK rule consequent generated. Thus, it will be ensured that this hyper-plane intersects with the swarm of points contained in E_θ, the most significant ones from E.

Since small angular values tend to cover larger regions in the search space, it seems adequate to generate small values for the parameters p_i in this third step. To do this, a modifier function is considered that picks smaller angular values with higher probabilities. For a possible definition for a function of this kind, refer to (Cordón and Herrera, 1997a, 1999b).

As regards the remaining vectors composing the individual, the components of $\vec{\sigma}$ are initiated to 0.001, and the ones in $\vec{\alpha}$, when considered, are set to *arctan (1)* by the authors (Cordón and Herrera, 1999b).

- *Fitness function*: The evolutionary learning is guided by a fitness function that locally measures the model error between the training data and the fuzzy rule output. The local error criterion shown in Eq. 8.1 combines two objectives, namely to find a TSK rule consequent that locally matches the data in the region defined by the antecedent and to induce cooperation in the fuzzy rule generation (as we shall see in Sec. 8.4.3).

8.2.2 *The iterative covering method*

The fuzzy rule generating method takes a set of training examples as input, then outputs a single rule that demonstrates a good performance. Given this generating method for learning individual rules, it becomes possible to learn an entire set of rules by invoking the generating method on the entire data set, to then remove the examples covered by the rule that was learned, and to repeatedly execute rule generation and example removal until an appropriate set of rules is obtained. This task is developed by the iterative covering method which incrementally generates a set of fuzzy rules of the desired kind that ultimately represents the information existing in the original training data set. This operation mode is the reason why this approach is called IRL.

Hence, at each iteration, the covering method invokes the fuzzy rule generating method to obtain the rule that best covers the remaining set of training examples. It then analyses the covering this rule causes on the training set, removing those examples that are covered to a sufficient degree in order not to generate more rules for that region in subsequent runs. The algorithm terminates when the fuzzy rule set generated appropriately covers the original training examples. Notice that since iterative covering method searches for rules in a greedy manner, it does not necessarily generate an optimal set of rules and a post-processing stage is usually needed to refine its composition.

The iterative covering method induces the formation of niches in the search space as it removes previously covered examples from the training set, thus eliminating the payoff associated to those regions in which these examples resided. This high-level modification of the fitness landscape directs the attention to previously unexplored regions in the search space.

Such a mode of operation ensures that fuzzy rules are generated in all regions that contain training examples. Moreover, in case frequency criteria are considered to guide the fuzzy rule generating method search, the covering method guarantees that the final fuzzy rule set generated in the first stage of GFRBSs based on the IRL approach satisfies the consistency and completeness properties.

Although the main aim of this first learning stage is to induce cooperation among individual rules to compose a preliminary fuzzy rule set, the iterative covering method also promotes *weak cooperation among rules*. The flexible search scheme, based on subsequently focusing on training examples not covered yet, tends to generate neighbour fuzzy rules cooperating softly.

The two following subsections present the iterative covering method in the multi-stage GFRBSs based on the MOGUL (Cordón and Herrera, 1997c, 2001; Cordón, del Jesús, Herrera, and Lozano, 1999; Herrera, Lozano, and Verdegay, 1998a) and the SLAVE (Castillo, González, and Pérez, 2001; González and Pérez, 1998a, 1999a) design guidelines.

8.2.2.1 *The iterative covering method of MOGUL*

In each iteration, the covering process invokes the generating method, obtaining the best fuzzy rule at the current moment. The relative covering that this rule causes on the example set is considered and accumulated to the global covering value caused by the whole fuzzy rule set derived till now on each example. Then, those training examples that have a covering degree larger than $\epsilon \in \mathbb{R}^+$ are removed from the training set.

Instead of an empty rule list, it is possible to initialise the iterative covering method with linguistic fuzzy rules obtained from expert knowledge.

The iterative covering method of MOGUL operates as follows:

(1) Initialisation:

    (a) Specify the values of ϵ and the other parameters considered in the different criteria used in the fitness function (k, ω, ...).

    (b) If linguistic rules are available from expert knowledge:

        i. Initialise the rule set B^g with those rules.

   ii. Compute the initial example covering degree $CV[l]$, $l = 1, \ldots, p$, caused by the expert rules in B^g.

   iii. Remove examples with $CV[l] \geq \epsilon$ from the training set E_p.

 (c) Else:

   i. Initialise B^g with the empty list.

   ii. Set the example covering degree to zero $CV[l] \leftarrow 0$, $l = 1, \ldots, p$.

(2) Invoke the generating method for the current set of examples E_p, and obtain the best fuzzy rule R_r.

(3) Add R_r to B^g.

(4) For every $e_l \in E_p$ do

 (a) $CV[l] \leftarrow CV[l] + R_r(e_l)$,

 (b) If $CV[l] \geq \epsilon$ remove it from E_p.

(5) If $E_p = \emptyset$ terminate else return to step 2.

As the iterative covering method terminates when no more training examples remain, it is guaranteed that the final fuzzy rule set covers every example in the original training data set. Moreover, when the fitness function considered in the fuzzy rule generation method includes any criterion regarding the consistency property, the final fuzzy rule set satisfies this property as well.

8.2.2.2 *The iterative covering method of SLAVE*

The iterative covering method of SLAVE distinguishes itself from the one in MOGUL by the following major differences:

- As SLAVE is mainly considered in classification problems, it follows a somewhat different learning approach. The concept (class or linguistic value) associated to the output variable is fixed and provided by the iterative covering method itself. Thus, the generating method only finds the optimal rule antecedent. Hence, the SLAVE covering method can only be used with descriptive fuzzy rules.
  The SLAVE covering method iterates over the set of concepts (possible problem classes). Once a rule is generated for the current concept, those examples matching it are removed from the training

set. The covering method continues searching for rules that refer to the same concept until it is exhaustively covered, and the iteration proceeds with the next concept.

- MOGUL accumulates the covering degree of the fuzzy rules subsequently being generated over every example and removes it when its covering value is larger than a threshold $\epsilon \in \mathbb{R}^+$. On the other hand, the SLAVE covering method removes the training examples in a different way, as for the removal decision it only considers the covering caused by the currently generated rule. When an example is covered to a degree $\lambda \in (0, 1]$ (λ-covered) by that rule, it is removed from the training set.

- SLAVE also employs a different termination criteria than MOGUL. In MOGUL, the iterative covering method terminates when all examples have been removed from the training set and thereby guarantees the completeness property. On the other hand, González and Pérez (1998a) noticed that the completeness property is difficult to satisfy when working with noisy training examples. Thus, the termination criterion in the SLAVE covering method is based on the so-called *weak completeness condition for a class*. This condition indicates whether the examples covered by the current set of rules for a particular class are representative enough to capture the class concept. Therefore, the SLAVE covering method might terminate generating fuzzy rules for a concept even though there remain class examples not yet covered by the current fuzzy rule set. According to González and Pérez (1998a), a set of rules T with output class C satisfies the weak completeness condition if and only if

$$
\begin{cases}
R_{s+1} \text{ is not } k\text{-consistent} \\
or \\
R_{s+1} \text{ is } k\text{-consistent and } E_\lambda(R_{s+1}) = \emptyset.
\end{cases}
$$

This definition, where $E_\lambda(R_{s+1})$ represents the examples λ-covered by R_{s+1}, assumes that the underlying algorithm ranks the set of rules generated for the concept $\{R_1, R_2, \ldots, R_s, R_{s+1}\}$ in decreasing order of optimality, i.e., R_i is a better rule than R_{i+1}. The iterative covering method terminates upon the first rule R_{s+1} that satisfies the weak completeness condition, without adding R_{s+1} to the final rule set $\{R_1, R_2, \ldots, R_s\}$.

The weak completeness combines the completeness and consistency conditions, thereby ensuring that the final fuzzy rule set is complete in a weak sense and k-consistent. For a formal proof of this assumption, refer to (González and Pérez, 1998a). It can also be shown that the algorithm does not generate contradictory rules, i.e., rules that suggest different consequents for the same antecedent.

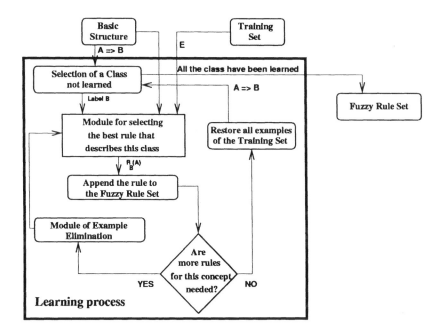

Fig. 8.3   Operation mode of the SLAVE covering method

The SLAVE covering method shown in Fig. 8.3 contains the following steps:

(1) Initialisation:

    (a) Specify values for λ, k and the other parameters in the generating method fitness function.

    (b) Initialise B^g with the empty set.

(2) For every possible value of the consequent variable Y (i.e., for every class B in the case of fuzzy classification rules or for every linguistic term, $B \in F$, in the case of Mamdani-type linguistic rules) do

(a) *Assign the set E_p to E.*

(b) *Invoke the fuzzy rule generating method, considering the data set E and B as inputs. Let R_r be the rule obtained as the output of this procedure.*

(c) *While the set of rules associated to the concept B does not satisfy the weak completeness condition, add R_r to B^g, eliminate the examples covered to a degree λ by R_r from E and go to step 2.b. Otherwise, proceed with the next concept of the consequent variable and continue with step 2.*

(3) *Terminate with the set B^g as the output if no concepts of the consequent variable are left.*

8.3 Post-Processing: Refining Rule Bases under Cooperation

The fuzzy rule generation stage of GFRBSs based on the IRL approach does not take into account the collaborative interaction between the fuzzy rules generated. Therefore, these kinds of genetic learning processes require a successive, post-processing process that refines the original rule set to ultimately improve the cooperation between the fuzzy rules.

The post-processing algorithm tries to address the CCP by promoting the cooperation among the fuzzy rules evolved during the fuzzy rule generation. The objectives are to improve the accuracy of the FRBSs and to simplify the fuzzy rule set either by removing redundant rules or by adjusting their membership functions.

The next two subsections introduce the two different post-processing algorithms based on the MOGUL and SLAVE paradigms.

8.3.1 *The post-processing algorithm of MOGUL*

The post-processing stage of MOGUL (Cordón, del Jesús, Herrera, and Lozano, 1999) develops the two possible refinement actions in two subsequent stages, each one dealing with the CCP in a different way. First, the *genetic simplification process* selects the most cooperative fuzzy rule set by removing the unnecessary fuzzy rules generated in the generation stage. Then, the *evolutionary tuning process* adjusts the membership function definitions for the selected rules.

Besides breaking down the post-processing stage into two processes to solve better the CCP, the following design aspects are to be considered (Cordón, del Jesús, Herrera, and Lozano, 1999):

- The GA in the simplification process employs the Pittsburgh approach evaluation scheme as it solves the CCP in the best possible manner. However, this method avoids the associated problem of a high-dimensional search space, by only working with those fuzzy rules generated in the first stage. Instead of adapting the membership function definitions, it only identifies those fuzzy rules that do not cooperate adequately and removes them. As a result, it selects the subset of fuzzy rules that in combination demonstrate the best performance.

  In addition, a genotypic sharing scheme (Deb and Goldberg, 1989) induces niches in the genetic population (see Sec. 2.2.3.4) with the objective to not only obtain a single best fuzzy rule set but rather a variety of potential solutions of comparable performance. The second stage is referred to as the *multi-simplification process*. However, we should notice that this idea is not new in the field but Krishnakumar and Satyadas (1995) also proposed a GFRBS generating different fuzzy rule sets by means of a niching GA.

- The third stage utilises an EA that deals with an even smaller search space because it only adapts the membership functions and not the fuzzy rule structure itself. Chapter 4 presented evolutionary processes for tuning the membership functions of a descriptive Mamdani, approximate Mamdani or TSK FRBS based on the MOGUL design guidelines.

  The evolutionary tuning process is applied to all the fuzzy rule sets generated by the multi-simplification process. Therefore, a set of fuzzy rules that was sub-optimal in the multi-simplification process may ultimately outperform the competing rule sets as its tuned membership functions allow the rules to cooperate in a better way.

The multi-simplification process incorporates the *sequential niche technique* (Beasly, Bull, and Martin, 1993) to induce niches, being the genetic simplification process proposed by Cordón and Herrera (1997c); Herrera, Lozano, and Verdegay (1998a), the basic optimisation technique iterated in each run of the multi-simplification process. The basic simplification algorithm and the particular aspects of multi-simplification are introduced in the next two subsections.

8.3.1.1 *The basic genetic simplification process*

The basic genetic simplification process was first proposed by Cordón and Herrera (1997c); Herrera, Lozano, and Verdegay (1998a). It is based on a binary coded GA, in which the individuals are selected using the stochastic universal sampling procedure (Baker, 1987) together with an elitist selection scheme. The offspring population is generated by using the classical binary two-point crossover and uniform mutation operator.

The coding scheme operates with fixed-length chromosomes. Assume the rules in the rule set B^g are enumerated from 1 to m. A subset of candidate rules B^s is represented as an m-bit string $C = (c_1, ..., c_m)$ in which each bit denotes the presence or absence of a particular rule from B^g in B^s:

$$If \ c_i = 1 \ then \ R_i \in B^s \ else \ R_i \notin B^s$$

The initial population is generated randomly except for a single individual that is initialised with the complete set of rules B^g by imposing $c_i = 1, \forall_i$.

The EA uses the fitness function $F(C_j)$ in Eq. 4.1 (Sec. 4.3.3.1) for MOGUL genetic tuning processes.

8.3.1.2 *The multi-simplification process*

The sequential niche algorithm requires a *distance metric* in order to define the similarity of two individuals (Beasly, Bull, and Martin, 1993). MOGUL is based on *genotypic sharing* as it uses the Hamming distance between two binary chromosomes $A = (a_1, ..., a_m)$ and $B = (b_1, ..., b_m)$ as the distance metric:

$$H(A, B) = \sum_{i=1}^{m} |a_i - b_i|$$

The *modified fitness function* used by the multi-simplification process penalises fuzzy rule sets that are similar to previously obtained solutions:

$$F'(C_j) = F(C_j) \cdot G(C_j, S)$$

where F is the original basic fitness function, $S = \{s_1, \ldots, s_k\}$ is the set of k solutions already found, and G is a similarity penalty function. Assuming that the fitness is to be minimised, Cordón, del Jesús, Herrera, and Lozano (1999) proposed the following penalty function:

$$G(C_j, S) = \begin{cases} \infty, & \text{if } d = 0 \\ 2 - (\frac{d}{r})^\beta, & \text{if } d < r \text{ and } d \neq 0 \\ 1, & \text{if } d \geq r \end{cases}$$

with d being the Hamming distance between C_j and the most similar individual in S, i.e., $d = Min_i\{H(C_j, s_i)\}$, with r being the *niche radius*, and β being the *power factor* that determines the concavity ($\beta > 1$) or convexity ($\beta < 1$) of the penalty function. The penalty becomes maximal if the individual C_j is identical to any of the solutions already found. Individuals that are further away from any previous solution than the niche radius r obtain no penalty.

The algorithm of the genetic multi-simplification process is shown below:

(1) Initialisation: Equate the multi-simplification modified fitness function to the basic simplification fitness function: $F'(C_j) \leftarrow F(C_j)$.

(2) Run the basic genetic simplification process, using the modified fitness function, adding the best individual found in the run to S.

(3) Update the modified fitness function to give a depression in the region near the best individual, producing a new modified fitness function.

(4) If S does not contain the desired number of fuzzy rule sets, return to step 2, else stop.

8.3.2   *The post-processing algorithm of SLAVE*

The post-processing algorithm of SLAVE uses a heuristic for the generalisation, specification, addition and elimination of fuzzy rules (González and Pérez, 1998b). Although this refinement algorithm can be applied to descriptive Mamdani-type KBs generated from any source, in general it is strongly associated to SLAVE. However, González and Pérez (1996a) successfully applied it to refine fuzzy rule sets obtained by learning algorithms different than SLAVE.

The basic elements of this post-processing algorithm are:

(1) *The rule model*: The descriptive DNF fuzzy rule considered is inherited from SLAVE and it is an important feature of this system because it permits to identify the relevant variables for each rule. Therefore, it is strongly associated to determine the structure of the rule and to obtain more comprehensive descriptions.

(2) *The heuristic*: The special way in which the process of generalisation, specification, addition and elimination of rules are combined.

(3) *The inference process*: The authors distinguish two different kinds of inference processes, one for a crisp consequent domain (fuzzy classification) and another for fuzzy consequent domain (fuzzy modelling and control). Both types of learning problems can be solved using SLAVE. For the first one, SLAVE uses a particular function for establishing the compatibility between an example and the rule set and a special procedure for solving the conflict problems. The mechanism for solving conflicts uses a criterion to sort the rules based on a measure of the accuracy on each rule. Hence, the refinement algorithm supposes that the rule set given as input is ordered using this criterion. For problems with fuzzy consequent domain, the order of the rule is not important, because the output is obtained by the combination of the active rules. In this case, a simple inference process and a defuzzification method are used.

This post-processing process has not only the aim of improving the cooperation level of the fuzzy rule set generated from SLAVE, but also it tries to reduce the dependency of the generation algorithm with the consistency parameters as well. In view of the experiments developed, González and Pérez (1998b) stated that rule sets obtained by SLAVE have a strong dependency on the parameters of the consistency conditions, being necessary a good estimation of these parameters for a correct learning. Furthermore, this module allows us to minimise the number of necessary rules to represent the problem knowledge, keeping the accuracy and improving the interpretability of these rules.

There are two different variants of the SLAVE post-processing algorithm, dealing with crisp and fuzzy consequent domains, respectively. Since they both are not based on GAs, we shall not describe them in this text. The interested reader can refer to (González and Pérez, 1998b).

8.4   Inducing Cooperation in the Fuzzy Rule Generation Stage

As mentioned previously, in the original proposal for the IRL approach, a weak cooperation is defined between neighbour fuzzy rules, being a consequence of the penalty of the rules already generated (usually, the example erasure). This cooperation level is sufficient when the concepts are exclusive and there is no noise or inconsistency in the example set. However, when working with databases affected by noise and inconsistency and when the concepts are not exclusive, a higher degree of cooperation must be established that permits good collaboration between rules from different values of the consequent variable.

There are two different ways to encourage this cooperation among the rules composing the fuzzy rule set in GFRBSs based on the IRL approach (Cordón, Herrera, González, and Pérez, 1998). On the one hand, it is possible to include a post-processing stage with the function of improving the cooperation among the rules in the preliminary fuzzy rule set obtained from the generation process, as we have seen in the previous section. On the other hand, the other possibility is to include any kind of cooperation within the first learning stage, the fuzzy rule generation process. Of course, both possibilities are not exclusive and may be used in the same GFRBS based on the IRL approach.

In this section, different extensions to the usual operation mode of GFRBSs based on the IRL approach involving inducing cooperation in the fuzzy rule generation stage will be introduced. To be precise, in the next three subsections, we shall present a different alternative for performing this task for each existing type of fuzzy rule set.

8.4.1   *Inducing cooperation in descriptive fuzzy rule generation processes: the proposals of SLAVE*

Usually, the degree of collaboration between rules from different concepts is measured using the FRBS inference process considered. However, in GFRBSs based on the IRL approach it is not easy to establish this cooperation between rules, since the first learning stage obtains the rules one by one, and the generation process has not got the whole fuzzy rule set for applying the inference process.

A way for defining this cooperation level involves applying the inference process partially on a subset of the examples and including this information in the evaluation of the rules that are going to be generated in the first learning stage. This subset contains the examples of the concepts that have been learnt by the fuzzy rule generation process. Hence, the cooperation between rules is measured by the error that the new rule produces in the outputs of the inference process, when it is included in the fuzzy rule set. This is the operation mode followed by SLAVE (Cordón, Herrera, González, and Pérez, 1998; González and Pérez, 1999a).

As mentioned previously throughout this chapter, SLAVE distinguishes two kinds of learning problems, crisp and fuzzy consequent domain ones. The difference between both problems is important for cooperation induction purposes since the inference process associated to each one of them is different. In the first case, the inference process can be usually seen as a competition between the rules for determining which of them will be the most appropriate for classifying a certain example, that is, the inference process selects only one rule each time. However, in the second case, the output of the FRBS is obtained by the combined action of all the fuzzy rules fired because of this the example, i.e., by means of an interpolative reasoning process.

In the first case, the authors establish competition relations between fuzzy rules having different consequent variable values for improving the overall behaviour of the fuzzy rule set generated (González and Pérez, 1999a). In the second case, SLAVE induces cooperation relationships between rules by including a new criterion in the fitness function of the fuzzy rule generating method for determining the cooperation degree of the new rule with respect to the fuzzy rule set already generated (Cordón, Herrera, González, and Pérez, 1998). Both approaches will be briefly described in the following two subsections.

8.4.1.1 *Cooperation between rules in SLAVE for crisp consequent domain problems*

The SLAVE proposal for inducing cooperation in the fuzzy rule generation process for crisp domain problems is based on keeping the information relative to the previously learnt rules and on redefining the concepts of positive and negative examples usually considered in SLAVE (González and Pérez, 1999a). Thus, the authors establish the concepts of maximum positive and negative covering degree of an example with respect to the fuzzy rule set already generated.

Let $E^B = \{e_1^B, e_2^B, \ldots, e_s^B\}$ be a set of s training examples that describes the B class, with $e_l^B = (ex^l, B)$ and $ex^l = (ex_1^l, \ldots, ex_n^l)$. Let $T = R_B \bigcup R_{\overline{B}}$ be the set of rules already generated, where R_B is the set of the rules learnt from the B class and $R_{\overline{B}}$ is the set of the rules learnt from the remaining classes.

Hence, noting by $A_i(ex^l)$ the matching degree between the antecedent of the fuzzy rule R_i and the input variable values of example e_l^B (see Sec. 8.2.1.1), the authors define the two following indices (González and Pérez, 1999a):

- *The maximum positive covering degree of an example over R_B:*

$$Q_B(R_B, e_l^B) = \max_{R_i \in R_B} \{A_i(ex^l)\}$$

- *The maximum negative covering degree of an example over $R_{\overline{B}}$:*

$$Q_{\overline{B}}(R_{\overline{B}}, e_l^B) = \max_{R_i \in R_{\overline{B}}} \{A_i(ex^l)\}$$

For each example from the training set, the previous definitions establish the maximum degree in which the rules of its class cover this example (Q_B) and the maximum degree in which the remaining rules cover it ($Q_{\overline{B}}$), respectively. GFRBSs based on the IRL modify the values of Q_B and $Q_{\overline{B}}$ in each iteration of the covering method, i.e., each time that a new rule is included in the fuzzy rule set being generated.

From these definitions, the authors redefine the concept of positive example for a fuzzy rule as follows: they only consider an example to be positive for a new fuzzy rule if the compatibility between the example and this rule is better than the maximum positive and the maximum negative covering degree of this example. In a similar way, they only consider an example as negative if the compatibility between the example from the other class and the new fuzzy rule is greater than or equal to the maximum positive covering degree for this example.

Let $E_p = E^B \bigcup E^{\overline{B}}$ be the training set, where E^B contains all the examples describing the B class and $E^{\overline{B}}$ contains all the examples describing the remaining classes. Let R_i^B be a new fuzzy rule for the B class that is not included in the R_B set.

Then, González and Pérez (1999a) defined the *fuzzy set of positive ex-*

amples for the fuzzy rule R_i^B as:

$$E_Q^+(R_i^B) = \{(e_l^B, Gr^+(R_i^B, e_l^B)) \mid e_l^B \in E^B\}$$

where

$$Gr^+(R_i^B, e_l^B) = \begin{cases} A_i(ex^l), & \text{if } A_i(ex^l) > Q_B(R^B, e_l^B) \\ & \text{and } A_i(e_x^l) > Q_{\overline{B}}(R_{\overline{B}}, e_l^B) \\ 0, & \text{otherwise} \end{cases}$$

and, consequently, the *fuzzy set of negative examples for the fuzzy rule R_i^B* as:

$$E_Q^-(R_i^B) = \{(e_l^{\overline{B}}, Gr^-(R_i^B, e_l^{\overline{B}})) \mid e_l^{\overline{B}} \in E^{\overline{B}}\}$$

where

$$Gr^-(R_i^B, e_l^{\overline{B}}) = \begin{cases} A_i(ex^l), & \text{if } A_i(ex^l) \geq Q_B(R_B, e_l^{\overline{B}}) \\ 0, & \text{otherwise} \end{cases}$$

Hence, the way to put into effect this cooperation induction procedure is based on obtaining both sets each time that a new fuzzy rule, R_i, is going to be evaluated in the fuzzy rule generating method and on computing their cardinalities $n^+(R_i)$ and $n^-(R_i)$ in the way shown in Sec. 8.2.1.1. These cardinalities will then be considered in the different criteria based on positive and negative examples composing the fitness function (see Sec. 8.2.1.1). This modification in the measurement of the positive and negative examples causes two improvements in the GFRBS behaviour:

(1) The examples are only considered to be positive or negative for a rule when causing a significant change over their classification, taking into account the fuzzy rules already generated.

(2) Its combination with the new genetic operators considered in the GA composing the fuzzy rule generating method in SLAVE (see Sec. 8.2.1.2) make it have a greater tendency to generalise the rules than the original one, thus causing a reduction in the number of rules necessary for the description of the system and an improvement in the readability of these fuzzy rules.

8.4.1.2 *Cooperation between rules in SLAVE for fuzzy consequent domain problems*

As mentioned previously, in this case SLAVE induces cooperation in the first learning stage by including an error measure that determines the cooperation degree of the new rule with respect to fuzzy rule set already generated.

To do so, the authors work with the crisp sets of positive and negative examples (see Sec. 8.2.1.1) instead of with the usual fuzzy ones considered in SLAVE because they state that, on the contrary that happens in crisp consequent problems, in fuzzy problems it is not necessary to grade the trust of the successes, since the rules work in cooperation for obtaining the output (Cordón, Herrera, González, and Pérez, 1998).

On the other hand, another modification is that, in this case, it is important the order in which SLAVE learns the concepts for the induction of cooperation between the fuzzy rules being generated. Hence, when inducing this kind of cooperation in the fuzzy rule generating method, the iterative covering method of SLAVE works by considering the concepts (the fuzzy labels in the output variable fuzzy partition) in increasing order. This operation mode ensures that when we address the generation of fuzzy rules to describe a new concept, we have previously obtained the subset of fuzzy rules describing the previous one, and thus we can consider how the new rules being generated affect the examples that match with them both. Notice that to achieve this, it is necessary not to consider the examples matching both a previous consequent variable and the antecedent part of the fuzzy rule currently being generated as negative examples for this rule (Cordón, Herrera, González, and Pérez, 1998).

To put all these ideas into effect, a new criterion is added to the fuzzy rule generating method fitness function with the aim of measuring the error that the new rule produces in the output when it is included in the rule set. Its value is computed by making inference, using the subset of learnt fuzzy rules and the one being currently generated, on the subset of the examples belonging to the current concept and to any of the previous ones. Anyway, the fitness function is based on an order between the different criteria (see Sec. 8.2.1.1), and the main criterion in this order is still the one considering both the maximisation of the number of positive examples and the verification of the soft consistency condition.

8.4.2 *Inducing cooperation in approximate fuzzy rule generation processes: the low niche interaction rate considered in MOGUL*

The weak rule cooperation induction performed in the first learning stage of multi-stage GFRBSs may even become a more significant problem when dealing with approximate FRBSs. As usually, each new fuzzy rule is generated without taking into account how it will cooperate with the previous ones obtained, but in this case this operation mode can have worse consequences due to the fact that the interaction level between neighbour approximate fuzzy rules is not fixed as in descriptive ones, but it changes from one rule to another. Hence, the newly generated fuzzy rule may interact insufficiently or excessively with the previous ones, making the approximate FRBS obtained perform badly.

Cordón and Herrera (1997c, 1999) proposed a solution for this bad behaviour, included in MOGUL assumptions, in the form of a criterion for the fuzzy rule generating method fitness function allowing us to generate the best possible approximate fuzzy rule in each covering method iteration taking into account both the goodness of this rule and its cooperation with the previous ones generated.

To put this criterion into effect, a second type of niching is induced in the generation process, apart from the one induced by the usual IRL operation mode (see Sec. 8.2.2). It will be based on a *phenotypic niching scheme* (Deb and Goldberg, 1989; Goldberg, 1989) (see Sec. 2.2.3.4) and will penalise the excessive proximity of the fuzzy rule being generated to the previously obtained ones, so obtaining a better cooperation level in the rule set. Cordón and Herrera (1996a, 1997c) presented a niche sharing function working in this way, the *low niche interaction rate* (LNIR).

Since this criterion is based on GA niching, some considerations must be taken into account to design it by means of a sharing function. On the one hand, we should note that, in this case, the payoff is not shared among all the individuals in the current genetic population, but between the individual being currently adapted (the rule being generated in this covering method iteration) and the rules previously generated. Therefore, the payoff associated to this individual will be lower when it is closer to a niche centre determined by the previously generated rules. Anyway, the role of the sharing function continues to be the sharing of the global payoff between the individuals located in the same niche, but in this case they have been generated in different runs of the EA composing the fuzzy rule generating method. Hence, the operation mode is quite similar to the one followed by the *sequential niche technique* (Beasly, Bull, and Martin, 1993) mentioned in Sec. 8.3.1. This operation mode allows us to work with this criterion even in case the EA considered is not a population-based one (e.g., an $(1 + 1) - ES$) (Cordón and Herrera, 1997c).

We should also note that one of the most important drawbacks associated to the classical sharing scheme is the need to know *where* each niche is and *how big* it is in order to allow the fitness sharing. Typically, this requirement is addressed by the assumption that if two individuals are close together, within a distance known as the *niche radius*, then their fitness must be shared. Although several methods have been proposed to determine this value (see (Deb and Goldberg, 1989)), the calculation of this radius is a very difficult task in the most cases.

Fortunately, in our case it is easy to determine the location and size of the different existing niches. As we are working in the phenotypic space, each individual represents an approximate fuzzy rule formed by n input fuzzy variables and an output one. Each variable takes as its value a triangular-shaped fuzzy number encoded in the string. Therefore, the centre of the niche in the solution space will be an $(n + 1)$-dimensional point, whose coordinates correspond to triangular membership function modal points. Two individuals will share their payoff if there is any interaction between the fuzzy sets composing them, i.e., if the fuzzy sets associated to the same variable in them both overlap each other. Hence the algorithm does not present a fixed niche radius value as in the classical sharing scheme, but rather the size of the niche depends on the membership function shapes encoded in the different individuals.

Considering that C encodes the approximate fuzzy rule being adapted

$$C \sim IF\ x_1\ is\ A_1\ and\ \dots\ and\ x_n\ is\ A_n\ THEN\ y\ is\ B$$

the LNIR penalises the fitness associated to C in the following way:

$$LNIR(C) = 1 - NIR(C)$$

with

$$NIR(C) = Max_i\{h_i\}$$

and

$$h_i = T(A(N_i x), B(N_i y)), i = 1, \ldots, d$$
$$A(N_i x) = T(A_1(N_i x_1), \ldots, A_n(N_i x_n))$$

being $N_i = (N_i x, N_i y)$ the centres of the rules R_i (niches) determined until now ($i = 1, \ldots, d$, where d is the number of iterations of the covering method developed).

Hence, $LNIR(C)$ penalises the excessive interaction between the fuzzy rules, which leads to bad cooperation between them. It is defined in $[0, 1]$ and gives the maximum value (no penalty) when the rule encoded in C does not interact with any of the rules generated until now. The minimum value (maximum penalty) is obtained when this rule is equal to one of those generated previously.

Figure 8.4 graphically shows a situation where there is interaction between the rule encoded by C and any of the rules generated until now.

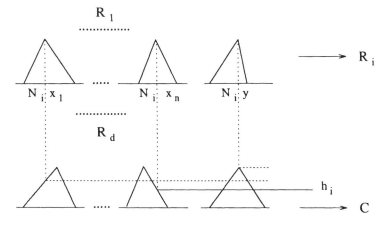

Fig. 8.4   Interaction between the current rule and the predetermined ones

Therefore, the combined action of the LNIR criterion and the iterative covering method modifies the fitness landscape in two different levels at each algorithm step. The purpose of these changes on the individual fitness payoff is to encourage the generation of individuals exploring new zones of the solution space in the subsequent runs while penalising the ones located in existing niches. The two different modification levels are analysed below:

- As mentioned in Sec. 8.2.2.1, the covering method removes examples from the training data set, eliminating the payoff associated to the space zones where these examples were located. This is a *high-level modification* because it translates the search focus to another space zone. This modification encourages adequate space exploration.
- When a niche has been located in a space zone and it continues to be the most promising one (i.e., the examples located in it have not been yet covered and they have associated a big payoff), new fuzzy rules will be generated in the same zone and they will interact with the ones generated until now. An adequate interaction rate (i.e., rule cooperation) is desirable to make the best use of the FRBS interpolative reasoning capabilities. This is obtained by using a *niche penalty function* that penalises the excessive proximity of the new rule to the previously generated ones.

  The frequency criteria existing in the generating method fitness function usually try to widen the supports of the generated fuzzy rules to extend their applicability and cover more examples, this way obtaining a larger reward. The niching criterion tries to narrow their support by penalising the excessive proximity. The combination of both kinds of criteria allows us to obtain a suitable interaction level among neighbouring approximate fuzzy rules.

  This is a *low-level modification* because the algorithm continues working in the same space zone but penalises excessive proximity to the niches located therein. This modification encourages adequate space exploitation.

8.4.3   *Inducing cooperation in TSK fuzzy rule generation processes: the local error measure considered in MOGUL*

Finally, there is also a proposal for inducing cooperation in the first learning stage of GFRBSs obtained from MOGUL for designing TSK FRBS (Cordón and Herrera, 1999b; Cordón, del Jesús, Herrera, and Lozano, 1999). It takes as a base an error criterion proposed by Yen and Gillespie (1995) to design TSK GFRBSs considering the local behaviour of the FRBS, which presents the expression shown in Eq. 8.1.

This criterion is analysed in different works (Cordón and Herrera, 1999b; Cordón, del Jesús, Herrera, and Lozano, 1999), concluding that it is really appropriate for inducing cooperation at the level of single fuzzy rules, due to the fact that it promotes the generation of fuzzy rules whose consequents will adjust better to the examples best covered by their antecedents, i.e., to the examples matching them to a higher degree. Working in this way, the adjustment of the examples matching them to a lesser degree will be done by means of the combined action of the different rules in the TSK KB, thus inducing the desired cooperation in the fuzzy rule generation process. Therefore, the fitness function may be designed by using this single criterion, due to the fact that it allows us to satisfy both goals to be achieved in this stage: generating TSK fuzzy rules with a good individual behaviour and with good cooperation between them. For an example of a MOGUL-based TSK GFRBS composed of a fuzzy rule generation process considering this single-criterion fitness function, refer to (Cordón and Herrera, 1999b).

8.5 Examples

The structure of this chapter is quite different from the two previous ones devoted to the Michigan and Pittsburgh genetic learning approaches, due to the fact that only two applications of the IRL approach to the design of multi-stage GFRBSs can be found in the specialised literature, MOGUL and SLAVE. They both have been introduced in a common framework throughout this chapter, but this last section is devoted to summarise them and to describe some of their particular aspects.

8.5.1 *MOGUL*

Cordón, del Jesús, Herrera, and Lozano (1999) proposed MOGUL (Methodology to Obtain GFRBSs Under the IRL approach), a methodology consisting of some design guidelines that allow us to obtain GFRBSs based on the IRL approach for designing different types of FRBSs (descriptive and approximate Mamdani-type, and TSK), applicable to a variety of problem domains, fuzzy modelling, fuzzy control, and fuzzy classification, among others.

MOGUL allows the user to build his own GFRBSs customised for his specific problem. Therefore, any user is free to add his particular requirements to MOGUL guidelines for designing any kind of FRBS to solve his problem in an adequate way. To do so, he only needs to design his own evolutionary process for each one of the GFRBS learning stages, as long as it complies with the MOGUL assumptions.

As seen throughout this chapter, MOGUL considers the two possible ways to deal with the cooperation between rules in the IRL approach, the induction of cooperation in the generation stage and the inclusion of a post-processing stage, which in itself is composed of the genetic simplification and the evolutionary tuning processes. Therefore, multi-stage GFRBSs obtained from MOGUL possess the following structure:

(1) *A fuzzy rule generation process* that generates a set of fuzzy rules that capture the information contained in the underlying training data set in an appropriate manner. According to Sec. 8.2, it rests upon two components: a *fuzzy rule generating method*, which induces competition at the level of individual rules with the objective to obtain the rule with best performance as regards the examples currently contained in the training set, and an *iterative covering method* that learns a set of rules by repeatedly invoking the generation step and removing those examples from the training set that are already covered to a sufficient degree.

(2) *A genetic multi-simplification process* for selecting rules, which simplifies the fuzzy rule set obtained in the previous stage to improve the cooperation among rules. This second process considers a genotypic niching scheme to generate different simplified fuzzy rule sets by means of a binary coded GA (see Sec. 8.3.1).

(3) *An evolutionary tuning process*, based on any kind of real-coded EA. It adjusts the membership functions of every fuzzy rule set resulting from the genetic multi-simplification process. Which exact type of tuning is used depends on the nature of the FRBS (refer to Chapter 4 for some examples). The most accurate fuzzy rule set obtained in this stage constitutes the final output of the entire learning process.

One of the key aspects of MOGUL is its ability to use previously available knowledge to refine the learning process. Since FRBSs are able to incorporate linguistic as well as numerical information into the design process (Wang, 1994), MOGUL should be able to utilise both types of information, but still must be able to derive the fuzzy rule set from a training data set alone, in case no expert information is available.

Hence, the GFRBSs obtained from MOGUL may work in different ways, depending on the kind and on the amount of information available. We describe the different possibilities in the following (Cordón, del Jesús, Herrera, and Lozano, 1999), starting with the case in which only a training data set is available, and concluding with the case in which the expert knowledge is sufficient to initially come up with a complete, preliminary fuzzy rule set.

(1) *Example set:* The whole multi-stage GFRBS is applied. The initial DB definition is acquired using normalised fuzzy partitions of the input and output domains (when designing a Mamdani-type FRBS) or of the input domain alone (when working with a TSK-type FRBS).

(2) *Expert fuzzy partitions (linguistic terms, with their associated membership functions) and example set:* When initial fuzzy partitions are provided for all the linguistic variables of the system, the learning process still includes all three stages, but the fuzzy sets defined by the expert are used to initialise the DB, rather than starting with normalised, uniform partitions.

(3) *Partial fuzzy rule set and example set (either with or without expert fuzzy partitions):* First, the iterative covering method incorporates the expert fuzzy rules into the fuzzy rule set to be generated in the first stage. In the case of TSK FRBS, the linguistic rule is first transformed into a TSK one by taking the modal point of the consequent fuzzy set (Wang, 1994). This partial fuzzy rule set is then completed by merging it with the fuzzy rules that the generation process learns from input-output data. In the next step, the genetic multi-simplification process produces a set of alternative, simplified fuzzy rule set definitions. Finally, the evolutionary tuning process is applied to adjust the membership functions, resulting in the ultimate rule set.

(4) *Preliminary fuzzy rule set and example set:* In case the expert was able to define a complete fuzzy rule set, only the evolutionary tuning process is invoked to improve the accuracy of the FRBS with respect to the training data by refining the original membership functions.

MOGUL is also distinguished by the following characteristics:

- The designer can customise the generation stage by using different techniques to generate single fuzzy rules rather than having to rely on a GA as in the other IRL design schemes. It is possible to either employ a non-evolutionary inductive algorithm (as in the descriptive GFRBS proposed in (Cordón and Herrera, 1997c)) or an ES (as in the soft constrained approximate GFRBS proposed in (Cordón and Herrera, 1997c)) instead of the usual GA. The GFRBS so obtained from MOGUL still operates in the usual IRL way, but the generation process itself becomes more efficient and less time-consuming.
- The fuzzy rule set has to satisfy several important statistical properties in order for the FRBS to demonstrate a desirable behaviour (see Sec. 1.4.6). MOGUL takes the *completeness* and *consistency* criteria into account, in order to improve the performance of the generated fuzzy rule sets.
- Genetic operators and coding schemes of the EA that are customised for each of the individual stage in the design process are beneficial to an efficient and robust exploration of the different search spaces. MOGUL encompasses a variety of relevant techniques to reduce the search space complexity and to perform an adequate, efficient exploration and exploitation of it (Cordón, del Jesús, Herrera, and Lozano, 1999).

For more information about some specific MOGUL-based multi-stage GFRBSs for designing different types of FRBSs refer to:

- *Descriptive Mamdani-type FRBSs* (Cordón, Herrera, and Lozano, 1996; Cordón and Herrera, 1997c; Cordón, del Jesús, and Herrera, 1998; Cordón, Herrera, and Sánchez, 1999).
- *Approximate Mamdani-type FRBSs* with hard constrained learning (Cordón and Herrera, 1996b, 2001), soft constrained learning (Cordón and Herrera, 1996a, 1997c), or unconstrained learning (Cordón and Herrera, 1997b; Herrera, Lozano, and Verdegay, 1998a).
- *TSK FRBSs* (Cordón and Herrera, 1999b; Cordón, Herrera, and Sánchez, 1999).

8.5.2 *SLAVE*

SLAVE (Structural Learning Algorithm in Vague Environment) was initially proposed by González, Pérez, and Verdegay (1993) and later refined in different papers (González and Pérez, 1997b, 1998a, 1999a, 1998c; Castillo, González, and Pérez, 2001). SLAVE is a GFRBS based on the IRL approach that starts with a simple description of the problem —the consequent variable (concept variable) and the set of possible antecedent variables— for generating the rules that describe the consequent variable. Based upon this description and a set of examples, the learning algorithm decides for each rule and each concept (each possible class in fuzzy classification or each possible linguistic value for the output variable in fuzzy control and fuzzy modelling), which variables are needed to describe this concept and which can be ignored (feature selection). The DNF fuzzy rule model introduced in Sec. 1.2.6.1 and further analysed in Sec. 8.1.1 constitutes the basic element in the SLAVE design process.

The key to this rule model is that each variable can take as a value an element or a subset of elements from its domain, i.e., the value of a variable is interpreted more as a disjunction of elements rather than just as a single element in its domain. Thus, as we have seen in Sec. 8.1.1, the DNF rule model allows SLAVE to select the relevant variables for each fuzzy rule in the following way: if the value of a variable coincides with its entire domain, it is obvious that this particular variable is of no relevance for describing the current concept and is therefore eliminated from the rule.

Associated to the feature selection, SLAVE selects a set of rules that describe the training example set. The initial fuzzy rule generating method in SLAVE is guided by a bi-criteria function, which takes as a base the concept of positive and negative examples for selecting the rule covering the maximum number of positive examples and simultaneously verifying the soft consistency condition to a high degree (see Sec. 8.2.1.1).

Another key aspect of SLAVE is its iterative covering method, based on the concept of λ-covering (see Sec. 8.2.2.2). This covering method focuses in each iteration on the examples associated to the current concept and on the fuzzy set of positive examples that match the fuzzy rule currently being generated for this concept. Training examples are removed from the data set if they are covered to a degree higher than $\lambda \in (0, 1]$ by the last rule generated. The learning for the current concept terminates if either one of the two following conditions is satisfied: there are no more examples associated to the concept in the training data set or the last rule obtained by the fuzzy rule generating method is not able to λ-cover any of the remaining examples associated to the concept in this set. This fact allows the fuzzy rule set generated from the first learning stage of SLAVE to satisfy the weak completeness condition.

Further extensions of SLAVE improved the definition of the fuzzy rule generating method fitness function to include additional criteria allowing the genetic learning process to obtain fuzzy rule sets that demonstrate superior behaviour. González and Pérez (1998c) and Castillo, González, and Pérez (2001) added two new criteria to improve the descriptive power of the rules, the maximisation of the rule variable and value simplicity, in addition to the two original ones (see Sec. 8.2.1.1). Cordón, Herrera, González, and Pérez (1998) and González and Pérez (1999a) defined a criterion to induce cooperation between rules in the first learning stage (see Sec. 8.4.1). In all cases, the fitness function is designed by establishing an order among the criteria, with the most important criterion being the bi-criteria function that integrates both the maximisation of the number of positive examples and the verification of the soft consistency property (see Sec. 8.2.1.1).

On the other hand, González and Pérez (1997a, 2001) presented a new variant of SLAVE to improve the way in which it performs the feature selection in the fuzzy rules. The new version was called *SLAVE-2* and it is based on a new fuzzy rule generating method, a GA working on two different levels: a *variable level* and a *value level*. In the first one, the evolution process identifies a set of currently active variables. In the second level, the value one, particular values for the active and non-active variables are encoded. The coding of a value associated to a non-active variable is ignored. This coding scheme was presented in Sec. 8.1.1.

As the authors enunciate, this separation into variables and their values may be helpful for large problems, since the fuzzy rule generation method restricts the search to a subset of active variables when searching for the best rule for a particular consequent. Moreover, in the evolution process, the subsets of active variables can be adapted, by eliminating or adding variables. In the initial SLAVE fuzzy rule generating method, a variable can change from active to non-active, but only on the local basis of individual genes. In SLAVE-2, the adaptation on the variable level is global, namely the entire variable is active or not. On the value level, the process is similar to the previous one.

Moreover, González and Pérez (1997b, 2001) improved SLAVE-2 by considering some information measures to generate the variable level of the chromosomes in the initial population of the fuzzy rule generating method according to the data collected in the training set.

The idea is to use a measure of the relevance of each variable with respect to a particular concept. At the beginning of the process, this measure will determine the capacity of the variable to represent the concept. Later on, the genetic process, using a set of operators, will modify these values and will therefore alter the criterion to determine the set of variables that must be activated to represent it.

To do so, an index measuring the dependence or independence degree between an input variable and a specific value of the consequent variable is considered. This measure is interpreted as an initial value for the relevance of each variable X_i with respect to class C (or linguistic label B_i) of the Y variable.

The values obtained from the latter information measure will be included in the initial codes for the variable chromosome on each individual of the population instead of a 0 or 1 default value. This way, the first part of the chromosome (variable chromosome) becomes real-coded, taking values in $[0,1]$, whilst the value chromosome keeps the same description. This new coding scheme allows SLAVE-2 to make smoother changes between relevance and irrelevance.

There are different possibilities (González and Pérez, 1997b) to interpret the previous values. González and Pérez (2001) considered the *activation threshold* interpretation by adding a new real-coded gene to the variable chromosome representing a threshold that the relevance value of a variable in the encoded fuzzy rule must exceed to be considered active. Of course, the new coding scheme with a real-coded component requires the use of specific genetic operators for these kinds of representations. The authors worked with BLX-α crossover (see Sec. 2.2.4.1) and Michalewicz's non-uniform mutation (see Sec. 2.2.4.2).

For more information about the different variants of this two-level GA, the interested reader can refer to (González and Pérez, 2001), which also provides an experimental analysis demonstrating how the new SLAVE version, SLAVE-2, outperforms the original one in accuracy, learning time and the number and interpretability of the obtained rules.

Chapter 9

Other Genetic Fuzzy Rule-Based System Paradigms

There are other GFRBS paradigms that do not clearly fit into the categories described in the previous chapters either because they are based on very specific techniques, such as GP for learning fuzzy rules, or because the genetic learning processes assume a particular structure, such as the selection of fuzzy rules from an available rule set, the genetic learning of DB components or other genetic based learning approaches (genetic integration of KBs, hierarchical evolutionary design, Nagoya approach, hybrid algorithms of Michigan and Pittsburgh approaches, etc.).

The following four sections describe these miscellaneous approaches.

9.1 Designing Fuzzy Rule-Based Systems with Genetic Programming

GP (see Sec. 2.3.3) is concerned with the automatic generation of computer programs. Proposed in the early 90's by J. Koza, this new field has rapidly grown in visibility as a promising approach to adapt programs to particular tasks by means of artificial evolution (Koza, 1999). GP improves a population of computer programs by evaluating their performance against a set of training cases. It has been applied to a remarkable variety of different domains, such as symbolic regression, electronic circuit design, data mining, biochemistry, robot control, pattern recognition, planning and evolving game-playing strategies.

Most GP systems employ a tree structure to represent the program that computes a function or executes an algorithm. The trees in GP are composed of primitives taken from the sets of *function* and *terminal* symbols.

Terminal symbols refer to nodes at the leaves of the tree and either provide a value, such as a constant or a reference to an input variable, or correspond to a primitive action such as a simple robot motion command. Function symbols correspond to non-leaf nodes in the tree and compute or process the arguments passed from its children nodes. Commonly used functions are arithmetic and logic operators, conditional statements and loop statements. Depending on the type of operation, unary, binary or n-ary, function nodes contain as many child nodes as arguments to the function.

Different proposals can be found for using the GP paradigm to evolve fuzzy rule sets, internally represented as type-constrained syntactic trees (Geyer-Schulz, 1995; Alba, Cotta, and Troya, 1999; Bastian, 2000; Hoffman and Nelles, 2000; Homaifar, Battle, Tunstel, and Dozier, 2000). Fuzzy GP, proposed by Geyer-Schulz (1995), combines a simple GA that operates on a context-free language with a context-free fuzzy rule language. Nowadays, one distinguishes among GPs that utilise a grammar for learning linguistic rules (Alba, Cotta, and Troya, 1999; Geyer-Schulz, 1995, 1996; Tunstel and Jamshidi, 1996) and approaches that use domain-specific knowledge to define the function and terminal set which constitute the building blocks for the fuzzy rules to be learned (Bastian, 2000; Battle, Homaifar, Tunstel, and Dozier, 1999; Hoffman and Nelles, 2000; Homaifar, Battle, Tunstel, and Dozier, 2000; Sánchez and García, 1999).

In the following we introduce both approaches: learning RBs with GP, and learning KBs with GP.

9.1.1   *Learning rule bases with genetic programming*

In the original approach, the GP operates on a grammar to learn linguistic rules (Alba, Cotta, and Troya, 1999; Geyer-Schulz, 1995, 1996; Tunstel and Jamshidi, 1996). This section analyses this approach according to the learning structure presented by Alba, Cotta, and Troya (1999). It is divided into three subsections that focus on the coding of RBs, the necessity of a typed-system and the generation of RBs.

9.1.1.1   *Encoding rule bases*

As each expression is supposed to contain a complete list of rules, the coding requires a representation of individual rules and a mechanism to combine them. A single rule is represented as a binary tree:

- The root node is an *IF* node representing an *"IF-THEN"* statement, of which the left and right branches represent the condition and the action part, respectively.
- Likewise, conditions and actions themselves are expressed as subtrees.

   - A fuzzy clause is a variable paired with a linguistic term, represented by an *IS* node whose left branch corresponds to the variable name, whereas the right branch carries the linguistic label associated.
   - A conjunction of multiple clauses is achieved by an *AND* node, with its two branches either containing further nested conjunctions or fuzzy clauses of the form *"variable IS linguistic term"*.

Figure 9.1 shows an example with input variables *X1* and *X2* and output variable *Y*.

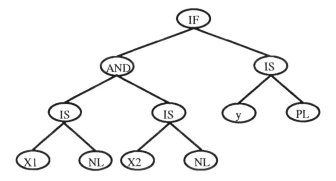

Fig. 9.1   A syntactic tree and the rule it represents: *IF X1 is NL and X2 is NL THEN Y is PL*

In principle, a list of rules could be represented as a list of trees. However, it is more appropriate to use a single tree representation because it allows rules to be tightly connected and therefore facilitates the propagation of desirable functional blocks. Two additional symbols are required in order to build rule list trees: the *EL* symbol represents an empty list whilst the *RLIST* symbol combines rules in the same fashion in which the *AND* symbol joins terms, i.e., a list of rules is a tree with an *RLIST* root-node and two branches either representing single rules or nested lists of rules.

This coding scheme provides is a very flexible and powerful representation of RBs. It takes into account the fact that a general RB should be able to encompass a variable number of rules of different structure.

9.1.1.2 *Necessity of a typed-system*

Due to the flexibility of the tree representation, crossover becomes a powerful tool because it exchanges genetic information at several levels: rules, terms or even variable names. However, the unconstrained, traditional GP crossover operator may generate trees that correspond to syntactically incorrect rules, for example in case a variable name is substituted by a whole rule. There are three possible remedies to that problem:

- Define closed functions, such that ill-defined rules are reinterpreted in some predefined way (Koza, 1992). For example, the *IS* function might consider anything different from a variable name in the left branch as a dummy variable with a fixed value.
- Repair the tree by deleting and adding sub-trees whenever necessary.
- Select crossover points such that the resulting trees are correct.

The first option produces a large number of useless trees mostly composed of dummy rules, whereas the second one is computationally expensive and does not preserve causality due to the repairing mechanism. Therefore, we restrict the presentation to the third option of constraining crossover to similar nodes.

Every function symbol contains a type attribute, according to the equivalence classes defined by the equivalence relation "is interchangeable with". For example, since an *EL* node may be swapped with an *RLIST* node, they share the same type attribute. Figure 9.2 shows all types.

The crossover operator first selects a random node in one parent. Then, a node of the same type is randomly selected in the other parent and the corresponding sub-trees are swapped, thus producing two valid offspring. However, although considering type attributes guarantees syntactically correct trees, crossover might still generate semantically incorrect trees in which the condition part accidentally contains output variables and/or the action part includes input variables (e.g., *IF Y is NL THEN X*1 *is ZE*). In order to avoid these detrimental combinations, two subtypes for both type II and III are defined. Every subtype refers to the set of variables that can

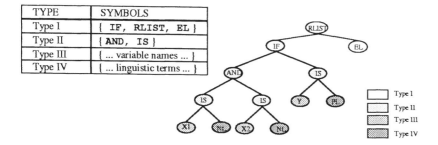

TYPE	SYMBOLS
Type 1	{ IF, RLIST, EL }
Type II	{ AND, IS }
Type III	{ ... variable names ... }
Type IV	{ ... linguistic terms ... }

Fig. 9.2   Basic types and symbols belonging to each type

appear in either the condition or the action part. For example, for type *VBLE* (type III), the subtypes *VBLEIN* and *VBLEOUT* are defined, representing input and output variables respectively. See Fig. 9.3 for a complete taxonomy of types and subtypes.

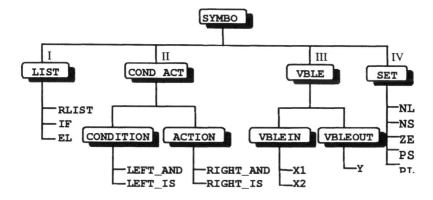

Fig. 9.3   Types and symbols used

9.1.1.3   *Generating rule bases*

An elitist generational GA is used to evolve trees. The tree sizes are bounded to a maximum depth of m_1 levels in the initial generation and m_2 in subsequent generations. The initial population is generated using a half-ramping mechanism (half of the population is composed of random complete trees of the given initial depth, and the other half is generated as random incomplete trees) (Koza, 1992). Trees are initialised according to the BNF grammar shown in Fig. 9.4 in order to guarantee their syntactical and semantic correctness.

\<TREE\>	::=	**EL** | *\<IF\>* | *\<RLIST\>*
\<RLIST\>	::=	*\<TREE\>* *\<TREE\>* **RLIST**
\<IF\>	::=	*\<COND\>* *\<ACT\>* **IF**
\<COND\>	::=	*\<L_IS\>* | *\<L_AND\>*
\<L_IS\>	::=	*\<VBLEIN\>* *\<SET\>* **LEFT_IS**
\<L_AND\>	::=	*\<COND\>* *\<COND\>* **LEFT_AND**
\<ACT\>	::=	*\<R_IS\>* | *\<R_AND\>*
\<R_IS\>	::=	*\<VBLEOUT\>* *\<SET\>* **RIGHT_IS**
\<R_AND\>	::=	*\<ACT\>* *\<ACT\>* **RIGHT_AND**
\<SET\>	::=	**NL** | **NS** | **ZE** | **PS** | **PL**
\<VBLEIN\>	::=	**X1** | **X2**
\<VBLEOUT\>	::=	**Y**

Fig. 9.4   BNF grammar of correct trees (trees are described and implemented as post-ordered strings)

Trees are generated in a top-down fashion, i.e., choosing a root-node and then recursively generating appropriate sub-trees. The trees are implemented as variable-length strings. To generate a node, a random type is chosen and subsequently a random symbol belonging to that type is assigned to that node. The selection of this random symbol depends on the level of the node (e.g. an *IF* node can not appear in levels below 3 because its branches extend at least two levels down —an *IS* sub-tree—). Figure 9.5 shows which symbols may be root of a sub-tree according to its parent and the maximal depth of the tree. Only bold nodes might serve as the root node of the entire tree.

In every generation, the two best individuals are copied to the next generation. The rest of the population is obtained through crossover operations between selected individuals. This is done selecting half of the population according to scaled fitness and the other half randomly. This selection procedure proved beneficial in other domains and represents a good trade-off between exploitation and exploration.

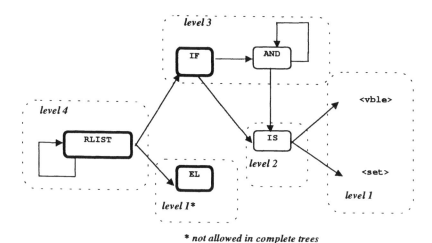

Fig. 9.5 Allowed argument types and the minimal level at which they may appear in a complete tree

9.1.2 *Learning knowledge bases with genetic programming*

Some GP approaches use domain-specific knowledge to define the function and terminal set for learning fuzzy rules: Hoffman and Nelles (2000) used GP for the structure identification of TSK FRBSs, whereas Battle et al. (1999) and Homaifar et al. (2000) considered a context-free fuzzy grammar and extended in order to accommodate membership functions evolution. Genes representing membership functions are added to the chromosomes of individuals in the population. On the other hand, Sánchez and García (1999) considered a GA-P algorithm (a hybrid method between GAs and GP (Howard and D'Angelo, 1995)) to design FRBCSs by coding the RB structure and the membership function parameters. In other work, Sánchez, Couso, and Corrales (2001) propose a hybrid GP-simulated annealing algorithm to design approximate FRBCSs. Bastian (2000) used the GP to identify the input variables, the RB and the involved membership functions of a fuzzy model.

In this subsection we present the GP approach to identify TSK FRBSs proposed by Hoffman and Nelles (2000, 2001). The GP tries to generate an optimal partition of the input space into Gaussian, axis-orthogonal fuzzy sets. A fuzzy rule constitutes a local linear model, that is valid for the region described by the premise part of the rule. For a given fuzzy input subspace (cluster), a local weighted least squares algorithm estimates the linear parameters in the rule conclusion. The non-homogeneous distribution of clusters enables a finer granulation in regions where the output depends in a highly non-linear fashion on the input.

Using a GP to partition the input space into clusters slightly differs from the conventional mode of computation. Instead of an algorithm or a function, each tree represents a possible distribution of cluster centres and widths. A node k from the function set $F = \{F_i\}$, $i \in \{1, \ldots, n\}$, where n is the dimension of the input space, corresponds to a Gaussian fuzzy set with centre c^k and standard deviation σ^k. The node function F_k denotes the dimension along which the fuzzy set is divided. The terminal set $T = \{L\}$ only contains one generic terminal symbol to close the branch with a leaf.

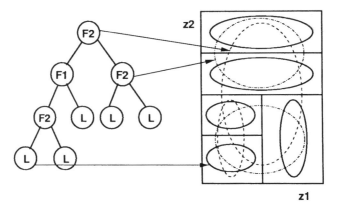

Fig. 9.6   GP tree and corresponding partition of input space. Solid lines denote clusters belonging to leaf nodes.

Figure 9.6 shows a possible tree and its correspondence to the resulting partition of the two-dimensional input space. Without loss of generality, we restrict the discussion to the two-dimensional case for sake of better illustration. Instead of considering the standard deviations σ_i, we focus our discussion on the widths δ_i of the hyper-rectangles which are related $\sigma_i = \alpha \cdot \delta_i$ by a single proportionality factor α.

The partition scheme starts with the root node, which describes a single cluster in the centre of the input space marked by a dashed ellipsoid. This cluster, or hyper-rectangle, is split into two hyper-rectangles along the dimension z_2 specified by the function symbol F_2 at the root node. Assume we split along dimension k for a parent node F_k with centre $(c_1, \ldots, c_k, \ldots, c_n)$ and widths $(\delta_1, \ldots, \delta_k, \ldots, \delta_n)$. The centre and width of the left child node become $(c_1, \ldots, c_k - \delta_k/4, \ldots, c_n)$ and $(\delta_1, \ldots, \delta_k/2, \ldots, \delta_n)$, accordingly the right child will be $(c_1, \ldots, c_k + \delta_k/4, \ldots, c_n)$ and $(\delta_1, \ldots, \delta_k/2, \ldots, \delta_n)$.

The partition process propagates through all child nodes which are function nodes until it encounters a leaf with the unique terminal symbol L. The leftmost node corresponds to the small lower left cluster. The centre c_i and width δ_i of the leaf node are added to the list of Gaussian clusters which define the next validity function $\Phi_l(\cdot)$. Notice that only leaf nodes contribute clusters (solid ellipsoids in Fig. 9.6), whereas inner tree nodes merely form the skeleton (dashed ellipsoids) of the input space partition.

Once the GP determined the cluster centres and standard deviations, a weighted local least squares algorithm (Nelles, 1999) computes the optimal linear parameters in the TSK rule conclusion by considering the mean square error between the data and the model output.

Obviously, the error becomes smaller as the number of clusters, and thereby the number of rules employed by the FRBSs increases. An FRBS with a large number of local models is more difficult to understand and computationally less tractable than a more comprehensive one with a fewer number of rules. This trade-off between model accuracy and complexity is taken into account by adding a penalty term for the number of local models $\#M$ to the fitness function. The overall fitness becomes

$$F = \frac{1}{NRMSE + p \cdot \#M}$$

where $NRMSE$ is the normalised root mean square error. The factor p defines the penalty per cluster relative to the size of the error. A small value of p results in a better approximation with the cost of an increased number of rules.

Mutation changes the split dimension of a randomly selected node. Crossover swaps two sub-trees among the parent trees.

9.2 Genetic Selection of Fuzzy Rule Sets

In high-dimensional problems, the number of rules in the RB grows exponentially as more inputs are added. A large rule set might contain many redundant, inconsistent and conflicting rules that are detrimental to the

FRBS performance.

Rule reduction methods aggregate multiple rules and/or select a subset of rules from a given fuzzy rule set in order to minimise the number of rules while at the same time maintaining (or even improving) the FRBS performance. Inconsistent and conflicting rules that degrade the performance are eliminated thus obtaining a fuzzy rule set with better cooperation.

Rule reduction methods have been formulated using NNs, clustering techniques and orthogonal transformation methods, and algorithms based on similarity measures, among others (Chiu, 1994; Halgamuge and Glesner, 1994; Rovatti, Guerrieri, and Baccarani, 1993; Setnes, Babuska, Kaymak, and van Nauta-Lemke, 1998; Setness and Hellendoorn, 2000; Yam, Baranyi, and Yang, 1999; Yen and Wang, 1999). Combs and Andrews (1998) proposed a different approach which attempts to reduce the growth of the RB by transforming elemental fuzzy rules into DNF-form.

Using GAs to search for an optimised subset of rules is motivated in the following situations:

- the integration of an expert rule set and a set of fuzzy rules extracted by means of automated learning methods (Herrera, Lozano, and Verdegay, 1998a),
- the selection of a cooperative set of rules from a candidate fuzzy *"IF-THEN"* rule set (Cordón, del Jesús, and Herrera, 1998; Cordón and Herrera, 1997c, 2000, 2001; Cordón, Herrera, and Zwir, 2000a; Ishibuchi, Nozaki, Yamamoto, and Tanaka, 1995; Ishibuchi, Murata, and Türksen, 1997; Krone, Krause, and Slawinski, 2000),
- the selection of rules from a given KB together with the selection of the appropriate labels for the consequent variables (Chin and Qi, 1998),
- the selection of rules together with the tuning of membership functions coding all of them (rules and parameters) in a chromosome (Gómez-Skarmeta and Jiménez, 1999), and
- the compact fuzzy model through complexity reduction combining fuzzy clustering, rule reduction by orthogonal techniques, similarity driving simplification and genetic optimisation (Roubos and Setnes, 2000).

In the following subsection, the basic algorithm for the genetic selection of rules from a set of candidate *"IF-THEN"* rules is introduced. Then we describe three of the said specific approaches that use the genetic selection:

- A specific selection approach integrating the use of linguistic modifiers in classification problems (Cordón, del Jesús, and Herrera, 1998).
- A specific GFRBS based on the ALM (Accurate Linguistic Modelling) methodology, that integrates the use of a special rule structure with two possible consequents and the genetic selection process (Cordón and Herrera, 1999a, 2000).
- A third approach based on a hierarchical KB and a learning methodology that uses the genetic selection for achieving cooperation via the integration of rules defined over linguistic partitions at different granularity levels (hierarchical DB) (Cordón, Herrera, and Zwir, 2000a).

9.2.1 *Genetic selection from a set of candidate fuzzy rules*

Let us denote a set of candidate fuzzy *"IF-THEN"* rules by S_{Cand}. Our task is to select a subset $S \in S_{Cand}$ that maximises a fitness function.

The genetic selection process is based on a binary-coded GA, using fixed-length chromosomes. Considering the rules contained in the fuzzy rule set S_{Cand} enumerated from 1 to m, an m-bit string $C = (c_1, ..., c_m)$ denotes the absence or presence of candidate rules from S_{Cand} in S, such that,

$$If\ c_i = 1\ then\ R_i \in S\ else\ R_i \notin S$$

The initial population is generated by introducing a chromosome representing the complete initial rule set S_{Cand}, that is, with all $c_i = 1$. The remaining chromosomes are initialised at random, with equal probabilities of 0 and 1.

According to the type of problem, the fitness function $E(\cdot)$ is based on different objectives. Each chromosome is first decoded into the corresponding fuzzy rule set, which is then evaluated according to the fitness criteria. In the following, we describe possible fitness functions, which assume an underlying training data set on which the rule subsets are evaluated.

- *Modelling problems:*
  Mean square error over a training data set, E_{TDS}, represented by

the following expression (Cordón and Herrera, 1997c, 2001, 2000):

$$E(C_j) = \frac{1}{2 \cdot |E_{TDS}|} \sum_{e_l \in E_{TDS}} (ey^l - S(ex^l))^2$$

where $S(ex^l)$ is the output value obtained from the FRBS using the fuzzy rule set coded in C_j, when the state variable values are ex^l, and ey^l is the known desired value.

A *τ-completeness property* to preserve the completeness of the initial fuzzy rule set can be also introduced (see Sec. 1.4.6).

- *Classification problems:*
  The classifier error rate over a training data set is defined as the ratio between the number of errors and the number of training examples (Cordón, del Jesús, and Herrera, 1998):

$$E(C_j) = \frac{\text{number of errors}}{\text{number of training examples}}$$

with the objective of minimising this function.

The maximum number of correctly classified patterns using a minimum number of rules (Ishibuchi, Nozaki, Yamamoto, and Tanaka, 1995) is computed as:

$$E(C_j) = W_{NCP} \cdot NCP(S) - W_S \cdot |S| \tag{9.1}$$

where W_{NCP} and W_S are constant positive weighting factors associated to the two objectives: the number of correctly classified training patterns $NCP(S)$ and the number of the linguistic classification rules contained in the fuzzy classification rule set $|S|$.

Constrained optimisation (Ishibuchi, Murata, and Türksen, 1997) could be generated by:

- Constraining the number of rules: For a desired upper bound on the number of rules N_{rules}, the fitness function becomes:

$$E(C_j) = W_{NCP} \cdot NCP(S) - W_S \cdot max\{0, |S| - N_{rules}\}$$

  with $W_{NCP} << W_S$ in order to penalise an excessive number of rules.

- Constraining the number of correctly classified patterns: Let $N_{patterns}$ be the desired number of correctly classified patterns:

$$E(C_j) = W_{NCP} \cdot max\{0, N_{patterns} - NCP(S)\} - W_S \cdot |S|$$

with $W_{NCP} >> W_S$ in order to penalise a small number of correctly classified patterns.

The learning is tackled as a two-objective optimisation problem using a multi-objective GA with two objectives in (Ishibuchi, Murata, and Türksen, 1997):

$$\text{Maximise NCP}(S) \qquad \text{Minimise } |S|$$

Finally, Ishibuchi, Nakashima, and Murata (2001) add a third objective, the minimisation of the total number of antecedent rule conditions, with the aim of designing compact FRBCSs with high classification ability.

- *Modelling as well as classification problems:*
Krone, Krause, and Slawinski (2000) proposed a fitness function, suitable for modelling and classification, combining the following three aspects:

  (1) Small modelling or classification error.
  (2) Minimum number of rules.
  (3) Maintenance of the number of covered examples by the original fuzzy rule set.

Since criteria (1) and (2) are partially competitive, the user selects a weighting factor $w \in [0, 1]$ that describes his preferred trade-off between both objectives in the fitness function:

$$E(C_j) = \begin{cases} w \cdot \frac{Err(C_j)^2}{Err(S_{Cand})^2} + \frac{(1-w) \cdot N_R(C_j)}{N_R(S_{Cand})}, & \text{if } Err(S_{Cand}) \neq 0 \\ w \cdot E(C_j)^2 + \frac{(1-w) \cdot N_R(C_j)}{N_R(S_{Cand})}, & \text{if } Err(S_{Cand}) = 0 \end{cases}$$

where S_{Cand} corresponds to the initial fuzzy rule set and C_j denotes the subset of rules in S_{Cand} that is currently evaluated. $Err(S_{Cand})$ is the error associated to the initial fuzzy rule set and $Err(C_j)$ the error of the current one. $N_R(S_{Cand})$ and $N_R(C_j)$ are the number of rules in each rule set. In system modelling, the error function Err is either the sum of the absolute errors or the mean square error. For classification problems, the error function Err becomes the number of wrong classifications.

Small values of the weighting factor w emphasise rule reduction in the fitness function, whilst large values of w emphasise on minimal modelling or classification error. The empty fuzzy rule set becomes optimal for $w = 0$, whereas the optimum for $w = 1$ is a fuzzy rule set with the smallest possible modelling or classification error, regardless of the number of rules.

To consider the third criteria, number of covered examples, the previous fitness $E(C_j)$ is multiplied by a penalty factor $p(C_j)$ of the form:

$$p(C_j) = \frac{N_{D,max} - N_{D,0}}{N_{D,max} - N_{D,i}}$$

where $N_{D,max}$ is the number of examples in the training set, $N_{D,0}$ is the number of examples not covered by the initial fuzzy rule set, and $N_{D,i}$ is the number of examples covered by the current rule set.

As described in Sec. 8.3.1.2, the *multi-simplification process* in MOGUL employs a sharing scheme (Cordón and Herrera, 2001) that allows the basic genetic selection process to generate a set of reduced fuzzy rule sets — instead of just one— by means of a sequential niche technique for multi-modal function optimisation (Beasly, Bull, and Martin, 1993).

9.2.2 Genetic selection of rule bases integrating linguistic modifiers to change the membership function shapes

9.2.2.1 The use of linguistic modifiers to adapt the membership function shapes

The definition of the membership function shapes is a difficult task, even for an expert familiar with the underlying problem. It is difficult to know in advance the number of fuzzy sets and the exact partition of the domain that achieves the optimal performance.

One way to improve the system performance without having to modify its original linguistic structure, is to use linguistic hedges in order to adapt the membership functions. Since Zadeh (1975) emphasised the utility of hedges for knowledge representation in approximate reasoning processes, several examples of application of linguistic modifiers to FRBSs have been developed (Bardossy and Duckstein, 1995; Chi, Yan, and Pham, 1996; Mitra and Pal, 1996).

A linguistic hedge is a function that alters the membership function of a fuzzy set in a way that increases or decreases the precision of the underlying concept. Two of the most common modifiers are: the concentration linguistic modifier "very", and the dilation linguistic modifier "more or less". The former reduces the membership degrees of points in the fuzzy set, whereas the latter increases them. They assume the following functional form:

$$\mu_{\text{Very } A_i^h}(x) = \left(\mu_{A_i^h}(x)\right)^2$$
$$\mu_{\text{More or Less } A_i^h}(x) = \sqrt{\mu_{A_i^h}(x)}$$

Their effects on a normalised fuzzy set with a triangular membership function are shown in Fig. 9.7.

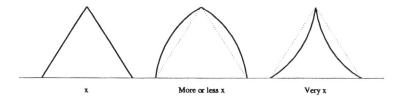

x   More or less x   Very x

Fig. 9.7   Linguistic Hedges

9.2.2.2   *A genetic multi-selection process considering linguistic hedges*

Cordón, del Jesús, and Herrera (1998) presented a multi-selection genetic selection process, which extends the one described in Sec. 8.3.1.2 that —at the same that eliminates unnecessary rules from the set of candidate rules S_{Cand}— refines KBs for classification problems by means of a linguistic hedge learning process. Although, in that paper, this process was proposed as a second stage of a GFRBCS based on MOGUL (see Chapter 8), it can also be used in isolation to refine a set of preliminary fuzzy classification rules. The two major differences with respect to the original approach are:

- Not only does the search process simplify the preliminary set of fuzzy classification rules but it also learns the best set of linguistic modifiers for the membership functions either defined in the DB or in the RB itself. Several subsets of rules that demonstrate the best cooperation are selected from the initial fuzzy rule set by sequentially applying the basic algorithm with niching.

- Each time a new KB definition (i.e., a fuzzy classification rule subset and a set of hedges) is selected in an iteration of the multi-selection process, a search process locally optimises this KB by either adding or eliminating a rule and/or modifying a linguistic hedge.

As said, the genetic learning of hedges may be carried out in two different ways:

- A hedge is obtained for each fuzzy set related to a linguistic label in the fuzzy partitions of the DB. In this case, this set of hedges is shared among all rules in the RB (*Hedges I*).
- The best set of hedges is selected for each single fuzzy classification rule in the RB (*Hedges II*).

Although in the second case the semantic is specific for each individual rule, the descriptive nature of the FRBCS is maintained in both approaches.

In the following section, the modifications compared to the original basic genetic selection process are briefly described. Section 9.2.2.4 introduces the algorithm of the genetic multi-selection process. Notice that the selection of different RBs along with their linguistic hedges associated is developed by means of the sequential niche technique (Beasly, Bull, and Martin, 1993) in the way shown in Sec. 8.3.1.2. Finally, Sec. 9.2.2.5 describes an extension of the algorithm considering arbitrary modifiers.

9.2.2.3 *The basic genetic selection method*

The main components of the basic selection process discussed in Sec. 9.2.1 remain the same. Only the coding scheme and the generation of the initial genetic population are different.

The *coding scheme* generates fixed-length chromosomes with two distinguishable parts, one related to the selected rules and the other referring to the linguistic hedges. Considering the rules contained in the initial rule set S_{Cand} enumerated from 1 to m, there are two different coding schemes depending on the hedge learning process that we want to carry out:

- *Hedges I*: The chromosome length is $h = m + \sum_{i=1}^{n} l_i$, with n being the number of input linguistic variables and l_i being the number of linguistic labels for the variable i. A chromosome $C_j = (c_1, \ldots, c_h)$

is divided into two parts. The first one maintains the previous coding to represent the subset of candidate rules to form the final RB. In addition, the second part contains as many genes as different linguistic terms are considered for each variable. Each gene can assume a possible set of integer values, with the allele '0' representing the absence of hedge and the remaining values standing for the set of possible hedges.

Figure 9.8 depicts this coding scheme and the resulting KB composed of fuzzy classification rules with a certainty degree for all classes in the consequent (see Sec. 1.5.3.2).

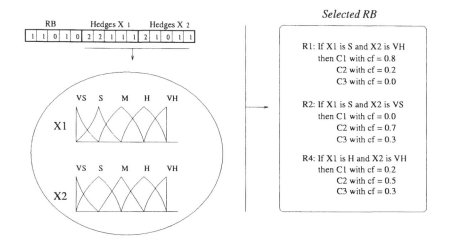

Fig. 9.8 Type I Hedges

- *Hedges II*: The chromosome length is $h' = m \cdot (n + 1)$. The chromosome is again divided into two parts. While the first m genes represent the same information as in the previous case, the $m \cdot n$ remaining genes represent the hedges for each of the individual rules, i.e., each linguistic variable in each rule has a gene associated to store the modifier related to it.

  Figure 9.9 shows the coding scheme as well as the form of the resulting KB.

The *initial population* is generated by injecting a single chromosome that corresponds to the complete initial rule set S_{Cand}, that is, with all $c_i = 1$,

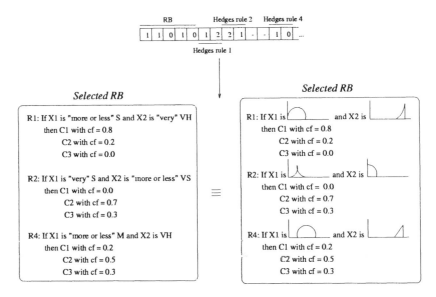

Fig. 9.9   Type II Hedges

$i \in \{1, \ldots, m\}$, and with all the genes encoding the hedge set initialised with 0. For each possible hedge considered, a chromosome representing the complete RB, and with all the genes in the hedge part initialised with the value representing the said hedge, is included in the population. The remaining chromosomes are generated at random.

The *fitness function* considers two criteria: a global classification error measure and a criterion penalising the violation of the completeness property (see Sec. 1.4.6). Hence, the final expression for the fitness function becomes:

$$F(C_j) = \begin{cases} E(C_j), & \text{if } TSCD(R(C_j), E_p) \geq \tau \\ 1, & \text{otherwise} \end{cases}$$

where

$$E(C_j) = \frac{\text{number of errors}}{\text{number of cases}}$$

Moreover, a *local search algorithm* individually optimises each of the KB definitions obtained, either inserting or eliminating a rule and/or changing a hedge, to invoke changes that potentially improve the KB behaviour. As may be observed, this is a straightforward and fast optimisation process.

The local search is carried out at the end of each multi-selection process iteration. To reduce the search space, the search process is divided into two phases: the RB part is optimised first and afterwards the search identifies the best set of hedges. First of all, the rule selection is optimised by means of a search in the RB space with a hamming distance of one to the current best solution, i.e., with either one rule more or one rule less in the RB. Finally, the search process is extended to the region of solutions within a Hamming distance of one to the hedge definition of the current KB.

9.2.2.4 *Algorithm of the genetic multi-selection process*

The algorithm of the genetic multi-selection process is shown below:

(1) Initialisation: Equate the multi-selection modified fitness function to the basic selection fitness function: $F'(C_j) \leftarrow F(C_j)$.

(2) Run the basic genetic selection process, using the modified fitness function.

(3) Run the local optimisation process to optimise the KB definition generated. Add the final KB obtained to the solution set Sol.

(4) Update the modified fitness function to penalise solutions similar to the current individual, producing a new modified fitness function.

(5) If Sol does not contain the desired number of fuzzy rule sets, return to step 2, else stop.

9.2.2.5 *Parametric extension*

Alternatively, the modifier applied to a membership function can be defined by a single parameter a according to

$$(\mu_{A_i^k}(x))^a$$

with different interpretations of this operator depending on the range of the parameter a. For values $a \in (1, \infty)$ the modified label is more precise than the original one. On the contrary, for values $a \in (0, 1)$, the effect is the opposite, that is, the modifier increases the imprecision of the original label.

González and Pérez (1999b) also analysed the use of this type of hedges in the SLAVE GFRBS introduced in Chapter 8.

9.2.3 *ALM: Accurate linguistic modelling using genetic selection*

ALM aims at designing fuzzy linguistic models from data, which are highly accurate and simultaneously afford an intuitive interpretation (Cordón and Herrera, 1999a, 2000). This approach is based on two aspects related to the interpolative reasoning scheme employed by FRBSs:

- The structure of the linguistic model is augmented by so-called double-consequent rules –explained in the following– which locally improve the model accuracy. This way, the usual single-consequent fuzzy rules will coexist with double-consequent rules in the KB.
- The KB generation process is designed in such a way that the cooperation among rules in the derived fuzzy rule set is addressed. To do so, a preliminary fuzzy rule set with a large number of single- and double-consequent rules is derived and then a genetic rule selection process is applied to obtain the subset of them best cooperating.

A double-consequent rule presents the usual linguistic rule structure but its consequent has two different linguistic labels associated. These rules locally improve the interpolative reasoning performed by the linguistic model by allowing the crisp output of a fuzzy rule fall in between the two modal points of the two linguistic labels in the consequent. Hence, they cause a "shift" of the fuzzy sets in the fuzzy partitions which allows the linguistic model accuracy to be increased in the same way than augmenting the granularity of the primary fuzzy partitions with the advantage of a lesser number of rules in the RB.

These kinds of rules are consistent with the linguistic structure of the model as they can be interpreted as:

IF X_1 is A_1 and ... and X_n is A_n THEN Y is between B_1 and B_2

Double consequent rules require to apply a defuzzification method based on the "first inference then aggregate" approach discussed in Sec. 1.2.2.3, that considers the matching degree of the rules fired in the aggregation of their outputs. Cordón, Herrera, and Peregrín (1997) propose the *centre of gravity weighted by the matching degree* defuzzifier, that computes the crisp output y_0 as:

$$y_0 = \frac{\sum_{i=1}^{T} h_i \cdot y_i}{\sum_{i=1}^{T} h_i}$$

with T being the number of single-consequent rules in the KB, h_i being the matching degree between the i-th rule and the current system input and y_i being the centre of gravity of the fuzzy set inferred from that rule. Notice that each double-consequent rule in the RB is decomposed into two single-consequent rules to make inference.

The generation process builds a preliminary fuzzy rule set composed of a large number of single- and double-consequent rules, depending on the complexity of the specific fuzzy region covered by each of them (no rules will be generated in the regions where the system is not defined). Subsequently, both types of fuzzy rules are treated as single-consequent ones, in that each double-consequent rule is decomposed into two single-consequent rules that share the same antecedent. Finally, the selection process identifies the subset of these candidate rules that demonstrates the best cooperation to form the final RB.

9.2.3.1 *Some important remarks about ALM*

ALM is not a specific learning method but a methodology to design RB learning processes. Any ALM process is composed of the two following steps:

(1) A *linguistic rule generation method*, which generates a preliminary rule set where single and double-consequent linguistic rules coexist according to the problem complexity.
(2) A *genetic selection process*, that selects the subset of rules cooperating best from the preliminary fuzzy rule set generated in the previous step.

We may draw two very important considerations:

- First, it is possible that, although the preliminary fuzzy rule set contains some double-consequent rules, the final RB does not include any such rule after the selection stage. In this case, the linguistic model benefited from the way in which the fuzzy rules were generated because many rule subsets with different cooperation levels were analysed. The rules in the final KB cooperate well, a property that can not be guaranteed in other inductive design methods, such as the Wang and Mendel method (Wang and Mendel, 1992b) which selects the consequent that locally best matches with the training data in the corresponding fuzzy input region.

- Second, it is possible that the final RB contains a smaller number of rules than RBs generated by other methods due to the two said extensions: the presence of two rules (a double-consequent rule) for the same input region and the generation of neighbouring rules with better cooperation may remove unnecessary rules from the final KB without sacrificing on approximation accuracy.

9.2.3.2 *A linguistic modelling process based on ALM*

This ALM process presents two stages: a preliminary linguistic rule generation method (taking as base the learning method proposed by Wang and Mendel (1992b)) and the basic genetic selection method introduced in Sec. 9.2.1.

The first stage assumes an input-output data set, E, from which the preliminary RB is generated by means of the following steps:

(1) *Consider a fuzzy partition of the input variable spaces.*
(2) *Generate a preliminary linguistic rule set.* This set is formed by the rules that best cover each example (input-output data pair) in E. The rule $R_l = IF\ X_1\ is\ A_1^l\ and\ \ldots\ and\ X_n\ is\ A_n^l\ THEN\ y\ is\ B_l$ —generated from the example $e_l = (x_1^l, \ldots, x_n^l, y^l)$— is obtained by assigning the linguistic label to the rule variable X_i whose fuzzy set best matches the corresponding example value x_i^l.
(3) *Assign a degree of relevance to each rule.* The degree of relevance associated to R_l is computed as:

$$G(R_l) = \mu_{A_1^l}(x_1^l) \cdot \cdots \cdot \mu_{A_n^l}(x_n^l) \cdot \mu_{B_l}(y^l)$$

(4) *Generate a preliminary RB.* The two most relevant rules in each input subspace —if they exist— become part of the preliminary RB.

Some combinations of antecedents may be associated to no rule (if there are no examples in that input subspace) or only to one rule (if all the examples in that subspace generated the same rule). Therefore, *the generation of double-consequent rules is only addressed when the problem complexity*, represented by the training set, *indicates that they potentially are able to improve the accuracy.*

The genetic selection process is then applied to the previous preliminary RB to obtain the final one.

Section 11.2 discusses various applications of the approach to model real-world systems. Other applications are to be found in (Cordón and Herrera, 1999a, 2000).

9.2.4 *Genetic selection with hierarchical knowledge bases*

Cordón, Herrera, and Zwir (2000a) extend the structure of the KB of FRBSs in a hierarchical way. Linguistic rules defined over linguistic partitions of different granularity levels provide additional flexibility, and thus improve the model accuracy in those regions in which the usual non-hierarchical models demonstrate poor performance.

This type of improvement is the point of departure for the development of different *hierarchical system of linguistic rules* (HSLR) learning methodologies, which are considered as refinements of the basic linguistic models. They preserve part of their descriptive power while they slightly change the model structure to increase the approximation accuracy. To do so, a learning method generates more specific rules in the regions of large modelling error and a genetic selection process is applied to obtain a set of hierarchical rules with good cooperation.

The following two subsections introduce the *hierarchical KB* (HKB) philosophy and different HSLR learning methodologies.

9.2.4.1 *Hierarchical knowledge base philosophy*

The inflexibility of the KB structure in fuzzy linguistic models is a consequence of the concept of linguistic variable (see Sec. 1.2.5). The more flexible HKB structure allows us to improve the accuracy of linguistic models while maintaining their interpretability at large.

The HKB is composed of a set of layers, that are defined by their components as follows:

$$layer(t, n(t)) = DB(t, n(t)) + RB(t, n(t))$$

with:

- $n(t)$ being the number of linguistic terms that compose the partitions of layer t.

- $DB(t, n(t))$ being the DB which contains the linguistic partitions of granularity level $n(t)$ of layer t.
- $RB(t, n(t))$ being the RB formed by those linguistic rules whose linguistic variables assume values in the former partitions.

Notice that, in the following, linguistic partitions with the same number of linguistic terms for all input-output variables are used, composed of triangular-shaped, symmetrical and uniformly distributed membership functions.

For the sake of simplicity, we are going to refer to the components of a $DB(t, n(t))$ and $RB(t, n(t))$ as *t-linguistic partitions* and *t-linguistic rules*, respectively.

This set of layers is hierarchically ordered according to the granularity level of the linguistic partition defined in each layer. That is, given two successive layers t and $t + 1$, the granularity level of the linguistic partitions of layer $t + 1$ is larger than the one of layer t. Each subsequent layer constitutes a refinement of the previous layer linguistic partitions. As a consequence of the previous definitions, the HKB is defined as the union of all layers up to level t:

$$HKB = \cup_t \ layer(t, n(t))$$

The remainder of this subsection studies the linguistic partitions and their extension to consider them as components of the $DB(t, n(t))$ of the *layer*$(t, n(t))$. Afterwards, we describe the relation between DBs from successive layers, and develop a methodology to build them under certain requirements. Finally, we explain how to relate these DBs with linguistic rules, i.e., to create RBs from them.

Hierarchical Data Base
Building up the hierarchical DB (HDB), one should bear in mind that the layers are ordered hierarchically, with increasing granularity level of the linguistic partitions.

To extend the classical linguistic partition, let us consider a partition P of the domain U of a linguistic variable A in the layer t:

$$P_A = \left\{ S_1, ..., S_{n(t)} \right\}$$

with S_k $(k = 1, \ldots, n(t))$ being linguistic terms which describe the linguistic variable A. These linguistic terms are mapped into fuzzy sets by the semantic function M, which gives them a meaning: $M_U : S_k \rightarrow \mu_{S_k}(u)$ (Zadeh, 1975).

Cordón, Herrera, and Zwir (2000a) extended this definition of P allowing the existence of several partitions, each one with a different number of linguistic terms, i.e., with a different granularity level. To do so, a parameter $n(t)$ is added to the definition of the linguistic partition P, which represents the granularity level of the partitions contained in the layer t where it is defined:

$$P_A^{n(t)} = \left\{ S_1^{n(t)}, ..., S_{n(t)}^{n(t)} \right\}$$

where $P_A^{n(t)} \in DB(t, n(t))$.

In order to build the HDB, the authors develop a strategy which satisfies two main requirements:

- To preserve all possible fuzzy set structures from one layer to the next in the hierarchy.
- To make smooth granularity transitions between successive layers.

In order to satisfy the first requirement, the membership function modal points, corresponding to each linguistic term, should be preserved throughout the higher layers in the hierarchy. The new *t+1-linguistic partition* is built by adding a new linguistic term located between each two consecutive terms of the previous *t-linguistic partition*. The support of these linguistic terms shrinks in order to create space for the new fuzzy set, located in between them. An example of the correspondence among an *1-*, a *2-* and a *3-linguistic partition*, with $n(1) = 3$, $n(2) = 5$ and $n(3) = 9$ respectively, is shown in Fig. 9.10.

Table 9.1 shows the number of linguistic terms needed in each *t-linguistic partition* in $DB(t, n(t))$ to satisfy the previous requirements. The values of parameter $n(t)$ represent the *t-linguistic partition* granularity levels and depend on the initial value of $n(t)$ defined in the first layer (e.g. 2 or 4 in the table).

In the general, the layer $t + 1$ DB is obtained from its predecessor by:

$$DB(t, n(t)) \rightarrow DB(t + 1, 2 \cdot n(t) - 1)$$

which means that a *t-linguistic partition* in $DB(t, n(t))$ with $n(t)$ linguistic terms becomes a *(t+1)-linguistic partition* in $DB(t + 1, 2 \cdot n(t) - 1)$.

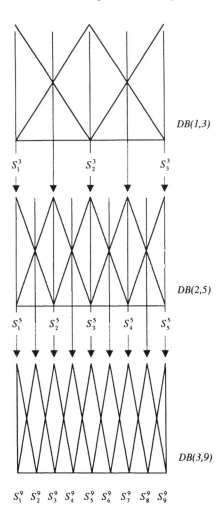

Fig. 9.10   Three layers of linguistic partitions which compose the HDB

In order to satisfy the previous requirements, each linguistic term $S_k^{n(t)}$ (term of order k from the *t-linguistic partition* in $DB(t, n(t))$) is mapped into $S_{2k-1}^{2 \cdot n(t)-1}$, preserving the former modal points, and a set of $n(t)$-1 new terms is created, each one between $S_k^{n(t)}$ and $S_{k+1}^{n(t)}$ ($k = 1, \ldots, n(t) - 1$). Table 9.2 shows this mapping and Fig. 9.10 depicts the corresponding HDB graphically.

Table 9.1   Hierarchy of DBs starting from 2 or 4 initial terms

DB(t,n(t))		**DB(t,n(t))**
DB(1,2)		DB(1,4)
DB(2,3)		DB(2,7)
DB(3,5) ·		DB(3,13)
DB(4,9)	or	DB(4,25)
$\vdots$		$\vdots$
DB(6,33)		DB(6,97)
$\vdots$		$\vdots$

Table 9.2   Mapping between terms from successive DBs

DB(t,n(t))		**DB(t+1,2·n(t)-1)**
$S_{k-1}^{n(t)}$	$\longrightarrow$	$S_{2k-3}^{2\cdot n(t)-1}$
		$S_{2k-2}^{2\cdot n(t)-1}$
$S_{k}^{n(t)}$	$\longrightarrow$	$S_{2k-1}^{2\cdot n(t)-1}$
		$S_{2k}^{2\cdot n(t)-1}$
$S_{k+1}^{n(t)}$	$\longrightarrow$	$S_{2k+1}^{2\cdot n(t)-1}$

This two-level successive layer definition can be generalised to $n(t)$, for all layers t, as follows:

$$n(t) = (N - 1) \cdot 2^{t-1} + 1$$

with $N = n(1)$, i.e., the number of linguistic terms in the initial layer partitions.

Hierarchical Rule Base

The *t-linguistic RB* structure $RB(t, n(t))$ is formed by a collection of linguistic Mamdani-type rules:

$$R_i^{n(t)} : IF\ X_1\ is\ S_{i1}^{n(t)}\ and\ \ldots and\ X_n\ is\ S_{in}^{n(t)}\ THEN\ Y\ is\ B_i^{n(t)}$$

with $X_1, \ldots, X_n$ and Y being the input linguistic variables and the output one, respectively; and with $S_{i1}^{n(t)}, \ldots, S_{im}^{n(t)}$, $B_i^{n(t)}$ being linguistic terms from different *t-linguistic partitions* of $DB(t, n(t))$, with the associated fuzzy sets specifying the underlying concept.

Hence, the fuzzy rules in $RB(t, n(t))$ take values in the linguistic partitions of $DB(t, n(t))$. As a consequence, rules contained in a higher layer are more specific than those in lower layers.

9.2.4.2 *System modelling with hierarchical knowledge bases*

This section introduces two different methodologies to develop an HKB:

- A local Two-Level HSLR Learning Methodology (HSLR-LM) proposed by Cordón, Herrera, and Zwir (1999), and its iterative extension (I-HSLR-LM) (Cordón, Herrera, and Zwir, 2000b), and
- a global approach (G-HSLR) proposed by Ishibuchi, Nozaki, Yamamoto, and Tanaka (1995).

A Local Approach: A Two-Level HSLR Learning Methodology

Cordón, Herrera, and Zwir (1999) introduced this methodology as a strategy to improve simple linguistic models preserving their structure and descriptive power, by refining the rule granularity in regions of large modelling error. To do so, the HSLRs generated are based on only two hierarchical levels, i.e., they are composed of an HKB of two layers.

In the following, the structure of the learning methodology and its most important components are briefly described:

(1) *HKB Generation Process:*

    (a) Generate the initial $RB(1, n(1))$ from the present $DB(1, n(1))$ using any inductive linguistic rule generating method (LRG-method), the initial *1-linguistic partitions* and a training data set.

    (b) Separate $RB_{bad}(1, n(1))$, the poor performance *1-linguistic rules*, subject to expansion from the good rules $RB_{good}(1, n(1))$, by comparing their modelling error with the error made by the entire rule set.

    (c) Generate the next finer granulised layer DB, $DB(2, 2 \cdot n(1) - 1)$.

    (d) For each $R_i^{n(1)} \in RB_{bad}(1, n(1))$:

        i. Select the *2-linguistic partition* terms which have a "significant intersection" with the terms in $R_i^{n(1)}$.

    ii. Join the previously selected term sets.

    iii. Extract *2-linguistic rules* from the combined selected *2-linguistic partition* terms using the LRG-method. These *2-linguistic rules* are the image of the expanded linguistic rule $R_i^{n(1)}$, i.e., the candidates to be in the HRB from rule $R_i^{n(1)}$ $(CLR(R_i^{n(1)}))$.

(e) Obtain a joined set of candidate linguistic rules, JCLR, as the union of the group of the new generated *2-linguistic rules* $CLR(R_i^{n(1)})$ and the former good performance *1-linguistic rules* $RB_{good}(1, n(1))$:

$$JCLR = RB_{good}(1, n(1)) \cup (\cup_i CLR(R_i^{n(1)}))$$

with $R_i^{n(1)} \in RB_{bad}(1, n(1))$.

(2) *HRB Selection Process:* Simplify the set JCLR by using the genetic selection process, in order to remove the unnecessary rules from it, and to generate an HKB with good cooperation:

$$HRB = Genetic\text{-}Selection(JCLR)$$

(3) *User Evaluation Process:* Evaluate the obtained model. If it is inappropriate, adapt the granularity of the initial linguistic partitions $n(1)$ and/or the threshold which determines whether an *1-linguistic rule* is expanded into a set of *2-linguistic rules*, and re-apply the algorithm in order to obtain a more accurate model.

Notice that this methodology was devised as a strategy to improve simple fuzzy linguistic models. Therefore, one can select any inductive LRG-method to build the HRB, based on the existence of a set of input-output data E_{TDS} and a previously defined $DB(1, n(1))$. Cordón, Herrera, and Zwir (1999) used two LRG-methods in order to illustrate this situation: those proposed by Wang and Mendel (1992b) and Thrift (1991).

This *Two-level HSLR-LM* was extended (Cordón, Herrera, and Zwir, 2000b) by considering it as an iterative process. While the former methodology was thought as a *simple descriptive refinement* of linguistic models, the *I-HSLR-LM* is viewed as an *accurate refinement* of those models, which preserves HSLR-LM features but looses description, having linguistic rules defined over more than two layers in the HRB, in order to improve the modelling accuracy performed by the learnt HSLR.

A Global HSLR Generation Method

Ishibuchi, Nozaki, Yamamoto, and Tanaka (1995) developed another approach along the same line. This method obtains an HSLR creating several hierarchical linguistic partitions with different granularity levels, generating the complete set of linguistic rules in each of these partitions, taking the union of all sets, and finally performing a genetic rule selection process on the entire rule set.

Hence, there is a main difference among both approaches: whilst the current proposal is a single method to derive HSLRs by *directly generating* fuzzy rules with different granularity levels, HSLR-LM is a methodology to *refine* linguistic models generated from any fuzzy rule generation method by *expanding* several of their bad performance rules into a set of fuzzy rules of higher granularity in order to increase the system accuracy. For the sake of simplicity, we refer to this method as a global HSLR (G-HSLR) generation method, in order to distinguish it from the local approach (HSLR-LM).

Although G-HSLR was designed to construct an FRBCS, whereas the main purpose of HSLR-LM was to perform linguistic modelling, both methods share some interesting similarities and differences:

- As said, while HSLR-LM takes a preliminary linguistic model and locally expands those rules in its KB which perform bad in the modelling of their respective regions, G-HSLR generates the HKB in a global way, i.e., it derives all rules at all granularity levels.
- Due to the said operation mode, G-HSLR allows the HSLR to contain several rules in different granularity levels in the same region of the space, thus resulting in a reinforcement of the more general rule (the rule in the lowest granularity level). Since the expansion process of HSLR-LM directly substitutes the expanded rule by its image, it does not initially perform a similar type of reinforcement.
- Both methods perform a genetic rule selection to extract the set of rules which best cooperates between them, i.e. the HRB, but on a different rule set.

Table 9.3 shows a common notation for both hierarchical methodologies when considering only two layers in order to clarify their similarities and differences. Remember that $CLR(R_i^{n(1)})$ stands for the image of the expanded bad linguistic rule $R_i^{n(1)}$, which joined to the former good performance *1-linguistic rules* constitutes the set of candidate linguistic rules to be included in the final HRB.

Table 9.3   Local and global HSLR learning methodologies

HSLR-LM	$HRB = Genetic - Selection$ $(RB_{good}(1, n(1)) \cup (\cup_i CLR(R_i^{n(1)})))$
G-HSLR	$HRB = Genetic - Selection$ $(RB(1, n(1)) \cup RB(2, n(2))))$

9.3   Learning the Knowledge Base via the Genetic Derivation of the Data Base

Genetic tuning (see Chapter 4) refines the components in the DB under the assumption that a predefined RB is used to evaluate the quality of the overall FRBS. In this type of *a posteriori DB learning* the DB is only slightly modified such that after tuning the mapping between input and output established by the rules in the original RB still captures the underlying problem.

In case the DB components are strongly modified, it becomes mandatory to derive a novel RB subsequent to the genetic modification of the DB. *A priori genetic DB learning* refers to a KB learning process in which a GA adapts the DB components such as scaling functions, membership functions and granularity parameters, whilst an additional fuzzy rule generation method derives the RB from the DB definition encoded in the chromosome.

Sections 9.3.2 and 9.3.3 describe two different approaches to *a priori* genetic DB learning:

- the genetic learning of membership functions, and
- the genetic learning of non-linear contexts.

First, we briefly introduce the basic aspects of learning the KB by means of DB derivation.

9.3.1   *Learning the knowledge base by deriving the data base*

As known, the KB learning problem is divided into two subtasks:

- *The DB learning*, that comprises the specification of the universes of discourse, the number of labels for each linguistic variable, as well as the definition of the fuzzy membership functions associated to each label.

- *The RB derivation*, involving the determination of the number of rules and the rule definitions themselves, namely the specific labels associated to each linguistic variable.

Most of the approaches proposed to automatically learn the KB from numerical information focus on RB learning, using a predefined DB. The usual way to define this DB involves choosing a number of linguistic terms for each linguistic variable (an odd number between 3 to 9, which is normally the same for all the variables). The membership functions associated to the linguistic terms are computed based on a uniform, equidistant fuzzy partition of the universe of discourse. Figure 9.11 graphically shows this type of KB learning. Examples for this learning approach using GAs can be found in Secs. 6.2 and 7.1 and in Chapter 8.

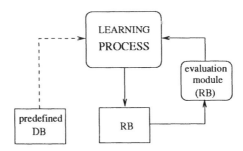

Fig. 9.11   Learning the RB

This way of defining the DB makes this component have a significant influence on the FRBS performance. Previous studies (Bonissone, Khedkar, and Chen, 1996; Zheng, 1992) indicated that, for example in fuzzy PI control, the system performance is much more sensitive to the choice of the semantics in the DB than to the composition of the RB. Considering a previously defined RB, the performance of the FLC depends on four KB components in decreasing order of relevance: scaling factors, peak values, width values and rules. As can be seen, the former three components belong to the DB.

Cordón, Herrera, and Villar (2000) studied the influence of fuzzy partition granularity (number of linguistic terms per variable) in the FRBS performance, showing that using an appropriate number of terms for each linguistic variable, the FRBS accuracy can be significantly improved without the need of a complex RB learning method.

Some approaches try to improve the preliminary DB definition once the RB has been derived. They are called *tuning processes* or *a posteriori DB learning processes*. Chapter 4 discussed genetic tuning processes and the GFRBSs based on MOGUL shown in Chapter 8 follows this learning approach. Fig. 9.12 graphically depicts this kind of learning.

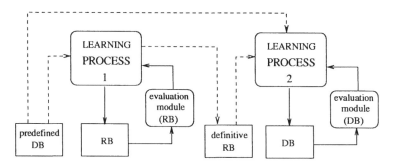

Fig. 9.12   Tuning process

Other approaches simultaneously learn both components of the KB. That way, they are able to generate superior solutions but as a drawback the larger search space complicates and slows down the learning process more. This approach is depicted in Fig. 9.13. For examples of this kind of learning processes using GAs refer to Secs. 6.3 and 7.2.

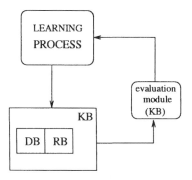

Fig. 9.13   Learning the KB

Finally, there is another way to generate the KB that considers two separate processes for deriving the DB and RB. In *a priori* DB learning, the DB generation process comprises an RB learning: each time the main process generates a new DB definition, it invokes the RB generation method to find the set of rules that in conjunction with the DB constitutes the optimal KB. This KB learning scheme shown in Fig. 9.14 follows a divide and conquer strategy, in that it decomposes the overall KB design task into two sub-processes that deal with simpler search spaces, namely the space of DB and RB components.

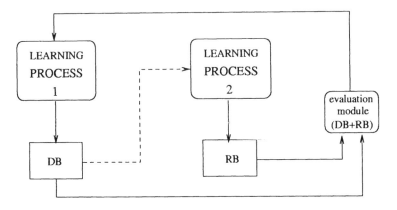

Fig. 9.14   Learning the KB via the derivation of the DB

The genetic processes presented in the two following subsections are examples of the latter KB learning approach.

9.3.2   *Genetic learning of membership functions*

Genetic learning of membership functions either assumes a constant granulation (Filipič and Juričić, 1996; Surmann, 1996) or adapts the granularity level in conjunction with the membership functions (Glorennec, 1996; Ishibuchi and Murata, 1996; Cordón, Herrera, and Villar, 2001).

9.3.2.1   *Genetic learning of isosceles triangular-shaped membership functions*

In the method proposed by Filipič and Juričić (1996), a GA is used to define the DB whilst the Wang and Mendel method (Wang and Mendel, 1992b) is considered to derive the RB. In particular, its chromosomes encode the numerical values of membership function parameters. For each set of membership functions, the Wang and Mendel procedure generates the fuzzy rules that locally best match with the training examples. Afterwards, the performance of the resulting FRBS considering the entire KB generated is evaluated on the given problem. That way, both the membership function definitions and the RB are improved in an iterative fashion.

Representation

The method employs isosceles, normalised, triangular-shaped membership functions. However, the fuzzy partition itself is not normalised as the left and right points of the triangle do not necessarily coincide with the centres of the two neighbouring fuzzy sets. However, the optimisation process is constrained such that for each point at most two fuzzy sets assume a non-zero membership degree. The example in Fig. 9.15 illustrates the parameterisation of the membership functions. In case the domain interval $[x^-, x^+]$ is partitioned into $M = 2 \cdot N + 1$ regions, $2 \cdot M - 2$ parameters are needed: x_{SN}^+ for region SN, x_A^- and x_A^+ for $A = S(N-1), \ldots, S1, CE, B1, \ldots, B(N-1)$, and x_{BN}^- for region BN.

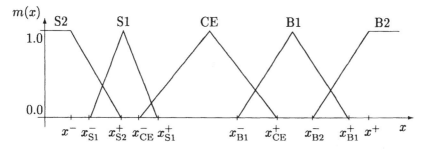

Fig. 9.15　Fuzzy regions for the domain interval of variable x and the corresponding membership functions

The genetic representation of membership functions is straightforward and therefore domain-independent genetic operators are easily applicable. A sub-string of $2 \cdot M - 2$ real-coded genes defines M membership functions for variable x. The allelic values are taken from the domain interval $[x^-, x^+]$. Then, $n + m$ of these sub-strings are concatenated to represent the membership functions for n input and m output variables. The number of membership functions may vary among different variables. For example, if there are three input variables, each partitioned into five fuzzy sets, and one output variable, partitioned into seven fuzzy sets, the corresponding chromosome has a length of $3 \cdot (2 \cdot 5 - 2) + (2 \cdot 7 - 2) = 36$. If the domain intervals of all variables are normalised to the interval $[0, 1]$, the GA is able to operate with the same set of alleles for every gene, when it encodes the membership function parameters.

In the following, we describe the genetic operators.

Genetic Operators

Four genetic operators are applied to alter chromosomes representing fuzzy membership functions in the GA: multi-point crossover, uniform mutation, whole arithmetical crossover, and non-uniform mutation.

- *Multi-point crossover* and *uniform mutation* are traditional operators, known from binary- and real-coded GAs, that exchange substrings between pairs of chromosomes and randomly alter selected gene values, respectively.
- The *whole arithmetical crossover* is defined as a linear combination of two chromosomes. Crossing $s_v^t = (v_1, \ldots, v_p)$ and $s_w^t = (w_1, \ldots, w_p)$, it gives the offspring $s_v^{t+1} = a \cdot s_w^t + (1 - a) \cdot s_v^t$ and $s_w^{t+1} = a \cdot s_v^t + (1 - a) \cdot s_w^t$, where $a \in [0, 1]$ is a parameter of the algorithm.
- The *non-uniform mutation* is described in Sec. 2.2.4.2.

9.3.2.2 *Genetic learning of membership functions with implicit granularity learning*

The method proposed by Glorennec (1996) operates with TSK FRBSs. A chromosome of real-valued genes represent a population of strong fuzzy partitions (see Sec. 4.3.2) on each input domain in such a way that each gene defines a modal value (i.e. the centre of a triangle or of a bell-shaped function in a fuzzy partition). A rule learning algorithm is invoked to generate the optimal TSK rule consequents prior to computing the fitness of an individual.

Representation

For the sake of simplicity, the following description is restricted to triangular membership functions. A typical chromosome C is of the form:

$$C = (c_{1,1}, \ldots, c_{1,m_1}, c_{2,1}, \ldots, c_{2,m_2}, \ldots, c_{n,1}, \ldots, c_{n,m_n})$$

for a simplified TSK FRBS with n inputs and m_i triangles by input, $i = 1$ to n. Moreover, the approach assumes an ordered, strong fuzzy partition (see Sec. 4.3.2 and Fig. 4.11), such that:

$$(\forall i) \ (c_{i,1} < c_{i,2} < \ldots < c_{i,m_i})$$

Rather than working with a fixed number of membership functions, the learning process itself determines the optimal number $(m_i)_{i=1}^n$. Therefore, the chromosome is composed of two separate parts (see Fig. 9.16) that encode:

- the number of membership functions for each input variable, and
- the parameters of the input membership functions.

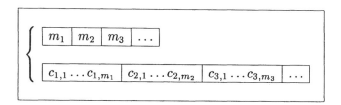

Fig. 9.16   Chromosome representation

Whereas the mutation and crossover operators modify the second string, only crossover alters the first string. In the following, we describe the genetic operators used and the method for generating the initial population.

Genetic Operators

The following common genetic operators generate new variants from the parent chromosomes:

Crossover: The exchange of genes between two parents concerns *all* the parameters for a selected input (see Fig. 9.17). Two individuals exchange their fuzzy partitions for a particular input as follows:

(1) Randomly select an input, j.

(2) Swap:

    (a) the number of membership functions for input j,

    (b) the parameters of the corresponding fuzzy partitions.

Non-uniform mutation: Mutation affects a random number of genes in the chromosome. Let $\theta_m \ll 1$ be a real number (e.g. $\theta_m = 0.3$) and let $c_{i,j}$ be a gene subject to mutation.

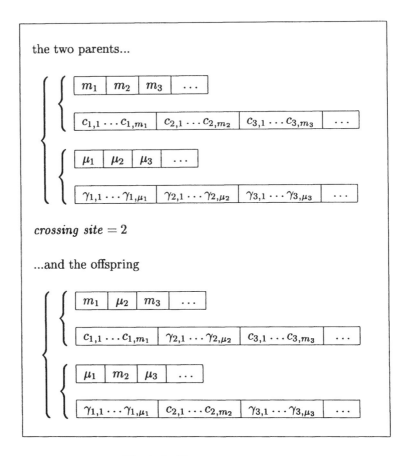

the two parents...

crossing site $= 2$

...and the offspring

Fig. 9.17   The crossover operator

(1) Select randomly $s \in \{-1, 1\}$.
(2) Select randomly $\lambda \in [0, \theta_m]$.
(3) $c_{i,j} \leftarrow (1 - \lambda) \cdot c_{i,j} + \lambda \cdot c_{i,j+s}$

Mutation is equivalent to shifting some of the modal values in the fuzzy partition.

Generation of the initial population

For each input i, $i = 1, \ldots, n$, let r^i_{min} and r^i_{max} be the minimum and the maximum number of membership functions respectively, specified by the user in advance. The chromosomes of the initial population are generated as follows:

(1) Randomly select $m_i \in [r^i_{min}, r^i_{max}]$, $i = 1, \ldots, n$.

(2) The variation domain $[x^i_{min}, x^i_{max}]$ of input variable X_i is uniformly divided into $m_i - 1$ parts to determine the initial values $c_{i,1}, \ldots, c_{i,m_i}$.

(3) Gaussian noise is added to the $c_{i,j}$.

9.3.2.3 *Genetic learning of the granularity and the membership functions*

The learning process proposed by Cordón, Herrera, and Villar (2001) automatically generates the KB of a descriptive Mamdani-type FRBS based on a learning approach composed of two methods that pursue different objectives:

- A genetic learning process for the DB that allows us to define:
  - The number of labels for each linguistic variable.
  - The characteristic points of each fuzzy membership function.

  Triangular membership functions are considered due to their simplicity.

- A quick *ad hoc data-driven method* that derives the RB considering the DB previously obtained. This method is invoked for each DB definition generated by the GA, thus allowing the proposed hybrid learning process to finally obtain the whole definition of the KB (DB and RB) by means of the cooperation between both methods. The fitness function of the GA computes the error of the complete KB over the entire training data set to measure the quality of the DB encoded in the current chromosome.

Hence, the major difference to the approach by (Filipič and Juričić, 1996) is that the granularity of the fuzzy partitions does not remain constant but is adapted during learning as well.

The next five subsections describe the main components of the genetic learning process.

Encoding the DB

Each chromosome is composed of two parts that encode the number of linguistic terms per variable (granularity) and the membership functions that define their semantics:

- Number of labels (C_1): For a system with N variables (including input and output variables), the granularity of each variable is stored in an integer array of length N. In (Cordón, Herrera, and Villar, 2001), the possible allelic values are taken from the set $\{3, \ldots, 9\}$.
- Membership functions (C_2): For triangular functions, an array of size $N \cdot 9 \cdot 3$ stores the membership functions parameters (N variables, with a maximum number of nine labels per variable, and three real-valued parameters per label). In case a chromosome employs fewer than the maximum number of labels in any variable, the surplus membership function parameters are ignored.

A complete chromosome C encodes the following information:

$$C_1 = (l_1, l_2, \ldots, l_N)$$
$$C_{2i} = (P_{i1}^1, P_{i1}^2, P_{i1}^3, \ldots, P_{il_i}^1, P_{il_i}^2, P_{il_i}^3)$$
$$C_2 = (C_{21}, C_{22}, \ldots, C_{2N})$$
$$C = C_1 C_2$$

in which l_i is the granularity of variable i, P_{ij}^1, P_{ij}^2, and P_{ij}^3 are the characteristic points of the fuzzy set j of variable i, and C_{2i} encodes the fuzzy partition of variable i in C_2 (all its labels).

In order to satisfy the order constraint on fuzzy sets, a variation interval limits the possible ranges of the characteristic membership function points as shown in Fig. 9.18. The intervals are calculated based on uniform fuzzy partitions for each variable. Thus, the variation intervals of the three characteristic points of fuzzy set j in variable i, (P_{ij}^1, P_{ij}^2, P_{ij}^3), are defined as:

$$P_{ij}^1 \in [L_{ij}^1, R_{ij}^1] = [V_{ij}^1 - \frac{V_{ij}^2 - V_{ij}^1}{2}, V_{ij}^1 + \frac{V_{ij}^2 - V_{ij}^1}{2}]$$
$$P_{ij}^2 \in [L_{ij}^2, R_{ij}^2] = [V_{ij}^2 - \frac{V_{ij}^2 - V_{ij}^1}{2}, V_{ij}^2 + \frac{V_{ij}^3 - V_{ij}^2}{2}]$$
$$P_{ij}^3 \in [L_{ij}^3, R_{ij}^3] = [V_{ij}^3 - \frac{V_{ij}^3 - V_{ij}^2}{2}, V_{ij}^3 + \frac{V_{ij}^3 - V_{ij}^2}{2}]$$

In case the number of fuzzy sets encoded in a chromosome for a variable changes, the uniform fuzzy partition is recalculated for the new level of granularity, and the variation intervals are adjusted accordingly.

Initial Gene Pool

The initial population is composed of four subsets with the same number of individuals, which are initialised separately:

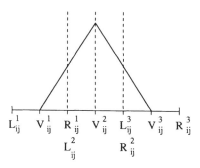

Fig. 9.18   Variation intervals for the membership function definition points

- Chromosomes in the first subset have the same number of labels for all variables and the membership functions are uniformly distributed across the variable domain.
- Chromosomes in the second subset acquire a different granularity per variable (different values in C_1) but the membership functions are still uniformly distributed as in the first group.
- Chromosomes in the third subset have the same number of labels for all variables. A uniform fuzzy partition is generated for each variable, from which the variation intervals of the characteristic points are calculated. Finally, values for the characteristic points are randomly chosen from the underlying variation interval.
- Chromosomes in the fourth subset have a different number of labels per variable and the membership functions are initialised by the same mechanism as in the third subset, namely drawing random values from the variation intervals.

Evaluating the chromosome
The chromosomes are evaluated in two steps:

- Invoke the RB generation method using the DB definition encoded in the chromosome to derive a complete KB.
- Calculate the mean square error (MSE) of the KB (genetically derived DB + RB) over the training set. In order to improve the generalisation capability of the final FRBS, the fitness function slightly penalises FRBSs with a large number of rules (NR) as done in Eq. 9.1 for the design of FRBCSs (Ishibuchi, Nozaki, Yamamoto,

and Tanaka, 1995):

$$F_C = \omega_1 \cdot MSE + \omega_2 \cdot NR \qquad (9.2)$$

The weighting factors ω_1 and ω_2 define the trade-off among the conflicting objectives of small model error MSE and small number of rules NR. In (Cordón, Herrera, and Villar, 2001), parameter w_1 is set to 1 whilst the MSE of the FRBS with the maximum granularity per variable and uniform fuzzy partitions (MSE_{max_lb}) and the number of rules of the corresponding RB (NR_{max_lb}) are used to estimate w_2 as follows:

$$\omega_2 = \alpha \cdot \frac{MSE_{max\_lb}}{NR_{max\_lb}}$$

with $\alpha \in [0, 1]$ being a weighting percentage.

Genetic operators
The coding scheme considered requires the design of particular genetic operators that maintain the integrity of the DB definition. Since the two chromosome parts are interdependent, the genetic operators have to process C_1 and C_2 in a cooperative and coherent manner.

Selection: The reproduction operator is Baker's stochastic universal sampling (Baker, 1987), which limits the actual number of offspring by the the floor and ceiling of the expected number of offspring together with an elitist reproduction.

Crossover: Two separate crossover operators recombine the parent chromosomes depending on whether their level of granularity is identical or different.

- *Crossover in case both parents define the same granularity level per variable:* If the chromosome parts C_1 of both parents are identical, namely both parents operate with the same number of labels for each variable, max-min-arithmetical (MMA) crossover operator (see Sec. 4.3.3.1) is applied to exploit the C_2 parts, whereas C_1 is passed directly to the offspring without modifications.
- *Crossover in case parents encode different granularity levels:* In this case, it makes no sense to try to recombine the genes in the C_2 parts to exploit the search space as they correspond to membership functions in fuzzy partitions of different granularity. Hence, standard

crossover operator is applied to both parts of the chromosome in order to use the information encoded in both parents to explore the search space looking for new promising zones. A single crossover point p is randomly chosen in C_1 and the tail segments in C_1 and C_2 to the right of the p-th variable are swapped between the parent chromosomes. That way, the levels of granularity in part C_1 of the descendent match with the encoding of their membership functions in C_2.

Let us look at an example in order to clarify the standard crossover application. Let

$$C_t = (l_1, \ldots, l_p, l_{p+1}, \ldots, l_N, C_{21}, \ldots, C_{2p}, C_{2p+1}, \ldots, C_{2N})$$
$$C_t' = (l_1', \ldots, l_p', l_{p+1}', \ldots, l_N', C_{21}', \ldots, C_{2p}', C_{2p+1}', \ldots, C_{2N}')$$

be the parent chromosomes subject to crossover at location p, the two resulting offspring are:

$$C_t = (l_1, \ldots, l_p, l_{p+1}', \ldots, l_N', C_{21}, \ldots, C_{2p}, C_{2p+1}', \ldots, C_{2N}')$$
$$C_t' = (l_1', \ldots, l_p', l_{p+1}, \ldots, l_N, C_{21}', \ldots, C_{2p}', C_{2p+1}, \ldots, C_{2N})$$

Mutation: Two separate mutation operators act on different chromosome parts:

- *Mutation on C_1*: The mutation operator for C_1 is similar to the one proposed by Thrift (1991) (see Sec. 7.3.1). A mutation on a gene in chromosome part C_1, either increases or decreases the number of labels by one, subject to the constraint that the number of labels does not fall below the minimum and does not exceed the maximum granularity limits specified in advance. After each mutation, the uniform fuzzy partition is recalculated for the new number of labels, and the corresponding part of C_2 is updated accordingly.
- *Mutation on C_2*: In the case of C_2 Michalewicz's non-uniform mutation (Sec. 2.2.4.2) is applied, since both parts are based on a real-coding scheme.

The entire recombination mechanism allows the GA to adequately balance the genetic search between exploration and exploitation. In the initial phase, standard crossover is applied most of the time and MMA only occurs in case two parents chromosomes coincidentally share the same levels of granularity. In the first stage, the GA mainly explores the search space, and identifies those regions that are likely to contain favourable solutions. As the population converges, exploitation of these promising regions begins to dominate the exploratory behaviour. At that point, MMA crossover occurs more frequently whereas the rate of standard crossover decreases. Notice that MMA crossover is only applied in the second part of the individuals as well as it exhibits a more exploitative behaviour than standard crossover. The MOGUL-based approximate GFRBS described in Sec. 8.5.1 successfully utilised this type of balanced exploration-exploitation strategy.

Restart

Due to the size of the search space and the particular coding in two separate entities, the genetic search might prematurely converge to local, suboptimal solutions. The authors stated that, in some preliminary experiments, the population rapidly converged to a particular granularity level (C_1), such that the GA no longer considered other possible partitions from there on. This phenomenon, known as *genetic drift*, is typical for multimodal search spaces (Goldberg, 1989). To avoid this problem, a restart operator increases the genetic diversity in the population whenever the relative difference in fitness between the best and the average individual becomes less than 5%.

The best individual is copied to the new population and the remaining chromosomes inherited 30% of their genes from it whereas the remaining 70% are reinitialised randomly (Eshelman, 1991). Re-initialisation takes into account that the effect of altering a gene is different in C_1 compared to C_2. Genes in C_2 simply acquire a new value randomly chosen from the variation interval of this point, whereas a new number of labels is randomly selected for a gene in C_1 and the corresponding segment in C_2 is initialised according to a uniform fuzzy partition.

9.3.2.4 *Genetic algorithm-based fuzzy partition learning method for pattern classification problems*

Ishibuchi and Murata (1996) introduced a genetic process that *a priori* determines the fuzzy partition of the pattern space for a classification problem. The resulting partition establishes, along with the process of obtaining the consequent described by (Ishibuchi, Nozaki, and Tanaka, 1992), the set of fuzzy classification rules that composes the FRBCS. The GA simultaneously determines the number of fuzzy rules and the membership function for each fuzzy set belonging to the antecedents.

The components of the method are briefly described in the following.

Encoding the Fuzzy Partitions

The coding scheme of the fuzzy partitions is based on the proposal of Nomura, Hayashi, and Wakami (1992b) for tuning the membership functions of an FLC. It employs binary chromosomes of fixed length, with a segment per variable. First, the domain of each variable is normalised. Then, a binary string with a predefined length of N_i bits, $S_i = (l_1, \ldots, l_{N_i})$, is associated to each variable X_i. Each bit l_j in the segment S_i corresponds to the fixed position $\frac{j-1}{N_i-1}$ in the [0,1] interval constituting the domain of the corresponding variable. If it is set to 1, both the centre of a triangular membership function, and the extremes of the neighbour membership functions will be located in that position.

Hence, the segment size N_i determines both the maximum number of fuzzy sets the fuzzy partition of variable X_i can have and the resolution in the tuning of that variable. Nomura et al.'s coding scheme required the two outer bits in each segment to be set to 1, thus restricting the possible values for the number of linguistic terms in each partition to $\{2, \ldots, N_i\}$.

Ishibuchi and Murata extend the previous coding scheme by removing the restriction on the outer bits, changing the coding scheme in two aspects:

- On the one hand, the range of possible granularities become $\{1, \ldots, N_i\}$. This way, a bit string with a single '1' is considered to represent a fuzzy partition with a single linguistic label for feature selection purposes. When a variable has a bit segment of this kind associated, it will not be considered in the fuzzy classification rule generation process.

- On the other hand, the values of both the leftmost and the rightmost bits of each segment take an additional meaning: when they are equal to 0, the membership function located in the corresponding edge of the fuzzy partition will be trapezoidal-shaped instead of triangular-shaped as usual. Therefore, individual rules become more descriptive, achieving the same classification accuracy with a smaller number of rules.

An example of this type of coding for a classification problem with two variables is shown in Fig. 9.19.

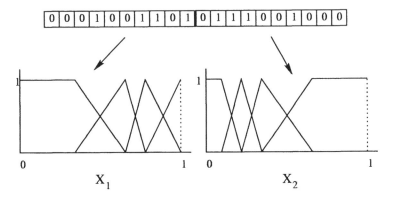

Fig. 9.19   Coding scheme

Genetic Operators

Multiple crossover operates between segments, which is equivalent to exchange the fuzzy partition of a variable between the parents. As mutation swaps adjacent bits inside a segment, it fine tunes the partition by slightly changing the support of the corresponding fuzzy set. An additional mutation operator randomly alters the value of a gene and thereby adds or removes a linguistic term from the partition of the variable. Figures 9.20, 9.21, and 9.22 illustrate the effect of these operators.

Fitness function

The fitness function has the objectives to maximise the number of correctly classified training examples and to minimise the number of rules. To do so, the DB definition encoded in the chromosome is considered to derive a complete FRBCS KB. The fuzzy classification rule generation method proposed by (Ishibuchi, Nozaki, and Tanaka, 1992) is invoked to determine the consequent for each rule in the RB and the correct classification rate NCP of the resulting FRBCS is measured over the training set. The final value of the fitness function for the chromosome is computed using Eq. 9.1.

Of course, the length of the segments N_i determines the tuning precision, and consequently, the classification accuracy achievable by the resulting FRBCS. However the increase on that length produces a deterioration in

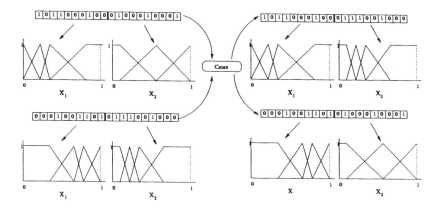

Fig. 9.20 Crossover operator

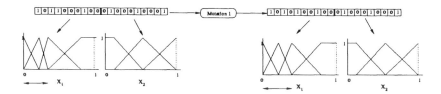

Fig. 9.21 Type 1 mutation operator

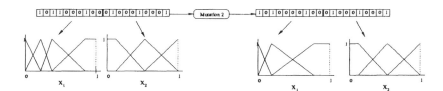

Fig. 9.22 Type 2 mutation operator

the GA efficiency, due to the expansion of the search space, highlighting the need of biasing the GA search with the available information to produce good results.

9.3.3   *Genetic learning of non-linear contexts*

Non-linear scaling (see Sec. 4.1.1.2) modifies the fuzzy distribution in a non-uniform manner and significantly changes the shape of the membership functions, thus requiring to redesign the RB.

Different approaches have been proposed to deal with the learning of non-linear contexts, considering tools such as NNs (Pedrycz, Gudwin, and Gomide, 1997) and, specially, GAs (Cordón, Herrera, Magdalena, and Villar, 2000, 2001; Gudwin, Gomide, and Pedrycz, 1997, 1998, Magdalena, 1997, Magdalena and Velasco, 1997). The following subsection presents one of these approaches which considers the learning approach studied in this section.

9.3.3.1   *Genetic learning process for the scaling factors, granularity and non-linear contexts*

Cordón, Herrera, Magdalena, and Villar (2000) proposed a genetic learning method for the DB of a descriptive Mamdani-type FRBS that allows us to define:

- The granularity for each linguistic variable.
- The linear scaling (working range).
- The non-linear scaling which, when applied on a fuzzy partition with triangular-shaped membership functions, defines regions in the domain for which the FRBS has a higher or a lower relative sensibility. These regions will maintain their triangular shape.

The function proposed by Magdalena and Velasco (1997) and described in Sec. 4.1.1.2, is employed to define the non-linear scaling. The parameter S used to distinguish between non-linearities with symmetric and asymmetric shape is always set to 0 by the authors, thus considering only the former possibility, i.e., lower granularity for middle or for extreme values (see Fig. 4.5). Hence, the scaling function mapping the normalised interval [-1,1] on itself takes the following form

$$f(x) = sign(x) \cdot |x|^a, \quad with \ a > 0$$

As said, triangular membership functions are considered due to their simplicity. The non-linear scaling function is only applied to the characteristic points, such that the scaled membership functions maintain their triangular shape. The transformed points are connected by piece-wise linear

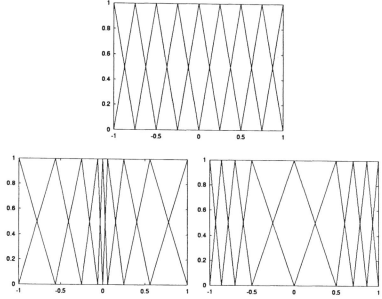

Fig. 9.23   Fuzzy partitions with $a = 1$ (at the top), $a > 1$ (down left), and $a < 1$ (down right)

functions. Depending on the value of parameter a, one obtains a uniform sensibility ($a = 1$), higher sensibility for centre values ($a > 1$), or higher sensibility for edge values ($a < 1$). Figure 9.23 shows a graphical representation for these three possibilities.

As usual in these kinds of learning processes, an ad-hoc data covering method is used to learn the RB based on the DB encoded in the chromosome prior to the chromosome evaluation.

In the following, we describe the main components of the genetic learning process.

Encoding the DB

The three main components of the DB are the number of linguistic terms per variable, the membership functions that define their semantics and the scaling factors.

As regards the membership functions, triangular-shaped functions, symmetrical and uniformly distributed across the variable working range are initially considered. Then, the said non-linear function is applied to the three characteristic points of each fuzzy set. Thus, the whole fuzzy partition is defined by the granularity and the sensibility parameter a for each variable, as well as by the working ranges. Accordingly, a chromosome is composed of three parts:

- Granularity (C_1): For a system with N variables (including input and output variables), the number of fuzzy sets per variable is encoded into an integer array of length N. The authors consider the possible values to be taken from the set $\{3, \ldots, 9\}$.
- Sensibility parameters (C_2): An array of real numbers of length N contains the sensibility parameter a for each variable. In (Cordón, Herrera, Magdalena, and Villar, 2000), the possible values for this parameter are the range $(0, 10)$.
- Working ranges (C_3): An array of $N \cdot 2$ real values stores the variable working range. If $[r_{min}^i, r_{max}^i]$ is the initial domain of a variable and d is the interval dimension ($d = r_{max}^i - r_{min}^i$), the range considered for the variable domain lower limit is $[r_{min}^i - (1/4 \cdot d), r_{min}^i]$, and the range for the upper limit is $[r_{max}^i, r_{max}^i + (1/4 \cdot d)]$.

A representation of the chromosome is shown next:

$$C_1 = (l_1, \ldots, l_N)$$
$$C_2 = (a_1, \ldots, a_N)$$
$$C_3 = (r_1^{inf}, r_1^{sup}, \ldots, r_N^{inf}, r_N^{sup})$$
$$C = C_1 C_2 C_3$$

Evaluating the DB

Each chromosome is evaluated in three steps:

- Generate the fuzzy partitions for all the linguistic variables using the information contained in the chromosome. First, each variable is linearly mapped from its working range (contained in C_3) $[r_i^{inf}, r_i^{sup}]$, $i = 1, \ldots, N$, to $[-1, 1]$. In a second step, uniform fuzzy partitions for all the variables are created considering the granularity per variable ($l_1, \ldots, l_N$, stored in C_1). Finally, the non-linear scaling function with the corresponding sensibility parameter ($a_1, \ldots, a_N$, in C_2) is applied to the definition points of the membership functions obtained in the previous step, obtaining the whole DB definition.
- Generate the RB, by invoking a fuzzy rule learning method considering the DB obtained.

- Calculate the fitness as the mean square error over the training set using the KB obtained (DB + RB).

Genetic operators

The genetic operators are designed appropriately according to the chromosome structure that reflects the various parts in the DB. Since there is a strong relationship between the three chromosome parts C_1, C_2 and C_3, the genetic operators are applied in a cooperative manner that considers the interdependencies between the different components.

Crossover:

- *Crossover when both parents have the same granularity level per variable:* If the two parents have the same values in C_1 (each variable has the same number of labels in the two parents), then the genetic search has located a promising region that has to be adequately exploited. This objective is accomplished by applying the MMA crossover, described in Sec. 4.3.3.1.
- *Crossover when the parents encode different granularity levels:* This second case suggests to explore the search space in order to discover new promising regions, considering that a suitable sensibility parameter or working range for a variable probably deteriorates if applied to a partition with a different level of granulation.

  Hence, when C_1 is crossed, the genes in C_2 and C_3 corresponding to the crossed variables are also exchanged between the two parents. In this way, a standard crossover operator is applied to the three parts of the chromosomes. This operator behaves similar to the one described in Sec. 9.3.2.3.

Mutation:

- *Mutation on C_1:* The mutation operator selected for C_1 is the one described in Sec. 9.3.2.3. It is important to notice that a mutation on C_1 does not require the redefinition of other parts of the chromosome, as it was the case in Sec. 9.3.2.3.
- *Mutation on C_2 and C_3:* Since both parts are based on a real-coding scheme, Michalewicz's non-uniform mutation operator is employed, which is described in Sec. 2.2.4.2.

Figure 9.24 shows the application scope of these operators.

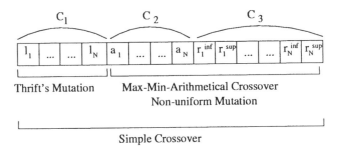

Fig. 9.24   Genetic representation and operators' application scope

In (Cordón, Herrera, Magdalena, and Villar, 2001), the authors propose a new *a priori* genetic DB learning process that extends the previous one in the following two aspects:

(1) The original non-linear scaling function shown in Sec. 4.1.1.2 is considered. This way, parameter S is not restricted to take value 0, thus allowing fuzzy partitions with symmetric and asymmetric non-linearities to be obtained.
(2) The fitness function in Eq. 9.2 is employed, thus allowing the GA to design FRBSs with the desired trade-off between accuracy and linguistic description.

9.4   Other Genetic-Based Machine Learning Approaches

There are other genetic-based machine learning (GBML) approaches to generate the RB/KB that do not fit into any of the previous categories, learning with GP, genetic selection and genetic learning of DBs, genetic tuning and genetic learning via the Michigan, Pittsburgh and IRL approaches. This section is devoted to describe several of these approaches.

9.4.1   *Genetic integration of multiple knowledge bases*

Wang, Hong, and Tseng (1998) proposed a GA-based fuzzy knowledge integration framework that integrates information from various fuzzy knowledge sources into a single KB. This is a special case of selection process where a final KB is obtained from the integration of different preliminary KBs.

The proposed framework integrates RBs and membership function sets (DBs). It maintains a population of fuzzy rule sets with their membership functions and uses the GA to automatically derive the resulting KB. It operates in two phases: fuzzy knowledge encoding and fuzzy knowledge integration.

Fuzzy knowledge encoding

The encoding phase first transforms each RB and its associated membership functions (DB) into a chromosome encoded as a variable-length string.

Each chromosome is divided into two parts: *RS-part* and *MF-part*. In the first part, each fuzzy rule is encoded into a fixed-length string. All substrings of intermediary rules are then concatenated to represent the entire fuzzy rule set. To effectively encode the associated membership functions, a two-parameter representation (centre and half of spread) is used. Thus, all pairs for every feature are concatenated to represent its membership functions. According to this representation, each chromosome thus consists of an intermediary fuzzy rule set and its associated membership functions.

This representation allows genetic operators to easily integrate multiple fuzzy rule sets and their membership function sets at the same time.

Fuzzy knowledge integration

The integration phase then chooses appropriate strings for "mating", gradually creating good offspring fuzzy rule and membership function sets. The offspring fuzzy rule sets with their associated membership functions then undergo recursive "evolution" until an optimal or near optimal set of fuzzy rules and membership functions has been obtained.

The fitness function evaluates the accuracy and the complexity of the derived KB. The evaluation function reflects this trade-off between high accuracy and low complexity of the rule sets as follows:

$$Fitness(RS) = \frac{[Accuracy(RS)]}{[Complexity(RS)]^\alpha}$$

The genetic fuzzy-knowledge integration framework uses a two-substring crossover and a two-part mutation operator.

9.4.2   *Flexibility, completeness, consistency, compactness and complexity reduction*

In the following two subsections we introduce the two approaches proposed by Jin, von Seelen, and Sendhoff (1999) and Jin (2000), related to the cited properties in the subsection title.

9.4.2.1   *Evolutionary generation of flexible, complete, consistent and compact fuzzy rule-based systems*

In an attempt to develop flexible, complete, consistent and compact FRBSs, an ES-based methodology is proposed by Jin, von Seelen, and Sendhoff (1999). It mainly focuses on the completeness, consistency and compactness of the FRBS. A distinctive feature of this approach is that, in addition to the KB itself, it also optimises the fuzzy inference operators, including t-norms and BADD defuzzification.

Flexibility, completeness and consistency are essential properties for any FRBS in order to exhibit good performance and to possess a clear, intuitive interpretation. Compactness is crucial when the number of the input variables increase as a means to limit the overall number of fuzzy rules. Briefly, the advantages of these properties are, according to Jin et al. (1999):

(1) The FRBS is compact and efficient because the number of fuzzy rules is largely reduced. A compact FRBS is desirable when the number of input variables increases.

(2) The FRBS is complete and contains no seriously conflicting rules, which improves the generalisation ability of the FRBS and guarantees that the knowledge acquired by the fuzzy rules is logically sound. Fuzzy rules are regarded as inconsistent if their premise parts are quite similar, but their consequents differ substantially, or they conflict with existing expert knowledge.

(3) The FRBS is expected to exhibit a larger degree of flexibility because soft fuzzy operators are incorporated to optimise the inference process.

ESs are used to optimise the coefficients of the soft t-norms considered as conjunctive and implication operators and BADD defuzzifier, the parameters of the fuzzy membership functions, and the structure of the fuzzy rules. The genetic coding of the variables for the fuzzy operators and fuzzy membership functions is straightforward:

- The coefficients for soft t-norms cover the range between 0.0 and 1.0.
- Although in principle the coefficient of BADD defuzzification can vary from 0 to ∞, the implementation imposes an upper bound on this parameter.
- Each membership function is specified by two parameters, the centre c and the width w of the Gaussian fuzzy set.
- A matrix encodes the rule structure. Every row is associated to a rule with a column per input variable and with an integer value associated to its respective label index. A '0'-entry indicates that the variable does not appear in the corresponding rule.

  Similarly, the structure of the consequents is encoded as a vector of positive integers of size N, identical to the number of rows in the premise matrix.

$$Struc_{consequent} = [c_1, c_2, ..., c_N]^T$$

where $c_j \in \{1, ..., K\}$, $j = 1, \ldots, N$, assuming that the domain of the consequent variable is partitioned in K fuzzy sets.

The chromosome is evaluated according to the said objectives: a completeness fuzzy similarity measure and a degree of inconsistency of a KB. No special means are taken to reduce the number of fuzzy rules. In practice, Jin, von Seelen, and Sendhoff (1999) stated that the EA tends to generate compact FRBSs.

9.4.2.2 *Genetic complexity reduction and interpretability improvement*

Jin (2000) proposed a hybrid data-based fuzzy modelling approach of high-dimensional systems with the following distinct stages:

- An initial fuzzy rule set is generated based on the conclusion that optimal fuzzy rules cover extrema (Kosko, 1995). In the rule generation process, a fuzzy similarity measure is adopted to check the similarity of each rule before incorporating it into the RB to avoid rule redundancy.

- Then, both the structure and parameters of the fuzzy rules are optimised using GAs and a gradient learning algorithm. The structure
  of the rule premise (i.e., the decision of which inputs should appear
  in the premise part of a fuzzy rule) is optimised using a GA based
  on a local cost function, which is consistent with the motivation to
  establish locally optimal fuzzy rules. It is shown that optimisation
  of the rule structure not only reduces the rule complexity and improves the system performance, but also reveals the dependencies
  between the system inputs and the system output. Subsequent to
  structure optimisation, a gradient learning algorithm optimises the
  parameters of the fuzzy rules.
  During learning, fuzzy rules with a very low firing strength are
  deleted, which plays an important role in reducing the number of
  fuzzy rules.
- Finally, interpretability of the FRBS is improved by fine tuning the
  fuzzy rules with regularisation.

The resulting FRBSs generated by this method have the following distinctive features:

(1) the FRBS is largely simplified;
(2) the FRBS is interpretable; and
(3) the dependencies between the inputs and the output are clearly
    shown.

Jin (2000) presented a successful application of this method to a system
with eleven inputs and one output with 20,000 training data and 80,000
test data. The final FRBS designed contains only 27 fuzzy rules and its
performance on both training and test data is satisfactory.

9.4.3 *Hierarchical distributed genetic algorithms: design-ing fuzzy rule-based systems using a multi-resolution search paradigm*

Kim and Zeigler (1996a, 1996b) employed multi-resolution search to design
optimal FRBSs in a variable structure simulation environment. The initial
search space is explored at a coarse resolution and some of the subspaces are
selected as candidate regions to contain the global optimum. Additional optimisation processes are invoked to investigate the candidate search spaces

in detail, processes which continue operating until a solution is found.

This search paradigm was implemented using hierarchical distributed GAs, search agents solving different degrees of abstract problems. Creation/destruction of agents is executed dynamically during the operation based on their performance. In the application to FRBSs, the hierarchical distributed GA investigates design alternatives such as different types of membership functions and the number of the fuzzy labels, as well as their optimal parameter settings, all at the same time.

9.4.4 *Parallel genetic algorithm to learn knowledge bases with different granularity levels*

Cheong and Lai (2000) proposed a new genetic learning approach that uses a parallel GA with three populations to optimise FRBSs with RBs generated from fuzzy partitions of three different granularities: 3×3, 5×5 and 7×7. The process also employs a novel method to create migrants between the three populations of the parallel GA to increase the chances of optimisation.

The algorithm operates as follows:

- *Coding of Chromosomes:* Real values are used to code the membership functions and integers to code the RB (consequent label index per rule). The number of parameters required to represent three-, five-, and seven-fuzzy set universes of discourse are one, three and five, respectively, per variable.
- *Initialisation of Chromosomes:* Chromosomes are initialised with random values or with values derived from expert knowledge. The latter way facilitates earlier convergence to the optimal solution.
- *Crossover:* To ensure that the membership functions of the offspring generated by the crossover operator always remain valid (i.e., ordered values within the permissible ranges in the universe of discourse), the crossover operator generates random values within the intervals represented by the values of the parents.
  Reproduction exchanges subsets of the RBs.
- *Mutation:* Mutation of the membership functions involves generating a random value within the permissible ranges, while mutation of the RBs is either purely random or random within a predefined range.

- *Parallel GA:* Since the approach optimises three types of FRBSs (generated from three, five, and seven-fuzzy set partitions) coded as chromosomes of different sizes, the incompatibility of the genetic operators does not permit to evolve the different types of FRBSs within a single population. Instead, the different sized chromosomes are evolved in parallel in three populations. Individuals can migrate between populations, which is beneficial to inject new genetic material into the populations and thereby avoids premature convergence to local minima. At the end of each generation, the five best individuals from each population are selected and migrated to the other populations.
  Due to the difference in size and composition, the chromosome structure is adapted to the population into which a migrant is injected. The following scheme is used to transform migrants from one type of chromosome to another.
- *Migrants:* To convert from three- to seven-fuzzy set membership functions and vice versa, five-fuzzy set membership functions have to be used as intermediary.
  When expanding a RB to a larger one, there are two problems:

  (1) Codes representing the output have to be mapped to a larger range.
  (2) New output codes have to be generated for the extra cells in the RB.

Then, different fuzzy rules must be introduced in order to obtain the output values of the membership functions and rules, obtaining values for the extra cells, via adjoining cell values.

The opposite process, i.e., the collapsing a large RB into a smaller one, is done by aggregating multiple cells of the larger rule matrix into single cells of the smaller one.

9.4.5   *Modular and hierarchical evolutionary design of fuzzy rule-based systems*

Regattieri-Delgado, Von-Zuben, and Gomide (1999, 2000a, 2000b) introduced a modular, hierarchical evolutionary method to design FRBSs. The method uses a GA working on different populations encoding information items of different levels, to finally evolve a population of complete FRBSs.

For this purpose, three hierarchical modules are defined namely fuzzy partition, individual rule and fuzzy rule set populations. The global evolutionary process finds the optimal set of fuzzy rules induced by the granularity required, the form of the membership functions and the inference procedure to be used with each individual fuzzy rule. This hierarchical configuration guides to implementation of an effective process of cooperation and competition between rules which allows fuzzy rule sets composed of a low number of rules to be obtained.

This approach considers simultaneously many important aspects that must be addressed when designing an FRBS:

- The complete system is automatically designed by means of GA techniques, providing a way to decide on: membership function parameters (type, shape and location), degree of overlap between fuzzy sets, fuzzy implication operator to be used to make inference with each individual rule, and number of rules in the FRBS.
- As said, the evolutionary process is structured in a hierarchical configuration that cooperatively evolves the following three populations: membership functions, individual rules, and fuzzy rule sets.
- The modules interact during the evolutionary process to obtain the optimal set of fuzzy rules and its associated parameters. The individuals in the membership function population provide the atomic elements (fuzzy clauses or membership functions) to be combined to produce a population of individual rules. The chromosomes in the population of single rules are then combined to form the individuals in the population of sets of rules.

A complete description of the representation and genetic operators can be found in (Regattieri-Delgado, Von-Zuben, and Gomide, 1999). The application of this hierarchical approach to design TSK FRBSs is to be found in (Regattieri-Delgado, Von-Zuben, and Gomide, 2001).

9.4.6 *VEGA: virus-evolutionary genetic algorithm to learn TSK fuzzy rule sets*

A learning algorithm based on virus theory of evolution is proposed by Shimojima, Kubota, and Fukuda (1996), the virus-evolutionary GA (VEGA).

The VEGA develops a horizontal propagation and a vertical inheritance of genetic information in a population. The main operator of the VEGA is a reverse transcription operator, which simultaneously plays the role of crossover and selection. Furthermore, a transduction operator generates a sub-string to be transmitted from one individual to others. Making use of the latter two operators, the VEGA is able to reduce the number of fuzzy rules in the derived fuzzy rule set.

In the following, we describe the VEGA algorithm and its application to learn simplified TSK KBs.

9.4.6.1   *Preliminaries: virus theory of evolution*

One of the most important evolutionary theories in biology is Darwinian theory of evolution, which was outlined in Sec. 2.1. However, recent progress of molecular biology results in proposals of other theories such as the virus theory of evolution, which is based on the view that virus transduction is a key mechanism to transport DNA segments across species. Here, the transduction means the genetic modification of a bacterium by genes from another bacterium carried by a bacteriophage. Many viruses in nature can easily cross species barriers and are often transmitted directly from individuals of one phylum to another as horizontal propagation. The incorporation of segments of host DNA into infective viruses and subsequent transfer to other cells is well known. Besides, whole virus genomes may be incorporated into germ cells and transmitted from generation to generation as vertical inheritance.

9.4.6.2   *Virus-evolutionary genetic algorithm architecture*

The VEGA is composed of two populations: a *host population* and a *virus population*. Whilst the host population is defined as a set of candidate solutions, the virus population contains a set of sub-strings of host individuals. A virus has the ability to transmit segments of DNA between species. Therefore, the virus infection develops the horizontal propagation of parts of genotypes in the host population. To incorporate the virus infection mechanism, a steady-state GA is adopted (see Sec. 2.2.3.3). In general, this GA operates with the crossover and mutation of only one pair of individuals in the population every generation.

An initialisation process randomly generates an initial host population, and then a virus individual is generated as a sub-string of a host individual. In the steady-state GA, the replacement scheme removes the currently worst individual from the population. Whereas the size of a host chromosome is constant, the size of the virus individual increases with the evolution of the host population. The procedure of the VEGA is as follows:

Initialisation
   repeat
      Selection
      Crossover
      Mutation
      Virus-infection
   until (Termination-condition = True)
end.

9.4.6.3   *Virus infection operators*

Shimojima, Kubota, and Fukuda (1996) assumed that the main process of a virus infection is a horizontal propagation of a sub-string among host individuals, though there are many different characteristics about virus infection in nature. Therefore, a virus transduces the genes from a host individual and transcribes to another host individual. The VEGA has two virus infection operators as follows:

- *Reverse transcription operator:* Virus transcribes its genes on the string of a host individual.
- *Transduction operator:* Virus transduces a sub-string from a host individual. As its fundamental operation, virus transduces a sub-string in addition/reduction to some genes on the host string.

Each virus has $fitvirus_i$ as a strength about the virus infection. The number of infection times of each virus is controlled under $fitvirus_i$. It is assumed that $fithost_j$ is the fitness value of a host individual j before the infection and $fithost'_j$ is the fitness value after the infection. The $fitvirus_{i,j}$ denotes the difference between $fithost_j$ and $fithost'_j$, which is equal to the value obtained by infecting to the host individual:

$$fitvirus_{i,j} = fithost'_j - fithost_j$$

$$fitvirus_i = \sum_{j \in S} fitvirus_{i,j}$$

where i is the virus number and S is a set of the host individuals which are infected by the virus i. The virus which can improve the performance of the host population has a high possibility of infecting to the host population at the next generation. Furthermore, each virus has a life power as follows:

$$life_{i,t+1} = r \cdot life_{i,t} + fitvirus_i$$

where t and r means the generation and the life reduction rate, respectively.

The initial value of $life_{i,0}$ is set to zero. First, a virus does the reverse transcription to a host individual selected randomly out of the host population. If $life_{i,t}$ takes a negative value, the virus individual transduces a new sub-string with the transduction operator from the host individual selected randomly. If not, the virus individual transduces a partially new sub-string from one of the transcribed host individuals with the transduction.

9.4.6.4   *The VEGA genetic fuzzy rule-based system*

Shimojima, Kubota, and Fukuda (1996) applied the VEGA to learn simplified TSK FRBs. In the following, the composition of this new GFRBS is described.

Coding
A binary coding is used to encode the information of the membership functions. An antecedent part is expressed by a binary number of $(3 \cdot n + 1)$ bits for every triangular-shaped membership function in this simulation. Each coefficient a, b, c needs n bits and one additional bit is used as a flag of the membership function's validity. A consequent part is encoded by an m bits binary number.

As can be seen, with the previous coding scheme, the antecedent part of the simplified TSK fuzzy rules presents an approximate nature. This is why we refer to the fuzzy rule set as TSK FRB instead of TSK KB.

Fitness Function
The fitness function involves the performance index and the number of membership functions and fuzzy rules in the way shown in the following equation:

$$F = \alpha \cdot P + \beta \cdot R_n + \gamma \cdot M_n$$

where P is the performance index, e.g. the approximation error between the output of the system and the desired performance, R_n is the number of rules, and M_n is the number of membership functions. The coefficients α, β, and γ are classified into two types: system performance (α) and system size $(\beta$ and $\gamma)$.

The objective is to acquire an accurate and compact FRBS without redundant fuzzy rules and membership functions. Therefore, the objective results in the minimisation problem of the fitness function F.

Selection

To perform selection, each individual of the host population is ranked based on the fitness value F, with lower fitness values obtaining a better ranking. The replacement scheme of VEGA removes the worst individual from the host population.

Depending on the values on the weighting factors α, β and γ, selection displays a preference for compact FRBSs $(\beta > \alpha, \gamma > \alpha)$ or high accuracy FRBSs $(\beta < \alpha, \gamma < \alpha)$.

Crossover Operator

In order to generate a new set of membership functions and rules, two-point crossover operator is applied, thus exchanging some rules between the two parents, which are randomly selected.

Mutation Operator

Two types of mutation operators are considered: (A) uniform distribution random set based mutation operator and (B) normal distribution random number based mutation operator. In both cases, the individual to be mutated as well as the mutation site are randomly selected. The mutation operator (A) changes some bits of the strings and is used for global and rough search. This operator can change the enable/disable flag of the membership function. The mutation operator (B) does not change the bits of the chromosomes, but adds (or subtracts) random values to (from) the parameters of the membership functions.

The random values are generated based on the age of the string. When the highest fitness value is improved, the age is reset to zero, otherwise the age is incremented. If the age is small, the random values are generated in a small region, and if the age is large, the random values are generated in a large region. Changing the region from small to large, expands the search space from narrow to wide. As in case of the mutation operator (A), this mutation operator also changes the validity of each membership function.

Virus Infection Operators

A virus population transmits sub-strings between individuals in a host population as horizontal propagation of genetic information. A virus has the information of some rules, membership functions, and a central point of the transduction/reverse transcription area on the input space. Virus infection operators applied to FRBS optimisation are as follows:

- *Transduction Operator:* A virus transduces some sub-strings from a host individual. This operator selects effective schemata (effective fuzzy rules in our case) to be transmitted between host individuals. A sub-string of a virus is generated by selecting some fuzzy rules from the string of a host individual. The fuzzy rules to be selected are the ones most frequently used in the host individual. At the same time, the central point of the virus is defined according to the transduced fuzzy rules.
- *Reverse Transcription Operator:* As said, in a reverse transcription operator, a virus transcribes its sub-string on the string of a host individual. Actually, a virus overwrites its sub-string on a host string.

  In the example shown by Shimojima, Kubota, and Fukuda (1996), the virus individual and the host individual are one fuzzy rule and three fuzzy rules, respectively. First, the virus deletes the fuzzy rules that include the central point of the virus. Next, the virus overwrites its fuzzy rules on the host string.

Shimojima, Kubota, and Fukuda (1996) applied the VEGA GFRBS to the well known cart-pole system (introduced in Sec. 11.3.1).

9.4.7 *Nagoya approach: genetic-based machine learning algorithm using mechanisms of genetic recombination in bacteria genetics*

A new approach to GBML —called *Nagoya approach* by the authors— was proposed by Furuhashi, Nakaoka, and Uchikawa (1994a, 1995). Furuhashi, Miyata, and Uchikawa (1997) associated this approach with the use of mechanisms of genetic recombination in bacterial genetics, which have several aspects in common with the VEGA introduced in the previous section. In the following, the Nagoya approach is described following this bacterial genetic interpretation.

9.4.7.1 *Bacterial genetics*

Bacterial genetics presents interesting mechanisms of genetic recombination. Bacteria can transfer DNA to recipient cells through mating. A male cell transfers a strand of genes into female cells. The female cells acquire characteristics of the male cells and then change to male cells. Characteristics of one bacterium can spread across multiple bacteria.

Bacteriophage carries a copy of the host gene across and incorporates it into the chromosome of the infected cell. The process is called transduction. Transduction is also possible to spread the characteristics of a bacterium among other bacteria.

These genetic recombinations lead to a mechanism of microbial evolution. Mutated genes are transferred from a bacterium to others and evolution of bacteria progresses rapidly.

9.4.7.2 *Nagoya approach: algorithm description*

Furuhashi, Miyata, and Uchikawa (1997) introduced the mechanisms of the previous bacterial genetics into the GBML. Multiple bacteria are reproduced and some genes of each chromosome are mutated and tested. The best genes are chosen and transferred to other bacteria. The new approach modified this mechanism in the extreme.

The algorithm acts in a classical GA way. The initial population, composed of n_{chr} individuals, is generated at random. Each chromosome is then evaluated and the genetic operators are applied to the individuals to generate a new population of chromosomes. The cycle is repeated until a termination condition is satisfied.

The following genetic operators are used.

Mutation and selection of genes
Assume there are n_p parts in a chromosome. A chromosome l is chosen and reproduces m clones. Parts i (i is randomly decided) of $m-1$ clones are mutated. Each chromosome is evaluated. The elite among the m chromosomes are selected and the rest are deleted. The preceding process: reproduction, mutation, evaluation, and selection, is repeated. The mutation is applied to a randomly chosen part rather than to the previously chosen parts. The new chromosome l is finally obtained. This genetic operation is applied to

all the n_{chr} chromosomes one by one.

The foregoing process can be interpreted as follows: A bacterium is reproduced to m clones and the same parts of their chromosomes are mutated. The elite part is selected and transferred to other $m-1$ bacteria and inferior genes are replaced with the elite genes. Then, other parts of the chromosomes are mutated, evaluated, selected and transferred. All the elite genes are aggregated to a chromosome of one bacterium. The n_{chr} bacteria evolve in this way.

Selection and reproduction of chromosomes
The selection and the reproduction are applied to chromosomes. Each chromosome has been evaluated under the foregoing operation. Those with lower fitness values are deleted. Some chromosomes randomly chosen from the remaining ones are reproduced.

Crossover
The crossover operator is applied to the newly generated chromosomes and the offspring are generated and evaluated. This is an operation of the conventional GA and is also efficient for improving the chromosomes.

The previous GA is efficient in local improvement of chromosomes, since the evolution is carried out at the level of genes of the chromosomes.

9.4.7.3 *Nagoya approach: extensions*

Lamarckian learning proposed by (Schaffer and Morishima, 1987) changes the probability of choosing crossover points according to the error distribution along the string. Also Davidor (1991) described the Lamarckian probability for mutations based on sub-goal fitness. Grefenstette (1991) proposed a Lamarckian learning for the Pittsburgh approach. The primary Lamarckian feature of this method is that a strength of a rule for conflict resolution is passed along when a rule is inherited to an offspring. In addition, crossover first cluster rules so that rules that fire in sequence within a high-payoff environment tend to be assigned to the same offspring. These methods were introduced to promote the inheritance of good strings by diminishing the counterproductive effects of crossover/mutation in destroying good schema. The Nagoya approach also promotes the inheritance of good schema. Furthermore, the new approach is aiming at improving local portions of chromosomes by using the *gene level search*.

Local search methods perform a local hill climbing in the vicinity of a current solution, bit by bit sweep. The new approach does a search at a gene level, i.e., mutations on sub-strings.

9.4.8 *Learning fuzzy rules with the use of DNA coding*

Yosikawa, Furuhashi, and Uchikawa (1995, 1996b, 1996a) proposed an interesting coding method based on biological DNA which is suitable to represent fuzzy rules. The DNA coding method has the following features:

(1) Flexible representation of fuzzy rules.
(2) Redundant and overlapped coding.
(3) Variable length of chromosome.
(4) No constraint on crossover points.

Furuhashi (1997) described the new coding method, a mechanism for the development from the artificial DNA chromosome to the fuzzy rules, and the way to find effective rules.

The DNA chromosome contains sets of fuzzy rules where an amino acid can be translated as an input variable or a form of membership function, and so on, and a sequence of amino acids makes a fuzzy rule.

Furuhashi applied the coding method to the discovery of effective fuzzy control rules based on chasing and avoiding actions of mobile robots, showing that the redundancy and overlapping of genes developed by the proposed method performed pretty well.

9.4.9 *Hybrid fuzzy genetic-based machine learning algorithm (Pittsburgh and Michigan) to designing compact fuzzy rule-based systems*

A hybrid GFRBS considering both fuzzy versions of Michigan and Pittsburgh approaches is proposed by Ishibuchi, Nakashima, and Kuroda (1999, 2000a, 2000b). The method is based on the Pittsburgh approach, where a set of fuzzy rules is coded as a string, and the Michigan approach is used as a mutation operation to partially modify each string (i.e., each fuzzy rule set) by generating new rules from existing good rules.

In this manner, this hybrid algorithm utilises the advantages of the two approaches. According to the authors (Ishibuchi, Nakashima, and Kuroda, 2000a):

- While the Michigan approach has high search ability to efficiently find good fuzzy rules in large search spaces for high-dimensional pattern classification, it can not directly optimise FRBSs.
- On the other hand, the Pittsburgh approach can directly optimise FRBSs while its search ability to find good fuzzy rules is not high.

To automatically adjust the number of fuzzy rules, the string length in the hybrid algorithm is not fixed. This means that each string includes a different number of fuzzy rules. The string length is changed when new strings are generated from parent strings by a crossover operation. To generate a new string from two parent strings (say, S_1 and S_2), some fuzzy rules are randomly selected from each parent to construct a new fuzzy rule set. The number of fuzzy rules to be inherited from each parent to the new rule set is randomly specified. Let N_1 and N_2 be the number of fuzzy rules to be inherited from S_1 and S_2, respectively. N_1 and N_2 are randomly specified in the sets $\{1, \ldots, |S_1|\}$ and $\{1, \ldots, |S_2|\}$, respectively, where $|S_i|$ is the number of fuzzy rules in the fuzzy rule set S_i. The generated new rule set includes $(N_1 + N_2)$ fuzzy rules.

For simultaneously maximising the classification ability of FRBCSs and minimising the number of fuzzy rules, the hybrid algorithm measures the fitness value of each rule set generated by means of Eq. 9.1.

The hybrid algorithm is structured as follows:

(1) Randomly generate N_{pop} rule sets.

(2) Calculate the fitness value of each fuzzy rule set in the current population.

(3) Generate $(N_{pop} - 1)$ fuzzy rule sets by the selection, crossover, and mutation. As the mutation operator, a single iteration of the Michigan-style algorithm is applied to all the generated rule sets after the selection and crossover operations.

(4) Add the best fuzzy rule set in the current population to the newly generated one.

(5) Terminate the algorithm if the pre-specified stopping condition is satisfied, otherwise return to Step 2.

Simulation results (Ishibuchi, Nakashima, and Kuroda, 2000a) demonstrate that the hybrid algorithm achieves the goal of simultaneously maximising the classification ability of FRBSs while minimising their number of fuzzy rules.

Chapter 10

Other Kinds of Evolutionary Fuzzy Systems

As discussed in Chapter 3, GFSs or Evolutionary FSs are FSs with added genetic/evolutionary components. They result from the hybridisation of GAs/EAs and FSs within the framework of soft computing. Although GFRBSs are the most prominent type of GFS, other kinds of GFSs have been proposed in the literature and successfully applied to real-world problems. This chapter is devoted to specialised approaches in GFSs, namely, genetic fuzzy neural networks, genetic fuzzy clustering and genetic fuzzy decision trees.

The reason behind this chapter is not only to complete the overview on GFSs but in particular that all three approaches are applicable and useful in various design stages of general FRBSs.

10.1 Genetic Fuzzy Neural Networks

Genetic fuzzy neural networks are the result of merging the domains of GAs and NNs previously described in Sec. 3.2. That section presented the integration of GAs with NNs, and of FL with NNs. At this point, we briefly discuss the possibility of augmenting neuro-fuzzy systems with additional genetic or evolutionary learning capabilities, into so-called genetic-neuro-fuzzy systems or genetic fuzzy neural networks*. This section describes those types of systems that belong to the white triangular area in Fig. 3.1 (page 81) that represents the intersection between FL, EC and NNs.

---

*Both terms refer to similar systems and we will use both of them without preferring one particular terminology to the other.

Genetic fuzzy neural networks are usually based on a fuzzy neural network (FNN), typically a fuzzy variant of a feed-forward multi-layered network, which is augmented by an evolutionary learning method. The evolutionary learning component employs any of the techniques described in Sec. 3.2.2, whereas the integrated fuzzy components are those described in Sec. 3.2.1. The resulting feed-forward multi-layered network is distinguished by two properties:

(1) The network incorporates some of the following fuzzy components:

    (a) fuzzy numbers that represent the connection weights,
    (b) fuzzy operations in the nodes of the network, and/or
    (c) fuzzy nodes that represent membership functions.

(2) The learning process integrates some of the following evolutionary techniques:

    (a) GAs for coarse-granularity followed by back-propagation for fine-granularity search,
    (b) GAs that obtain the weights of the NN,
    (c) GAs that adapt the transfer functions of the nodes, and/or
    (d) GAs that optimise the topology of the net.

To illustrate the general ideas, we briefly consider some particular cases.

10.1.1　*Genetic learning of fuzzy weights*

One possible combination—described by Krishnamraju, Buckley, Reilly, and Hayashi (1994)—is an FNN whose connection weights are fuzzy numbers that are adapted by a GA. A three-layered feed-forward NN computes a single output, from either a real-valued input or a symmetric triangular fuzzy number. The system is described by an input fuzzy set $\overline{X}$, a set of k weights ($\overline{W}_i$, $i = 1, \ldots, k$), that connect the single input neuron with the k neurons in hidden layer, and a second set of k weights ($\overline{V}_i$) for the connections between the k neurons in hidden layer and the output neuron. The structure is illustrated in Fig. 10.1.

The connection weights are symmetric triangular fuzzy numbers. The output is computed similar to the signal propagation in standard NNs except that due to the fuzzy weights, real arithmetic is replaced by fuzzy

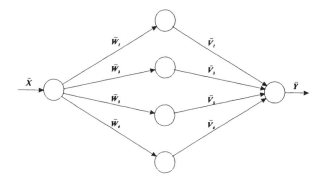

Fig. 10.1   Krishnamraju et al.'s fuzzy neural net

arithmetic. Consequently, the output of the net is computed as:

$$\overline{Y} = f_o \left(\sum_{i=1}^{k} f_h \left(f_i(\overline{X}) \cdot \overline{W}_i \right) \cdot \overline{V}_i \right) \tag{10.1}$$

where f represents the transfer function of the output (f_o), hidden (f_h) and input (f_i) neurons.

The objective of the genetic fuzzy neural system is to determine the fuzzy numbers $\overline{W}_i$ and $\overline{V}_i$ that minimise the error between the output $\overline{Y}_l$ and the desired target value $\overline{T}_l$, over a set of training examples ($l \in \{1, \ldots, L\}$).

The first question is how to define the error measure over the fuzzy sets $\overline{Y}_l$ and $\overline{T}_l$. This approach utilises α-cuts of $\overline{Y}_l$ and $\overline{T}_l$ ($\alpha \in \{0, 0.1, \ldots, 0.9, 1\}$), and computes two error bounds that correspond to the lower and upper extremes of the eleven α-cuts considered:

$$E_1 = \frac{1}{2} \sum_{\alpha} \sum_{l} (y_{l1} - t_{l1})^2 / L$$

$$E_2 = \frac{1}{2} \sum_{\alpha} \sum_{l} (y_{l2} - t_{l2})^2 / L$$

The crisp overall error is computed as the aggregated value $E = E_1 + E_2$ of the lower and upper bound.

The second question is how to code the fuzzy numbers (weights). As in this case the applied fuzzy numbers are symmetric and triangular, it is sufficient to code the extreme values of the support of the fuzzy numbers (two real numbers per weight in the net). Consequently, the GA represents a triangular fuzzy number $(\overline{W}_i)$ by the left and right point (w_{i1}, w_{i3}) (due to the symmetry of the fuzzy numbers, the central point is computed from $w_{i2} = w_{i1} + w_{i3}/2$) and the coding scheme generates chromosomes as:

$$Z = (w_{11}, w_{13}, \dots, w_{k1}, w_{k3}, v_{11}, v_{13}, \dots, v_{k1}, v_{k3}) \quad (10.2)$$

Based on this coding scheme, the network in Fig. 10.1 is defined by sixteen binary-coded real numbers that are optimised through genetic learning.

In a subsequent work, Reilly, Buckley, and Krisnamraju (1996) added a fuzzy bias term to the neurons in the hidden and output layers. With these new terms, Eq. 10.1 is transformed into

$$\overline{Y} = f_o \left(\left(\sum_{i=1}^{k} f_h \left(f_i(\overline{X}) \cdot \overline{W}_i - \Theta_i \right) \cdot \overline{V}_i \right) - \Theta_o \right)$$

Due to the additional elements, the genetic code in Eq. 10.2 is accordingly extended to:

$$Z = (w_{11}, w_{13}, \dots, w_{k1}, w_{k3}, \Theta_{11}, \Theta_{13}, \dots, \Theta_{k1},$$
$$\Theta_{k3}, \Theta_{o1}, \Theta_{o3}, v_{11}, v_{13}, \dots, v_{k1}, v_{k3})$$

10.1.2 Genetic learning of radial basis functions and weights

A Radial Basis Function (RBF) network is, simply stated, a three-layer network that implements a mapping $f : \mathbb{R}^n \to \mathbb{R}^m$ described by:

$$f_k(u) = \sum_{j=1}^{N} \pi_k^j \cdot \phi^j(\|u - \omega^j\|)$$

where N is the number of hidden units, $u \in \mathbb{R}^n$ is the input vector, $\omega^j \in \mathbb{R}^n$ is the centre of the j-th hidden unit, ϕ^j is the radial basis function (transfer function) of the j-th hidden node, and π_k^j is the weight of the connection from the j-th hidden unit to the k-th output node. The most common transfer functions ϕ are Gaussian:

$$\phi^j(u) = \exp\left[-\frac{\|u - \omega^j\|^2}{(\sigma^j)^2} \right]$$

in which the second parameter (σ^j) represents the width of the *receptive field* of the neuron.

Due to the local nature of the neuron activation function, RBF NNs are closely related to FRBSs. In fact, Nie and Linkens (1993, 1995) implemented a simplified TSK FLC using an RBF NN, considering the function ϕ^j as the *IF* part of the j-th rule, and the weight vector (π^j) as the *THEN* part of the rules. Consequently, each neuron in the hidden layer represents a rule premise while the connection weight (from hidden to output neurons) defines the action (consequent) of a singleton rule. An RBF NN is equivalent to an approximate FRBS in which each rule contains the definition of its membership functions.

Several authors (Linkens and Nyongesa, 1995; Seng, Khalid, and Yusof, 1999) take this type of fuzzy neural system as a point of departure and employ a GA to tune the parameters of the net. Linkens and Nyongesa (1995) tuned the n-dimensional receptive fields (represented by its centre and width) of each node, and its output weights. In addition, scaling factors are also coded. These parameters are encoded by concatenating the 31-bit strings that encode an individual neuron. Assuming N neurons in the hidden layer and k neurons in the output layer (output variables), the resulting code has a length of $31 \cdot (2 \cdot n + k) \cdot N$ bits, since it represents the n-dimensional centre and widths of each hidden neuron plus the k weights that connect it to the output neuron.

Seng, Khalid, and Yusof (1999) proposed a slightly different approach where each neuron in the hidden layer constitutes a fuzzy rule whose premise part refers to an n-dimensional vector received from n neurons in the input layer. The major difference with respect to the previous approaches is that a neuron no longer operates with a unique, locally defined receptive field but that different nodes share a common receptive field for the same input variable. That way, the *approximate FRBS* is transformed into a *descriptive FRBS* that employs a common, global term set definition for each input variable. Consequently, the number of neurons in the hidden layer is the product of the number of linguistic terms associated to each input variable[†], whereas the number of parameters to be coded is proportional to the sum of the number of linguistic terms[‡].

The FNN employs Gaussian membership functions and singleton output

---

[†]This is equivalent to the number of rules in the decision table of an FRBS using those linguistic terms.

[‡]The system codes the parameterised linguistic terms, having each of them one or several parameters to be coded.

fuzzy sets. The output of a single rule is computed as the product of the hidden neuron activation multiplied by the connection weight between the hidden neuron and the corresponding output neuron. Similar to the aggregation method in TSK FRBSs, the overall output of the FNN is computed as the weighted sum of the individual rule outputs.

Each Gaussian RBF is represented by the two parameters centre and width. As there exists only one RBF definition per linguistic term, the genetic representation only encodes the common term sets of the input variables. In case of an FNN with two input variables and five linguistic terms per variable, the genetic code contains twenty $(2 \cdot 2 \cdot 5)$ parameters to represent the ten Gaussian RBFs. The connection weights between the hidden and output layer are subject to learning as well. The number of connections equals the product of the number of neurons in hidden and output layers. In the previous case, there are twenty-five connections for a single output network.

The overall genetic code contains the parameters of the Gaussian functions and the connection weights from hidden to output neurons, which are linearly mapped into an eight-bit binary code. The overall FNN is functionally equivalent to the one proposed by Lee and Takagi (1993c), except that the later one uses triangular rather Gaussian membership functions and first order TSK rules rather than simplified ones with singleton outputs. The length of the code, considering L linguistic terms per input variable, becomes $8 \cdot (2 \cdot L \cdot n + k \cdot L^n)$, where L^n is the number of neurons in the hidden layer.

10.1.3 *Genetic learning of fuzzy rules through the connection weights*

A quite common structure for FNNs is the five-layered NN (Narazaki and Ralescu, 1993; Nauck and Kruse, 1992) which is composed of:

(1) input variable layer,
(2) input set term layer,
(3) rule layer,
(4) output set term layer, and
(5) output variable layer.

Figure 10.2 shows an example of this architecture. The nodes in the second and fourth network layer represent the membership functions asso-

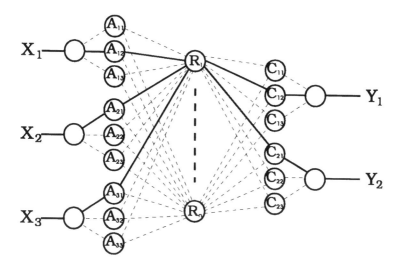

Fig. 10.2 Rule-like fuzzy neural net

ciated to the input and output linguistic variables, whilst the nodes in the third layer encode rules that associate linguistic input terms with output variable terms. The nodes in this layer aggregate their inputs through logical conjunctions as the minimum or product operator. The connections leading into and out of the node R_1, correspond to the rule

R_1 : IF X_1 is A_{12} and ... and X_2 is A_{21} and X_3 is A_{31}
     THEN Y_1 is C_{12} and Y_2 is C_{21},

in which connections in bold lines have a weight value of 1, and connections in dashed lines carry a zero weight.

Chung, Lin, and Lin (2000) employed an FNN of the same type, called Falcon (Fuzzy Adaptive Learning COntrol Network), with trapezoidal membership functions. The connection weights from the third to the fourth layer only assume the values 0 or 1. The learning process involves a three-step hybrid learning algorithm that comprises fuzzy ART, GAs and back-propagation.

The first step in the learning process is the definition of the input and output fuzzy partitions by means of a fuzzy clustering process (see Sec. 10.2.3) using the fuzzy ART algorithm (Carpenter, Grossberg, and Rosen, 1991). In a second step, a GA is applied to learn the weights for the connections between the third and fourth layers, i.e., the consequents of the rules. Considering that those weights are restricted to 0 or 1, a binary code is applied working with classical binary genetic operators. Finally, fine tuning of the membership functions is performed using the back-propagation algorithm.

10.1.4   *Combination of genetic algorithms and delta rule for coarse and fine tuning of a fuzzy-neural network*

The previous section presented a learning approach that generates fuzzy rules through connection weights, but in addition is based on the interesting idea to utilise GAs and back-propagation as subsequent, complementary steps of the learning process. This concept was introduced to the field of FNNs as a remedy to the problem that back-propagation due to its gradient descent search tends to converge to local minima of the error function. The problem of local minima does not only occur in conventional NNs but is inherent to FNNs as well. Consequently, the same idea of augmenting the gradient descent search with an evolutionary global search technique is applicable to FNNs as well.

The general idea is to start the learning process by applying the multipoint search techniques of GAs, to obtain a good initial solution, which is further locally optimised by the back-propagation delta rule. Using common terminology of GAs, this approach reflects the idea to divide the search process into an exploration and an exploitation phase (see Sec. 2.2.1). The role of the GA is to perform a global, exploratory search for a suitable initial starting point, while the back-propagation delta rule is concerned with exploitation of that preliminary solution.

In the work by Chung, Lin, and Lin (2000), the *exploration* and *exploitation* phase are clearly distinguishable since GAs and back-propagation are applied in different stages of the learning process. On the other hand, Ishigami et al. (1995) presented an approach in which the two methods operate in an intertwined rather than purely sequential fashion. The inner loop fine tuning stage that uses the delta rule is an integral part of the outer loop genetic optimisation process. The overall learning process is based on a loop that contains the following steps:

(1) Create the initial population.

(2) Apply genetic crossover and mutation in the coarse tuning stage.

(3) Apply the delta rule in the fine tuning stage.

(4) Evaluate the fine tuned solution according to the fitness function.

(5) Select individuals according to their fitness.

(6) If the desired target error is achieved or a maximum number of generations elapsed, terminate the algorithm; otherwise continue with step (2).

We omit a thorough analysis of the genetic code or genetic operators applied by Ishigami et al. (1995), as the main purpose of this section is to illustrate the idea of a learning process that operates on two levels of granularity: coarse granularity in the evolutionary stage and fine granularity in the gradient descent stage.

10.2 Genetic Fuzzy Clustering

10.2.1 *Introduction to the clustering problem*

Cluster Analysis involves the partitioning of objects into groups (*clusters*) according to a similarity measure or inherent structure of the data. The objective of cluster analysis is to group the sample in such a way that the objects within a cluster are more similar to each other than to those in other clusters (Anderberg, 1973).

Such grouping can be developed by clustering techniques, which are among the *unsupervised* learning methods since the class to which each object belongs is *a priori* unknown. Cluster Analysis has many applications. Not only can it be used for classification and pattern recognition (for example, for image segmentation), but also provides a powerful tool for the reduction of complexity in modelling, exploratory data analysis and optimisation (Babuska, 1998).

From a formal point of view, the clustering problem involves obtaining a partition P of a set of objects (patterns) $E = \{x_1, \ldots, x_l, \ldots, x_N\}$, $x_l = (x_l^1, \ldots, x_l^j, \ldots, x_l^M)$ into c groups. As previously said, an element $P_i \in P$ is called a cluster which is usually characterised by a *prototype* v_i.

There are many possible definitions for a cluster, depending on the clustering objective. The most accepted and generic one is the view that a cluster is a group of objects that are more similar to one another than to members of other clusters (Babuska, 1998), a criterion that is directly derived from the basic objective of clustering. The term *similarity* is usually understood in a mathematical way, thus often defined by means of a *distance norm*. The composition of this norm conditions the cluster shapes and, consequently, the prototype structure. For example, the usual Euclidean distance defines hyper-spherical clusters in the M-dimensional space. Many other shapes such as ellipsoids, lines, rectangles or planar boundaries are obtainable through other distance norms.

Depending on the structure of the partition P, two different kinds of clustering can be distinguished (Bezdek, 1981; Chi, Yan, and Pham, 1996; Babuska, 1998):

- *Hard clustering*: The partition P is a set of disjoint subsets of E in such a way that each object belongs to exactly one cluster.
- *Fuzzy clustering*: Each object belongs to a cluster to a certain degree according to the membership function of the cluster. The partition is fuzzy in the sense that a single object can simultaneously belong to multiple clusters.

The following two subsections are devoted to both types of clustering. The third section deals with the application of EAs to fuzzy clustering, which results in the so-called genetic fuzzy clustering (GFC).

Clustering techniques can also be classified into three groups according to the algorithmic approach (Bezdek, 1981):

(1) *Hierarchical clustering methods*: The clusters are subsequently created in an hierarchical structure by reallocating memberships of one object at a time with respect to the similarity measure. Within this family, one distinguishes between *agglomerative* and *splitting* clustering. Agglomerative clustering starts with a set of N clusters, whose prototypes are identical to the N objects in the sample, and the clustering algorithm iteratively groups them together to form bigger clusters. Splitting clustering starts with a single cluster composed of the entire sample that is iteratively partitioned into smaller clusters.

(2) *Graph-theoretic clustering methods*: The sample is regarded as a weighted graph, where each object comprises a node and the weight associated to an edge that connects two objects represents the similarity among them.

(3) *Objective function-based clustering algorithms*: An objective function is considered to measure the desirability of the partition and

non-linear optimisation algorithms are used to optimise this objective function.

The clustering methods reviewed in the following belong to the latter group, regardless if hard or fuzzy clustering is considered.

10.2.2 *Hard clustering*

Hard clustering involves finding a hard partition of a sample into c disjunctive clusters. A hard partition is defined as a family of c classical subsets $P_i \in P(E)$, $i = 1, \ldots, c$, with the following properties (Bezdek, 1981):

$$\bigcup_{i=1}^{c} P_i = E$$

$$P_i \cap P_j = \emptyset, \quad 1 \leq i \neq j \leq c$$

$$\emptyset \subset P_i \subset E, \quad 1 \leq i \leq c$$

Hence, when considering a hard partition, the hard clustering problem presents the following number of different possible solutions —which is known as the Stirling approximation— (Anderberg, 1973):

$$\frac{1}{c!} \cdot \sum_{i=1}^{c} \left\{ (-1)^{c-i} \cdot \binom{c}{i} \cdot i^N \right\}$$

There are different objective function-based hard clustering methods proposed in the literature, which consider many different optimisation techniques: iterative algorithms, adaptive vector quantisation, self-organising maps, simulated annealing and EAs, among others. In the following, we introduce one of the best known methods, the *hard c-means (HCM) clustering algorithm*, which follows the iterative approach.

(1) *Partition the sample E into c clusters $P_1, \ldots, P_i, \ldots, P_c$ (c is known a priori) following any rule. If no specific rule exists, the patterns are randomly partitioned.*

(2) *Compute the set of cluster prototypes (centres) $V = \{v_1, \ldots, v_i, \ldots, v_c\}$, $v_i = (v_i^1, \ldots, v_i^j, \ldots, v_i^M)$, as follows:*

$$v_i^j = \frac{\sum\limits_{x_l \in P_i} x_l^j}{|P_i|}$$

with $|P_i|$ being the cardinality of P_i, i.e., the number of patterns in
the i-th cluster.

(3) Redistribute the sample by associating each pattern $x_l \in E$ to the
closest cluster P_i according to its distance to the respective centre
v_i. Similarity is measured by a distance norm, commonly using the
Euclidean metric:

$$x_l \in P_i \text{ if and only if } d(x_l, v_i) < d(x_l, v_j), \ j \in \{1, \ldots, k\}, \ i \neq j$$

(4) If the cluster centres do not change, stop; else go to step 2.

Often a cluster validation index is considered to measure the quality of
the sample partition obtained. The most common clustering criterion is:

$$J(V) = \sum_{l=1}^{N} \sum_{x_l \in P_i} d^2(x_l, v_i)$$

The operation mode of the HCM clustering algorithm can be regarded as
a local search procedure that tries to minimise this index. In each iteration,
patterns are reassigned to their closest cluster centre, thus obtaining a new
partition that decreases the value of the index $J(V)$. The termination
condition in step 4 is equivalent to the observation that the value of $J(V)$
does not change from one iteration to the next.

For the final partition, one can compute the variances σ_i, $i = 1, \ldots, c$,
of the cluster centres as follows:

$$\sigma_i^j = \sqrt{\frac{\sum_{x_l \in P_i} (x_l^j - v_i^j)^2}{|P_i|}}$$

These two components, centres and variances, uniquely define an input
space partition.

10.2.3 *Fuzzy clustering*

Fuzzy clustering is based on the fact that objects do not entirely belong to
one particular class but rather acquire a partial degree of membership —
between 0 an 1— to a cluster (Bezdek, 1981; Babuska, 1998). In many cases,
fuzzy clustering provides a more intuitive alternative to hard clustering as it
naturally reflects the notion of an object that is equally similar to multiple
cluster prototypes rather than a unique single cluster. The membership

values play an important role in the clustering process since they allow the classification procedure to be more flexible and robust when dealing with noisy and uncertain data.

Fuzzy clustering is based on the concept of a fuzzy partition, a generalisation of the hard partition. A fuzzy partition U is represented by a fuzzy relation μ_{il} —the fuzzy partition matrix—, $i = 1, \ldots, c$ and $l = 1, \ldots, N$, defined between the set of clusters P and the sample of patterns E, with the following constraints:

$$\mu_{il} \in [0, 1], \qquad 1 \leq i \leq c,\ 1 \leq l \leq N$$
$$\sum_{i=1}^{c} \mu_{il} = 1, \qquad 1 \leq l \leq N$$
$$0 < \sum_{l=1}^{N} \mu_{il} < N, \quad 1 \leq i \leq c$$

There are three main categories of fuzzy clustering, namely: fuzzy clustering based on fuzzy relation, fuzzy clustering based on objective function and generalised k-nearest neighbour rule, which are summarised by Yang (1993). Objective function-based clustering allows the most precise formulation of the clustering criterion. The algorithms that follow this paradigm aim at minimising the objective function

$$J(U, V) = \sum_{i=1}^{c} \sum_{l=1}^{N} \mu_{il}^{m} \cdot d^2(x_l, v_i)$$

with μ_{il} being the membership degree of l-th data x_l to the i-th cluster P_i, $d(x_l, v_i)$ being the distance between the prototype of that cluster, v_i, and the object x_l, and with $m \in [1, \infty)$ being a weighting exponent which determines the fuzziness of the resulting clusters. As m approaches 1, the fuzzy partition moves closer to a hard partition with ($\mu_{il} \in \{0, 1\}$). On the other hand, as $m \to \infty$, the partition becomes maximally fuzzy ($\mu_{il} = \frac{1}{c}$). Typically, one chooses a value of $m = 2$.

The fuzzy clustering algorithms treated in this section are based on an objective function. Perhaps, the best known and most widely used fuzzy clustering algorithm of this type is the Fuzzy C-Means (FCM), developed by Dunn (1974) and extended by Bezdek (1981). The FCM algorithm is an extension of the HCM in that it iteratively minimises a fuzzy version of the HCM cost function. It uses the said index J, based on the Euclidean norm:

$$d(x_l, v_i) = \sqrt{\sum_{j=1}^{M} (x_l^j - v_i^j)^2}$$

Hence, the fuzzy clusters generated by FCM are hyper-spheres centred at the M-dimensional cluster prototypes v_i.

The composition of the FCM algorithm is shown as follows:

(1) *Initialise the fuzzy partition matrix U at random in a way that satisfies the said constraints.*

(2) *Compute the set of cluster centres $V = \{v_1, \ldots, v_i, \ldots, v_c\}$ as follows:*

$$v_i = \frac{\sum_{l=1}^{N} (\mu_{il})^m \cdot x_l}{\sum_{l=1}^{N} (\mu_{il})^m}$$

(3) *Update the membership values to the previously computed cluster centres:*

$$\mu_{il}' = \frac{1}{\sum_{j=1}^{c} \left(\frac{d(x_l, v_i)}{d(x_l, v_j)} \right)^{2/(m-1)}} \qquad i = 1, \ldots, c, \; l = 1, \ldots, n$$

(4) *If $\max\limits_{i,l} | \mu_{il} - \mu_{il}' | \leq \epsilon$ (with ϵ being a predefined tolerance threshold), stop; else continue with step 2.*

As the weighting exponent m approaches 1, the centroids v_i become identical to the mean of the clusters (like in HCM), whilst as m tends to ∞, all centroids converge toward the overall mean of the entire sample E.

Although FCM has been widely applied and performs relatively well, it suffers from the same limitations as HCM (Velasco, López, and Magdalena, 1997):

(1) Since FCM is based on the Euclidean distance, it only identifies hyper-spherical clusters.

(2) The algorithm is not able to calculate the optimal number of clusters, c, which is fixed and must be defined in advance.

(3) Although the FCM algorithm is guaranteed to converge, in essence it remains a local search algorithm that searches for the optimum by using a hill-climbing technique. As all hill-climbing methods, it

might get trapped in local optima and is very sensitive to the initial conditions.

In regard to restriction to hyper-spherical clusters, there are different variants of the FCM algorithm —usually called FCM-type algorithms— to generate clusters of different shapes. As we shall see in the following, the change in the fuzzy cluster shapes is achieved by using a different distance norm d and a more complex representation of the cluster prototypes. In the remainder of this subsection, we analyse different kinds of FCM-type algorithms, depending on the cluster shapes, and describe some examples.

In the next subsection, the applications of EAs to the field of fuzzy clustering are described, which are usually aimed at solving the three said types of problems.

Compact clusters versus shell-type clusters

A fuzzy cluster is *"compact"* (or *"filled"*) if both the cluster boundary as well as the points enclosed by the boundary belong to the cluster prototype. In contrast, *"shell-type"* clusters are only comprised of the boundary structure (circle, ellipse, etc.) itself whereas the interior points do not belong to the cluster prototype.

In view of the previous definitions, the usual FCM algorithm generates compact hyper-spherical clusters. As seen when describing this algorithm, distances to compact clusters are measured between points and cluster centres, whereas distances to shell-type cluster prototypes are measured between points and shells.

FCM-type algorithms for compact clusters

In addition to the compact hyper-spherical clusters generated by the basic FCM, there are other variants of this fuzzy clustering algorithm dealing with compact clusters of different shapes. They can be classified into two groups (Babuska, 1998):

(1) *Algorithms using an adaptive distance norm*, which are based on replacing the classical Euclidean distance considered in FCM by a *transformed Euclidean distance*. The inner-product norm is modified this way. Let the generic squared distance function of index J denoted by:

$$d_A^2(x_l, v_i) = ||x_l - v_i||_A^2 = (x_l - v_i)^T A(x_l - v_i)$$

with A being an $M \times M$ variance matrix for the different directions of the coordinate axes in the M-dimensional sample space.

If A is the identity matrix, one obtains the usual Euclidean distance: $d_A^2(x_l, v_i) = (x_l - v_i)^T(x_l - v_i)$.

However, other variance matrices A will result in different cluster shapes. For example, if all but the diagonal matrix elements in the matrix are set to zero, and the diagonal is set to the variance in the corresponding coordinate axis,

$$A = \begin{bmatrix} (\frac{1}{\sigma_1})^2 & 0 & \cdots & 0 \\ 0 & (\frac{1}{\sigma_2})^2 & \cdots & 0 \\ \cdots & \cdots & \cdots & \cdots \\ 0 & 0 & \cdots & (\frac{1}{\sigma_m})^2 \end{bmatrix}$$

a *diagonal norm* is induced in $\mathbb{R}^M$, and the fuzzy clusters become axis-parallel hyper-ellipsoids.

On the other hand, if A is defined as the inverse of the covariance matrix of the sample E, $A = R^{-1}$, with

$$R = \frac{1}{N} \sum_{l=1}^{N} (x_l - \bar{x})(x_l - \bar{x})^T$$

with $\bar{x}$ being the sample data mean, the *Mahalanobis norm* is induced (Bezdek, 1981), and the clusters become non-axis-parallel hyper-ellipsoids.

The two most known fuzzy clustering algorithms belonging to this family are the *Gustafson-Kessel (GK)* (Gustafson and Kessel, 1979) and the *maximum likelihood estimation* algorithms (Gath and Geva, 1989).

The GK algorithm searches for a set of hyper-ellipsoidal clusters specified by their centre, size and orientation. Each cluster P_i employs its own norm-inducing matrix A_i, which yields to the following adaptive distance norm:

$$d_A^2(x_l, v_i) = (x_l - v_i)^T A_i (x_l - v_i)$$

Hence, cluster prototypes are composed of the cluster centres v_i and the variance matrices A_i, which are also subject to optimisation in the FCM functional J. The variance matrices are thus optimised by the algorithm by means of the Lagrange multiplier method resulting in the iterative update rule:

$$A_i = [\rho_i \, det(F_i)]^{\frac{1}{N}} \, F_i^{-1}$$

with $\rho_i = |A_i|$ ($\rho_i > 0$, $\forall i$) and F_i being the fuzzy covariance matrix of the i-th cluster P_i defined by:

$$F_i = \frac{\sum\limits_{l=1}^{N} \mu_{il}^m (x_l - v_i)(x_l - v_i)^T}{\sum\limits_{l=1}^{N} \mu_{il}^m}$$

Apart from the additional update rule for the variance matrices A_i, the GK algorithm is otherwise identical to the basic FCM. The variance matrices update in the GK algorithm takes place after the new cluster centres are computed (step 2 of the FCM), and before the fuzzy partition matrix is updated (step 3 of the FCM).

(2) *Algorithms based on hyper-planar or functional prototypes.* In this case, prototypes are r-dimensional linear or non-linear subspaces, with $0 \leq r \leq M - 1$. Notice that, in contrast to the compact FCM-type algorithms, the prototypes are defined over a different manifold than the underlying sample space.

Several examples of these algorithms are: Fuzzy C-elliptotypes (Bezdek, Coray, Gunderson, and Watson, 1981), Fuzzy C-varieties (Bezdek, 1981), and fuzzy regression models (Hathaway and Bezdek, 1993). We omit a discussion of these approaches since, to our knowledge, no EA has been applied to fuzzy clustering algorithms of this type. We refer the interested reader to (Babuska, 1998) for an overview on these kinds of FCM-type techniques.

FCM-type algorithms for shell-type clusters

Among the different FCM-type algorithms designed for shell-type clusters, the most prominent ones are the *Fuzzy C-Shells (FCS)* (Davé, 1990) and the *Fuzzy C-Spherical Shells (FCSS)* (Krishnapuram, Nasraoui, and Frigui, 1992). In both cases, the prototypes are composed of the cluster centre v_i and the radius r_i.

The squared distance norm considered in FCS is:

$$d^2(x_l, (v_i, r_i)) = (\|x_l - v_i\| - r_i)^2$$

with $|| \cdot ||$ being the Euclidean norm. Hence, the clusters generated from this algorithm have a circular shell-shape.

As Klawoon (1997, 1998) enunciated, the problem of the FCS algorithm is that this distance norm leads to a set of coupled non-linear equations for v_i and r_i that can not be solved in an analytical way. Thus, an additional numerical iteration procedure to solve non-linear equations —such as the Newton method— is invoked in each iteration of the fuzzy clustering algorithm, which increases the computational burden.

FCCS solves this problem by using the modified distance norm:

$$d^2(x_l, (v_i, r_i)) = (||x_l - v_i||^2 - r_i^2)^2$$

which generates hyper-spherical shells and avoids the need to deal with non-linear equations. However, the problem remains for non-hyper-spherical shell shapes utilised by other fuzzy clustering algorithms of this type.

To minimise the FCM index J, the FCSS algorithm rewrites the previous distance norm as (Krishnapuram, Nasraoui, and Frigui, 1992):

$$d^2(x_l, (v_i, r_i)) = p_i^T \cdot M_l \cdot p_i + z_l^T \cdot p_i + b_l$$

where

$$p_i = \begin{bmatrix} -2v_i \\ v_i^T \cdot v_i - r_i^2 \end{bmatrix}$$

and

$$b_l = (x_l^T \cdot x_l)^2, \qquad z_l = 2(x_l^T \cdot x_l) \cdot y_l, \qquad M_l = y_l \cdot y_l^T$$

with

$$y_l = \begin{bmatrix} x_l \\ 1 \end{bmatrix}$$

Thus, the FCSS algorithm contains the following steps:

(1) Initialise the fuzzy partition matrix U at random in such a way that the constraints are satisfied.

(2) Compute H_i and w_i for each cluster P_i as follows:

$$H_i = \sum_{l=1}^{N} \mu_{il}^m \cdot M_l, \qquad w_i = \sum_{l=1}^{N} \mu_{il}^m \cdot x_l$$

where M_l and z_l are calculated according to the previous expressions.

(3) *Compute p_i for each cluster P_i as follows:*

$$p_i = -\frac{1}{2}(H_i)^{-1}w_i$$

(4) *Update the membership values of the patterns to the previously computed cluster centres in the usual way:*

$$\mu'_{il} = \frac{1}{\sum_{j=1}^{c}(\frac{d(x_l,v_i)}{d(x_l,v_j)})^{2/(m-1)}}, \qquad i = 1,\ldots,c\,;\, l = 1,\ldots,N$$

(5) *If $\max\limits_{i,l} \mid \mu_{il} - \mu'_{il} \mid\, \le \epsilon$, terminate; else continue with step 2.*

For an overview on fuzzy shell clustering, we refer the interested reader to (Krishnapuram, Frigui, and Nasraoui, 1995).

10.2.4 *Different applications of evolutionary algorithms to fuzzy clustering*

There are several references in the literature proposing the use of EAs in fuzzy clustering. The majority is devoted to improve the results of FCM-type algorithms by using the EA to optimise some parameters of the FCM algorithm itself. However, there are other approaches in which the EA directly substitutes the FCM algorithm as the fuzzy clustering problem-solving technique. In the following we classify the existing approaches. The remaining subsections briefly introduce some of the methods.

A) Use of the EA to optimise the parameters of an FCM-type algorithm

These GFC processes are characterised by solving the fuzzy clustering problem in combination with any kind of FCM-type algorithm. Two main approaches can be distinguished according to the FCM-type algorithm parameter subject to optimisation:

- *Prototype-based GFC* encodes the fuzzy cluster prototypes V, which are evolved by an EA guided by a centroid-type objective function, i.e., a measure that computes the quality of the clustering performed from the cluster centres and the sample. This objective function is usually some variant of the FCM index, J, introduced in

Sec. 10.2.3. The value of the index for a specific set of prototypes is computed based on the fuzzy partition U, which is calculated from the cluster centres according to the expression in step 3 of the FCM algorithm. The majority of the GFC approaches belongs to this group.

- *Fuzzy partition-based GFC.* In this case, the individuals encode the fuzzy partition matrix U, and the cluster prototypes V are directly computed according to the expression in step 2 of the FCM algorithm. This approach is less popular than prototype-based GFC as the number of parameters grows linearly with the number of patterns in the sample.

Notice that, similar to the basic FCM algorithm, the GFC approach assumes that the number of clusters is known a priori. The EA provides a remedy to the problem of standard FCM algorithms, the possibility of getting trapped in local optima and the sensitivity of the obtained solution on the initial conditions. Since EAs are global search techniques, the GFC algorithm is less sensitive to initial conditions and is more likely to avoid local minima.

On the other hand, the GFC algorithm generalises to other type of fuzzy clustering algorithms such as FCS that do not employ an Euclidean distance norm without a need of a significant change in the EA. In this case, the resulting GFC algorithm is able to generate non hyper-spherical fuzzy clusters for which the index J depends in a non-linear fashion on the cluster prototype parameters.

B) Use of the EA to define the distance norm

Methods in this group consider an adaptive distance metric such as those analysed in the previous section. An EA learns the parameters of the distance function in order to achieve an optimal performance of the FCM-type algorithm for a given data sample.

As seen in Sec. 10.2.3, a parameterised distance norm changes the cluster shapes. Therefore, these approaches provide a solution to the first and third problem within FCM, namely global search and non hyper-spherical clusters.

C) Use of the EA to search for a complete solution to the fuzzy clustering problem regardless of the FCM-type algorithm

The genetic approaches belonging to this group directly solve the fuzzy clustering problem without utilising an FCM-type algorithm for computing the new cluster partition. The cluster parameters are directly encoded and the role of the EA is to evolve the optimal fuzzy partition.

These methods are distinguished by two characteristic features: they are able to automatically determine the optimal number of clusters and they deal with clusters of arbitrary shape. Therefore , they simultaneously address all of the three said FCM problems.

For the sake of clarity, we refer to approaches in this group as *"pure GFC"*, in contrast to methods that are integrated with a conventional FCM-type algorithm and therefore go by the name *"hybrid GFC* approaches".

10.2.4.1 *Prototype-based genetic fuzzy clustering*

Prototype-based GFC is the most extended variant of GFC. To our knowledge, it was originally proposed by several authors independent of each other (Babu and Murty;Bezdek and Hathaway;Hall, Bezdek, Boggavarapu, and Bensaid;Schulte) around 1994.

Prototype-based GFC algorithms usually employ the classical FCM algorithm, i.e., they deal with hyper-spherical clusters (Babu and Murty, 1994; Bezdek and Hathaway, 1994; Hall et al., 1994, 1995, 1998, 1999; Schulte, 1994; Liu and Xie, 1995; Van Le, 1995; Nascimiento and Moura-Pires, 1996, 1997; Egan, Krishnamoorthy, and Rajan, 1998). In this case, the EA evolves chromosomes that encode the c cluster centres v_i. Each individual cluster centre is represented by an M-dimensional binary-coded (Bezdek and Hathaway, 1994; Hall et al., 1994; Van Le, 1995; Egan et al., 1998), Gray-coded (Hall et al., 1995, 1999; Liu and Xie, 1995) or real-coded (Babu and Murty, 1994; Schulte, 1994; Nascimiento and Moura-Pires, 1996, 1997; Klawoon, 1997, 1998) vector. The partial representations of the c prototypes are joined to form the chromosome encoding the entire prototype set.

However, several approaches utilise other fuzzy cluster shapes and thus integrate one of the other FCM-type algorithms introduced in the previous section. In these cases, the genetic representation becomes more complex, as it encodes additional parameters of the cluster prototypes. For example, the proposals introduced by (Schulte, 1994; Klawoon, 1997, 1998) operate with circular shell clusters as they employ the FCS algorithm. Therefore, the prototype representation includes both the cluster centre v_i and the radius r_i. On the other hand, Klawoon (1998) also consider compact rectangular clusters, in which case the genetic code in addition incorporates the diagonal matrices B_i.

In most approaches, the fitness function minimises the FCM index J in one way or another. In order to compute the index J, the cluster centres are decoded from the chromosome and the corresponding fuzzy partition matrix U is derived according to step 3 of the FCM-type algorithm (see Sec. 10.2.3).

In the following, we describe two different approaches within this group. The former, proposed by Hall, Ozyurt, and Bezdek (1998, 1999), serves as a prototypical example, whereas the proposal of Nascimiento and Moura-Pires (1996, 1997) contains several unique features that distinguish it from other GFC algorithms in the same class.

Hall et al.'s prototype-based GFC algorithm
The GFC algorithm proposed by Hall, Ozyurt, and Bezdek (1999) comprises the following components:

- *Representation:* A chromosome encodes a set of c cluster centres v_i, $i = 1, \ldots, c$, each one of them encoded in an M-dimensional Gray-coded array.
- *Fitness function:* The adaptation function

$$R_m(V) = \sum_{l=1}^{N} \left\{ \sum_{i=1}^{c} d(x_l, v_i)^{\frac{1}{1-m}} \right\}^{1-m}$$

  decreases the runtime of the GFC algorithm as it directly computes the fitness of a set of cluster centres and thereby bypasses the intermediate step of first computing the fuzzy partition matrix as in other FCM-type algorithms.
- *Selection mechanism:* Tournament selection is considered as presented in Sec. 2.2.3.3.
- *Crossover operator:* A particular crossover operator, the c-fold (with c being the number of clusters) two-point crossover, is specifically designed for this problem. One crossover operation is applied to each binary sub-string which represents an individual cluster.

Hall, Ozyurt, and Bezdek (1999) claimed that the algorithm converges faster since each cluster centre is adapted independently of the others.

- *Mutation operator:* The standard uniform bit-flip mutation operator is used.

Nascimiento and Moura-Pires' prototype-based GFC algorithm

Nascimiento and Moura-Pires (1997) proposed a prototype-based GFC algorithm with the following characteristics:

- *Representation:* A real-coded chromosome represents the M components of each cluster centre.
- *Fitness function:* The authors proposed two different fitness functions: the usual centroid-type one, i.e., the minimisation of the index J (implemented as the maximisation of $\frac{1}{J}$), and the cluster validity measure used in the GFC method proposed by Yuan, Klir, and Swan-Stone (1995), which is entirely based on the fuzzy partition itself rather than to rely on the prototypes:

$$F(V) = \frac{1}{N} \cdot \sum_{l=1}^{N} \sum_{i=1}^{c} \mu_{il}^2$$

Notice that, in the latter case, a non centroid-type fitness function is used in a prototype-based GFC method. This is only possible due to several special characteristics of this genetic FCM algorithm, which are described next.

- *Selection mechanism:* The standard roulette wheel selection is used.
- *Crossover operator:* The simple one-point crossover operator is used. Moreover, an additional reordering operator is applied to the parent strings prior to the crossover. The reordering operator aligns the cluster centres encoded in the parent chromosomes according to a distance norm. Therefore, cluster centres encoded by the same segment in the two parent chromosomes are located nearby in sample space. This way, the operator provides a remedy to the familiar "competing conventions" problem known in genetic-neuro systems, namely that the same set of clusters (weights in the NN case) is encoded in a different order by the two parent chromosomes.

- *Mutation operator:* An application-specific mutation operator, the *Distance-Based Mutation*, is considered. This operator is distinguished by the fact that it modifies the fuzzy partition matrix rather than the genotype that encodes the cluster centres. Three steps comprise it:

  (1) The fuzzy partition matrix U is computed from the cluster centres encoded in the chromosome.
  (2) A cluster and a data point are randomly picked for mutation. The membership value of the data point is randomly changed depending on its distance to all cluster centres. The probability of changing the membership of a pattern belonging to a certain cluster increases with the proximity of the pattern to the corresponding cluster centre.
  (3) A new cluster centre list is computed according to the mutated fuzzy partition U'.

- *Genetic local search:* As mentioned in Chapter 8, genetic local search processes (Suh and Gutch, 1987; Jog, Suh, and Gutch, 1989) refine the individuals obtained in each generation of the GA by means of a local search procedure.

  The proposal by Nascimiento and Moura-Pires (1997) follows this concept as the underlying FCM algorithm constitutes the local search technique. The GA not only optimises the FCM parameters but the approach in fact constitutes a truly hybrid genetic FCM algorithm. Crossover and mutation generate a new set of cluster centres, which is used to initialise the FCM algorithm, which then iteratively improves the cluster centre configuration and the resulting fuzzy partition.

10.2.4.2 *Fuzzy partition-based genetic fuzzy clustering*

Coding the fuzzy partition matrix requires substantially more parameters than prototype coding. Hence, the dimension of the search space and thereby the complexity of the optimisation problem are significantly increased. Due to this reason, only a few contributions have been made in this group of GFC algorithms. In the following, we briefly review two of them.

Van Le's fuzzy partition-based GFC algorithm

Van Le (1995) proposed an EP algorithm (see Sec. 2.3.2) to generate fuzzy partitions with hyper-spherical fuzzy clusters in combination with the classical FCM algorithm. The representation scheme directly encodes the $c \times N$ fuzzy partition matrix U in an $(c \cdot N)$-dimensional real-coded array.

The algorithm contains the following steps:

(1) Randomly generate the individuals $U^k = \mu_{il}^k$, $k = 1, \ldots, \alpha$, in the initial population. Determine the set of centres $V^k = \{v_1^k, \ldots, v_c^k\}$ from the fuzzy partitions U^k and evaluate each individual according to the FCM index J.

(2) For each $k = 1, \ldots, \alpha$, generate one offspring $U^{\alpha+k}$ mutating each individual component μ_{il}^k, $i = 1, \ldots, c$, $l = 1, \ldots, N$, as follows:

$$\mu_{il}^{\alpha+k} = \frac{1}{S} \cdot \mu_{il}^k \cdot e^{-d(x_l, v_i^k)}$$

with $S = \sum_{h=1}^{c} \mu_{hl}^k \cdot e^{-d(x_l, v_h^k)}$.

(3) Evaluate each offspring by determining the set of centres $V^{\alpha+k}$ and computing the value of J.

(4) Select the α best individuals from the $2 \cdot \alpha$ candidates (parents + offspring) to form the new population.

(5) If the termination condition is satisfied, stop; else go to step 2.

As usual, the algorithm minimises the FCM index J. However, Van Le suggests an alternative objective function that maximises the following fuzzy partition-based measure:

$$\sum_{l=1}^{N} \sum_{i=1}^{c} \mu_{il}^2$$

Notice that the average of this index over the number of patterns in the sample N has been also used in the prototype-based hybrid genetic FCM algorithm proposed by Nascimiento and Moura-Pires (1997) (see Sec. 10.2.4.1) and in the distance-based one introduced by Yuan, Klir, and Swan-Stone (1995) (that will be presented in Sec. 10.2.4.3).

Zhao et al.'s fuzzy partition-based GFC algorithm

The approach by Zhao, Tsujimura, and Gen (1996) employed a real-coded GA, using the same representation scheme as the previous EP algorithm.

The basic FCM is considered, the fitness function is composed of the index J and fitness scaling is used.

The chromosomes in the initial population are generated in such a way that they satisfy the boundary constraints associated to fuzzy partitions. The initialisation process is shown as follows:

> *(1) Generate c random numbers $r_{1l}, \ldots, r_{cl}$ in [0,1] for the l-th membership array —the one associated to object x_l— of the chromosome.*
>
> *(2) Calculate $\mu_{il} = \frac{r_{il}}{\sum_{j=1}^{c} r_{jc}}$ for $i = 1, \ldots, c$. The obtained μ_{il} satisfies the first two conditions of a fuzzy partition (see Sec. 10.2.3).*
>
> *(3) Repeat steps 1 and 2 n times for $l = 1, \ldots, n$ in order to generate the entire chromosome.*
>
> *(4) If the generated chromosome does not satisfy the third condition (i.e., it contains empty clusters, namely clusters to which every pattern has a membership degree of 0), return to step 1.*
>
> *(5) Repeat steps 1 to 4 until enough chromosomes are generated for the initial population.*

The approach uses the standard roulette wheel selection and the real-coded arithmetical crossover. The mutation operator is specific in order to generate meaningful descendents that satisfy the usual constraints imposed on a fuzzy partition. A column vector of the fuzzy partition matrix —the cluster membership values of a particular pattern— is randomly selected. The membership vector is replaced by a set of new values generated by means of step 1 and 2 of the above initialisation routine.

10.2.4.3 *Genetic fuzzy clustering by defining the distance norm*

The EA in GFC processes within this family optimises the parameters that define the distance norm utilised by the fuzzy clustering algorithm. In particular, the EA adapts the variance matrix A that defines the non-hyper-spherical distance norm described in Sec. 10.2.3. To evaluate a chromosome, the FCM algorithm is invoked using the parameterised distance norm encoded in the chromosome. The fitness of the encoded set of distance parameters is computed from the fuzzy partition generated by the FCM algorithm.

Yuan et al. (1995) proposed an (1+1)-ES (see Sec. 2.3.1) and a diagonal variance matrix A with $a_{ij} = 0$ for $i \neq j$. Therefore, the chromosome encodes the M diagonal matrix elements in a real-coded vector

$w = (w_1, \ldots, w_M)$ resulting in the distance norm:

$$d(x_l, v_i) = \sqrt{\sum_{j=1}^{N} w_j \cdot (x_l^j - v_i^j)^2}$$

The EA minimises the index measure

$$1 - \frac{1}{N} \cdot \sum_{l=1}^{N} \sum_{i=1}^{c} \mu_{il}^2$$

computed from the fuzzy partition U generated by the FCM algorithm.

The (1+1)-ES operates in the usual manner. The offspring is generated from the parent by adding a normally distributed random variable to the objective vector. The better of the two individuals becomes the new parent in the next iteration of the ES.

10.2.4.4 *Pure genetic fuzzy clustering*

Pure GFC processes directly solve the fuzzy clustering problem by searching the space of possible fuzzy cluster sets, rather than to invoke an FCM algorithm to identify the fuzzy partition.

Moreover, pure GFC approaches usually deal with arbitrary cluster shapes (such us triangles, trapezoids or clusters with an infinite support) and automatically learn the optimal number of clusters, which is also encoded in the chromosome.

Some GFC algorithms in this group are to be found in (Fogel, 1993; Turhan, 1997; Velasco, López, and Magdalena 1997, 1998, 2000; Buckles et al., 1994, 1995; Burdsall and Giraud-Carrier, 1997). The remainder of this section briefly illustrates the behaviour of three of them.

Turhan's pure GFC algorithm

Turhan (1997) considered hyper-triangular clusters obtained from the combination of individual triangular membership functions. The membership of a pattern x_l to a hyper-triangular cluster, $\mu_i(x_l)$, is computed as the average of the membership degrees of its elements to the corresponding individual fuzzy set, $\mu_i^j(x_l^j)$:

$$\mu_i(x_l) = \frac{1}{M} \sum_{j=1}^{M} \mu_i^j(x_l^j)$$

Each individual fuzzy set is described by three parameters, (m_l, m_r, D), with D being the centre of the triangle, and m_l and m_r being the left and right points, respectively. These parameters are binary-coded and the entire chromosome for an M-dimensional triangular fuzzy cluster is obtained by joining the partial chromosomes of the M individual fuzzy sets. Consequently, the entire chromosome of a set comprised by c fuzzy clusters is obtained by joining the partial representations of the c M-dimensional fuzzy clusters. Due to the fixed-length coding scheme, this GFC approach is not able to automatically obtain the number of clusters.

The approach employs a binary GA with a steady-state replacement scheme. In each generation, new offspring are generated, which replace those members of the current population that have a lower fitness than the offspring. Uniform mutation and n-point crossover are employed to generate new variants.

The fitness function to be maximised uses the same fuzzy partition-based measure considered in (Van Le, 1995):

$$\sum_{l=1}^{N} \sum_{i=1}^{c} \mu_{il}^2$$

subject to the usual constraints of fuzzy partition.

Moreover, a *penalty* criterion penalises chromosomes that encode so-called "underdeveloped clusters", which contain a small number of patterns.

López et al.'s pure GFC algorithm

López, Magdalena, and Velasco (1997, 1998) proposed a GFC algorithm to derive trapezoidal-shaped fuzzy clusters for one-dimensional problems. In a subsequent work (López, Magdalena, and Velasco, 2000) the authors introduced an extension dealing with multi-dimensional clustering.

The coding scheme only encodes the cut points between two successive clusters and an additional parameter specifying the fuzziness of the fuzzy partition that varies from triangular to rectangular fuzzy membership functions. All parameters are stored in a real-coded vector. The three restrictions associated to fuzzy partitions are easily satisfied.

The GFC algorithm also identifies the optimal number of clusters. A niching GA (see Sec. 2.2.3.4) induces the formation of competing sub-populations (species) in the genetic population, each one of them containing individuals with a different number of clusters.

Initially, the population is sampled with the same number of chromosomes in each sub-population. During evolution, sub-populations with above average better fitness are allocated a larger number of individuals, at the cost of below average sub-populations for which the number of individuals is reduced. The overall population size remains fixed and a lower and an upper thresholds for the number of individuals per sub-population are considered to avoid the extinction or domination of sub-populations. The number of individuals in a sub-population is computed according to the following steps:

(1) The fitness of the best individual in each sub-population measures the species strength.

(2) The mean of the species strengths is computed.

(3) The number of individuals allocated to a sub-population in the next generation is calculated. Sub-populations of above (below) average strength acquire a larger (smaller) number of individuals.

The fitness function considered involves maximising:

$$\frac{1}{1 + g(c) \cdot J}$$

in which c is the number of clusters in the fuzzy partition encoded in the chromosome, $g(c)$ is a functional that penalises an excessively large number of clusters and J the usual FCM index.

The following GA is applied to each sub-population:

(1) First, a group with the best individuals in the population, the elite, is preserved.

(2) An intermediate population of size the number of individuals in the population less the number of elite individuals is obtained from the current population by means of the classical roulette wheel selection scheme.

(3) An offspring population is generated by applying uniform crossover (each descendant takes genes from his two parents with probability 0.5) and soft mutation (operator described in Sec. 6.3.2) to the individuals in the intermediate population.

(4) The individual in the intermediate population that is substituted by the new offspring is chosen by means of similarity corrected roulette

wheel selection in order to preserve the diversity of the genetic material in the population.

(5) *A hard mutation operator, which changes an existing chromosome by a randomly generated one, is applied to the intermediate population with a very low probability in order to introduce diversity.*

(6) *Finally, the new population is built by joining the said elite individuals to the current intermediate population.*

Buckles et al.'s pure GFC algorithm

The GFC algorithm introduced by Buckles et al. (1994, 1995) operates with hyper-ellipsoidal clusters. A variable-length binary GA evolves the hyper-ellipsoids, which are encoded by the following set of parameters:

- An M-dimensional vector for the ellipse centre.
- An M-dimensional vector for the axis lengths.
- An $\frac{M \cdot (M-1)}{2}$-dimensional array for the orientation of each axis.

For the particular two-dimensional problems presented in the papers, the chromosome contains five parameters that encode the hyper-ellipsoidal cluster: two two-dimensional vectors for the centre and the axis lengths, and another parameter for the orientation of the ellipsoid axis.

In principle, these parameters define a crisp cluster. In order to determine the fuzzy membership value of a pattern x, the intersection between the ellipse P_i and the line connecting its centre and the pattern, noted as p_e, is computed. Let r_o and r_e be the distances of the pattern x and the intersection point p_e to the centre of the ellipse. The membership value of the pattern x to the i-th cluster is computed as:

$$\mu_i(x) = \begin{cases} 1, & \text{if } r_o \leq r_e \\ e^{\frac{-(r_e-r_o)^2}{\beta^2}}, & otherwise \end{cases}$$

in which β is a parameter that defines the slope of the membership function.

The entire chromosome is composed of a variable number of binary-coded hyper-ellipsoidal cluster representations. Initially, every individual contains a single cluster, number which is subsequently increased by the following set of genetic operators:

- *Crossover:* The crossover operator (*module crossover*) is adopted to be compatible with variable-length chromosomes. Two chromosomes of lengths l_1 and l_2 —both lengths are multiples of the code length for a single cluster— are crossed. A random integer number r is drawn with uniform probability from $\{1, \ldots, l_{max}\}$ —with $l_{max} \gg l_1$ and $l_{max} \gg l_2$— and both parents are crossed, each in a different crossover site, with $c_1 = (r \mod l_1)$ being the crossover point in the first parent is and $c_2 = (r \mod l_2)$ in the second.

- *Mutation:* The common uniform mutation operator is applied to the variable-length chromosomes.
- *Insertion:* The role of the insertion operator is to augment a chromosome with additional clusters. The selection mechanism determines a parent (donor), from which a chromosome segment containing several complete cluster representations is randomly chosen. This donor segment is inserted at a randomly chosen location in the original chromosome.
- *Deletion:* The deletion operator removes a randomly chosen segment from a chromosome.

The proposed approach is targeted at supervised classification problems. Hence, the fitness function is a specific classification error measure and the clusters encoded in each chromosome carry a class label. An additional penalty term promotes the formation of ellipsoids of small size.

10.3 Genetic Fuzzy Decision Trees

10.3.1 *Decision trees*

Decision trees are suitable for classification problems in which instances are represented by attribute-value pairs and the target concept has discrete values. A decision tree classifies an example by propagating it along a path from the root node down to a leaf node which contains the classification for this example. Each node tests a particular attribute, each of its branches corresponds to a particular value of this attribute.

Figure 10.3 shows a typical decision tree that predicts if a customer qualifies for credit based on its annual income and employment status. Each instance is classified according to the attributes *income*={*small, medium, large*} and *employment*={*none, part time, full*}. The example *employment=full, income=large,* is sorted down the two right branches and categorised as a positive example according to the label at the leaf node. Notice that a path to a leaf node does not necessarily use every attribute to ob-

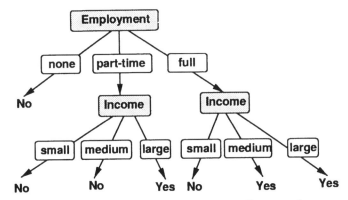

Fig. 10.3   Decision tree for customer credit approval

tain a classification. All customers with no employment are classified as negative examples, regardless of their income.

The next question is how to generate a decision tree from a set of training examples. Most approaches are based on the ID3 or C4.5 learning algorithm proposed by Quinlan (1986, 1993). ID3 constructs decision trees in a top-down manner, at each node selecting the attribute that best discriminates the training examples. It then creates a descendant node for each possible value of the attribute. The training examples are distributed among the nodes according to their value. This procedure is repeated until either all examples at a node carry the same target classification, or the attributes are exhausted.

What is the best attribute to test at each node? Intuitively, it makes sense to choose an attribute that separates positive and negative examples well. This quality can be measured by a statistical property, the so-called *information gain*. The algorithm selects the attribute that provides the maximal information gain. Information gain is based on the entropy of a set of examples, a concept from information theory that measures how cluttered the examples are. Assume a set of examples S that contains instances of N possible target values. The entropy of S relative to the target concept is defined by

$$E(S) = -\sum_{i=1}^{N} p_i \cdot \log_2 p_i \qquad (10.3)$$

where p_i is the proportion of examples with classification i. Notice that the entropy has a maximum when each target value appears equally often $p_i = 1/N$. The information gain measures the reduction in entropy when the examples are partitioned based on an attribute. The information gain for an attribute A with values a_j is defined as

$$G(S, A) = E(S) - \sum_j \frac{|S_j|}{|S|} \cdot E(S_j)$$

where S_j is the set of examples with attribute value a_j, $|S_j|, |S|$ are the number of instances in S_j and S, and $E(S_j)$ is the entropy of the examples in S_j. The next split occurs along the attribute A^* which maximises the information gain

$$A^* = \text{argmax}_A G(S, A)$$

Suppose our training set contains the seven examples of customer credit approval in Table 10.1.

Table 10.1 Training examples for decision tree learning

#	employment	income	credit
D1	none	medium	no
D2	none	large	no
D3	part-time	large	yes
D4	part-time	medium	no
D5	full	medium	yes
D6	full	small	no
D7	full	large	yes

The original entropy in regard to the target concept credit is

$$
\begin{aligned}
E(S) &= -p_{yes} \cdot \log p_{yes} - p_{no} \cdot \log p_{no} \\
&= -3/7 \cdot \log_2 3/7 - 4/7 \cdot \log_2 4/7 = 0.985
\end{aligned}
$$

The information gains for the attributes *income* and *employment* are computed as

$$
\begin{aligned}
G(S, \text{income}) &= E(S) - \sum_{j=\text{small,medium,high}} |S_j|/|S| \cdot E(S_j) \\
&= 0.985 - 1/7 \cdot 0 - 3/7 \cdot 0.91 - 3/7 \cdot 0.91 = 0.205
\end{aligned}
$$

$$G(S, \text{employment}) \;=\; E(S) - \sum_{j=\text{none,part-time,full}} |S_j|/|S| \cdot E(S_j)$$

$$= \; 0.985 - 2/7 \cdot 0 - 2/7 \cdot 1.00 - 3/7 \cdot 0.91 = 0.309$$

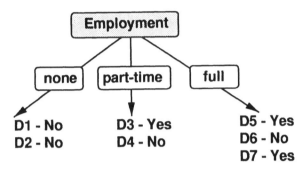

Fig. 10.4   Root node using employment as the attribute to split the tree

The attribute employment has the higher information gain and is therefore used at the root node as depicted in Fig. 10.4. For node with attribute value *none*, both examples $D1, D2$ carry the same target classification. Therefore, it becomes a leaf node and is assigned the target value *no*. The other two nodes have no unique classification and the examples are split using the remaining attribute *income*. The learnt tree is identical to the one shown in Fig. 10.3. There exists no example for the attribute values *employment=part-time, income=small*. In that case, the leaf node acquires the most common classification of its parent node. Since the parent node contains one positive and one negative example, the tie is resolved by a coin flip, in this case with the classification *no*. For more information on decision tree learning, the interested reader is referred to (Quinlan, 1993).

10.3.2   *Fuzzy decision trees*

The basic idea of fuzzy decision trees is to combine example-based learning in decision trees with approximate reasoning of FL (Janikow, 1998). This hybridisation integrates the advantages of both methodologies, the compact knowledge representation of decision trees with the ability of FSs to process uncertain and imprecise information. The learning and inference mechanism of standard decision trees are modified in order to reflect the fuzziness of instances.

In standard decision trees, instances are partitioned in a binary fashion, whereas a node in a fuzzy decision tree distributes instances across multiple branches. Instead of a sharp decision boundary, attribute values become fuzzy and overlap. For example, an applicant with an income of $20000 might belong to the category *income=small* with a membership degree of $\mu_{small} = 0.3$. The target classification also becomes fuzzy, for instance a training example might look like *income=$35000, employment=30, credit=0.6* stating that a person with an annual income of $35000 who is employed 30 hours a week, qualifies for credit to the degree 0.6. We first describe the process to generate a fuzzy decision tree from data and then discuss the inference mechanism.

The major difference to binary decision trees is that an example belongs to a node to a certain degree. Instead of counting the number of examples at a particular node, fuzzy decision trees aggregate their degree of membership. The proportion P_i^N of examples $D_j \in S^N$ with classification i at node N is computed as

$$P_i^N = \sum_{S^N} \min \left(\mu_N(D_j), \mu_i(D_j) \right) \tag{10.4}$$

where $\mu_N(D_j)$ is the membership degree of an example D_j at node N and $\mu_i(D_j))$ is the membership degree of the example in regard to the target value i. We like to point out that the min-operator used for conjunction in Eq. 10.4 can be replaced by any t-norm. Further, let $P^N = \sum_i P_i^N$ denote the overall example count at node N. Standard decision trees calculate the proportion of examples p_i with classification i. In a similar way, fuzzy decision trees measure the membership degree P_i^N/P^N of examples at a particular node. Therefore, the entropy of the examples S at node N is defined by analogy with Eq. 10.3 by

$$E(S^N) = -\sum_i P_i^N/P^N \cdot \log_2 P_i^N/P^N \tag{10.5}$$

where the ratio P_i^N/P^N counts the examples with classification i over the overall number of examples. The next step of the fuzzy decision tree building algorithm computes the entropy for a split along attribute A with values a_j

$$E(S^N, A) = -\sum_j P^{N|j}/P^N \cdot E(S^{N|j})$$

where $N|j$ is the successor node of N following the attribute value a_j. The algorithm selects the attribute A^* which maximises the information gain

$$G(S^N, A) = E(S^N) - E(S^N, A)$$
$$A^* = \operatorname{argmax}_A G(S, A) \qquad (10.6)$$

The node N is split into as many sub-nodes $N|j$ as there are attributes a_j. The degree of membership of example D_k at node $N|j$ is computed incrementally from node N as

$$\mu_{N|j}(e_k) = \min\left(\mu_{N|j}(D_k), \mu(D_k, a_j)\right) \qquad (10.7)$$

where $\mu(D_k, a_j)$ indicates how well D_k matches the attribute a_j. A sub-node $N|j$ is removed if no example has a degree of membership larger than zero. The algorithm terminates either when all examples at a node have the same classification, or when all attributes are already used for splits.

Figure 10.5 shows a fuzzy decision tree for the examples listed in Table 10.2 and the fuzzy sets for the two variables *employment* and *income*. The numbers indicate to which degree an example belongs to a particular node and to which degree examples are classified as *no* or *yes*.

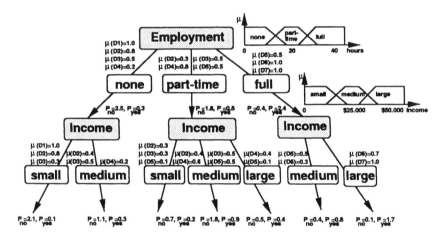

Fig. 10.5 Fuzzy decision tree for customer credit approval

The examples D_1, D_2, D_3, D_4 match the node with the attribute *none* for the criteria *employment* to the degrees $\mu_{none}(D_1) = 1.0$, $\mu_{none}(D_2) = 0.8$, $\mu_{none}(D_2) = 0.5$, $\mu_{none}(D_3) = 0.2$. The proportion of positive and

Table 10.2 Training examples for fuzzy decision tree learning

#	employment in hours	income in $	credit
D1	0	$10.000	0.0
D2	10	$15.000	0.0
D3	15	$20.000	0.1
D4	20	$30.000	0.3
D5	30	$25.000	0.7
D6	40	$35.000	0.9
D7	40	$50.000	1.0

negative examples at this node is computed as:

$$P_{no} = \min(1.0, 1.0) + \min(0.8, 1.0) + \min(0.5, 0.9) + \min(0.2, 0.7) = 2.5$$
$$P_{yes} = \min(1.0, 0.0) + \min(0.8, 0.0) + \min(0.5, 0.1) + \min(0.2, 0.3) = 0.3$$

The degree of membership of an example D_i for the other nodes is calculated in the same way. Since no example matches the classifications *(employment=none, income=large)* and *(employment=full, income=small)* to a degree larger than zero, the corresponding nodes are discarded in the tree. The proportions P_{yes} and P_{no} indicate to which degree this node favours approval or denial of customer credit. The node *(employment=none, income=small)* for example strongly favours *no*, the node *(employment=full, income=large)* supports *yes*, and the node *(employment=part-time, income= large)* shows no clear preference for either decision.

The fuzzy representation requires a modified inference mechanism for fuzzy decision trees. The inference result is a number in the interval $[0, 1]$, rather than a crisp decision for *no* or *yes*. According to its risk willingness, the creditor chooses a threshold to decide upon credit approval. Assume, we want to classify a new example D_j with the attribute values *employment=25, income=$30.000*. The inference scheme first computes the membership of the example within each leaf node as shown in Table 10.3.

Now we compute the support for both possible decisions *no* and *yes*. In general the inference mechanism utilises the centre of gravity χ_k of output fuzzy set k. In our particular case, the decision fuzzy sets are singletons with $\chi_{no} = 0.0$ and $\chi_{yes} = 1.0$.

There are multiple suggestions how to aggregate the information con-

Table 10.3 Degree of membership to leaf nodes for the example *employment=25, income=$30.000*

employment	income	μ
none	small	0.0
none	medium	0.0
part-time	small	0.0
part-time	medium	0.6
part-time	large	0.4
full	medium	0.3
full	large	0.3

tained in the leaf nodes (Janikow, 1998). In the following we describe the two most common inference procedures. The first method only considers the most frequent classification within a leaf. In the second method, all classifications contribute to the decision process. In addition, one can weight the contribution of a leaf node l by its example count P_k^l, such that the inference pays more attention to leaf nodes that have a large number of examples. That way, the classification becomes more reliable in case of a few noisy or atypical examples.

In the first approach, each leaf node l contributes the centre χ_{max} of the fuzzy set with the most common classification. In our example, the two leaf nodes with *employment=part-time* vote for *no* since $P_{no} > P_{yes}$. On the contrary, the two nodes with *employment=full* favour a *yes*-decision. Each vote is weighted by the matching of the example $\mu_l(D_j)$ with the particular node as computed by means of Eq. 10.7 and shown in Table 10.3.

$$\delta_j = \frac{\sum_l \mu_l(D_j) \cdot \chi_{max} \cdot \mu_l(D_j)}{\sum_l \mu_l(D_j)}$$

For the given example, the classification results in

$$\delta_j = \frac{(0.6 \cdot 0.0 + 0.4 \cdot 0.0 + 0.3 \cdot 1.0 + 0.3 \cdot 1.0)}{(0.6 + 0.4 + 0.3 + 0.3} = 0.6/1.6 = 0.375$$

Assume a threshold of 0.7 for approval, the bank ultimately declines credit to the customer. In order to take the number of examples into account, the proportion each leaf is weighted by the proportion P_{max}^l of training examples with the dominant classification.

$$\delta_j = \frac{\sum_l P^l_{max} \cdot \mu_l(D_j) \cdot \chi_{max} \cdot \mu_l(D_j)}{\sum_l P^l_{max} \cdot \mu_l(D_j)}$$

The classification including example counts becomes almost the same as before.

$$\delta_j = \frac{(1.8 \cdot 0.6 \cdot 0.0 + 0.5 \cdot 0.40.0 + 0.8 \cdot 0.3 \cdot 1.0 + 1.7 \cdot 0.3 \cdot 1.0)}{(1.8 \cdot 0.6 + 0.5 \cdot 0.4 + 0.8 \cdot 0.3 + 1.7 \cdot 0.3)}$$
$$= 0.75/2.03 = 0.369$$

The second approach reflects more the notion of FL, in the sense that all classifications at a leaf node contribute to the outcome, rather than just the most frequent classification. In this case the contribution of each leaf node is proportional to the matching degree of the example $\mu_l(D_j)$ and the proportion P^l_k of training examples with classification k.

$$\delta_j = \frac{\sum_l \sum_k P^l_k \cdot \mu_l(D_j) \cdot \chi_k}{\sum_l \sum_k P^l_k \cdot \mu_l(D_j)} \tag{10.8}$$

In our example, the classification is computed according to Eq. 10.8 as

$$\delta_j = \frac{0.9 \cdot 0.6 \cdot 1.0 + 0.4 \cdot 0.4 \cdot 1.0 + 0.8 \cdot 0.3 \cdot 1.0 + 1.7 \cdot 0.3 \cdot 1.0}{(0.9 + 1.8) \cdot 0.6 + (0.4 + 0.5) \cdot 0.4 + (0.8 + 0.4) \cdot 0.3 + (1.7 + 0.1) \cdot 0.3}$$
$$= 1.45/2.88 = 0.503$$

Notice that we omitted the contribution of *no*-examples in the numerator which do not contribute since $\chi_{no} = 0$. The confidence for credit approval is higher since the nodes with *employment=part-time* have a significant proportion of positive examples, which the previous method disregards. Notice also that both methods come to the same conclusion in case each leaf node has a unique classification.

10.3.3 *Optimising Fuzzy Decision Trees*

The underlying idea of optimising fuzzy decision trees is to improve the distinguishability of examples. According to Eqs. 10.4 and 10.5, the entropy at a node decreases as the target classifications become more unique. Therefore, the entropy of node can be reduced by properly adjusting the fuzzy sets that partition the attribute.

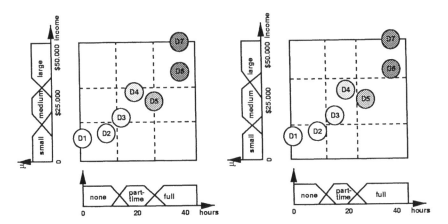

Fig. 10.6   Static optimisation of the fuzzy input partition, left initial partition, right optimised partition

Figure 10.6 compares the initial fuzzy sets on the left with the ones optimised on the right. Notice that the grey value of a data point corresponds to the target classification, the darker the more confidence into credit approval. The optimisation procedure mimics the genetic tuning process described in Chapter 4. The GA encodes the centre and width of the fuzzy membership functions subject to tuning. Janikow (1996) proposed to impose additional constraints on the membership parameters in order to improve the search efficiency. The constraints impose certain properties of the fuzzy partition such as set symmetry, non-containment or a predefined overlap among neighbouring sets. The constraints limit the number of free parameters, reduce the dimension of the search space and that way improve the speed and quality of optimisation.

Fuzzy decision trees can be optimised in two different ways. Static optimisation takes place before the actual tree is constructed, whereas dynamic optimisation is an integral part of the tree construction process itself. Static optimisation aims to minimise the entropy of the data. Instances are partitioned into elementary cells ν_r according to their attribute values a_j. The examples in Fig. 10.6 are partitioned by three fuzzy sets for each attribute, resulting in nine fuzzy cells ν. The example count for instances with target classification k at a cell ν_r is computed similar to Eq. 10.4 as

$$P_k^{\nu_r} = \sum_{e_j} \mu_{\nu_r}(e_j) \cdot \mu_k(e_j)$$

The total count is the sum over all possible target classifications is

$$P^{\nu_r} = \sum_k P_k^{\nu_r}$$

With these preliminaries, we can compute the entropy of an individual cell ν_r given the data S according to Eq. 10.5

$$E^{\nu_r}(S) = -\sum_k P_k^{\nu_r}/P^{\nu_r} \cdot \log P_k^{\nu_r}/P^{\nu_r}$$

The overall entropy E

$$E(S) = \frac{\sum_r P^{\nu_r} \cdot E^{\nu_r}}{\sum_r P^{\nu_r}}$$

is computed as the sum over the individual entropies E^{ν_r} of a single cell ν_r weighted by its example count P^{ν_r}. The GA is supposed to minimise the entropy. Therefore, the fitness function becomes the reciprocal value of the entropy

$$F = \frac{1}{E(S)}$$

The optimised fuzzy partition in Fig. 10.6 on the right discriminates the examples D_3, D_4, D_5 better than the original partition and thereby reduces the entropy.

Dynamic optimisation employs a similar fitness criterion but only optimises the entropy of the attribute for the next node. In principle, the same attribute can be optimised in a different way at each node it occurs. That way one improves the accuracy for the price of loosing the comprehensibility of the tree. This trade-off among accuracy and comprehensibility is another instance of the descriptive versus approximate representation problem previously discussed in Sec. 1.2.6.2.

Prior to tree construction, the dynamic algorithm optimises the initial fuzzy sets in the same way as static optimisation by minimising the entropy. In dynamic optimisation, the standard tree building algorithm and the GA optimisation step are interleaved. First, we define a set of open nodes, which includes those nodes without a successor node. Notice that in the beginning this set only contains the root node. From this set of open nodes, the algorithm selects the node N with the highest event count P^N. Therefore, the algorithm proceeds in a best-first traversal rather than a depth-first traversal as the standard tree-building method. The GA optimises the entropy E based on the training examples at the selected node, subject to the constraint that attributes already used in a previous node remain constant. Expand the node in the usual way by splitting along the attribute a_j that provides the maximal information gain according to Eq. 10.6. Finally, add the newly generated nodes to the open set and remove their predecessor. The algorithm terminates when the examples at each leaf node carry a unique classification or if all attributes have been used for splits.

Chapter 11

Applications

The last chapter describes some real-world applications of GFRBSs. The following four sections present examples how GFRBSs solve problems in the domains: *Classification*, *System Modelling*, *Control Systems* and *Robotics*.

11.1 Classification

Pattern recognition constitutes an important application of FRBSs as FRBCSs provide a suitable mean to classify incomplete and imprecise data (Chi, Yan, and Pham, 1996; Kuncheva, 2000).

In this domain, GFRBSs employ an EA in order to learn or tune fuzzy classification rules. GFRBCSs have been evaluated on a number of classical benchmark problems, such as the Iris, Pima, or Wine data sets, which are available from the UC Irvine repository at ics.uci.edu (Merz and Murphy, 1996).

Section 11.1.1 briefly reviews GFRBSs for classification problems. Sections 11.1.2 and 11.1.3 present two applications of GFRBCSs to the diagnosis of myocardial infarction and breast cancer, respectively.

11.1.1 *Genetic fuzzy rule-based systems to learn fuzzy classification rules: revision*

Fisher's Iris database (Fisher, 1936) constitutes the most prominent benchmark for GFRBCSs. The data set contains 150 examples of iris flowers with four attributes (petal and sepal width and length) and three classes (setosa, versicolor and virginica).

The design paradigms MOGUL and SLAVE (see Chapter 8) are used to design GFRBCSs based on the IRL approach which are evaluated on the Iris data set (Cordón, del Jesús, and Herrera, 1998; González and Pérez, 1998a, 1999a). One of the important characteristics of SLAVE is the high linguistic description level of the obtained RB. SLAVE generates the following RB composed of three rules for the Iris data set:

R_1 : *IF Petal_l is Very_low THEN Class is Setosa*
also
R_2 : *IF Petal_w is High or Very_high THEN Class is Virginica*
also
R_3 : *IF Petal_l is Medium and Petal_w is less or equal to Medium*
        THEN Class is Versicolor

The different proposals by Ishibuchi et al. (the GFRBCS based on the Michigan approach presented in Sec. 6.2.2, the different genetic selection algorithms introduced in Sec. 9.2.1, the design process based on the genetic learning of the DB introduced in Sec. 9.3.2.4 and the hybrid GFRBCS following both the Pittsburgh and Michigan approaches shown in Sec. 9.4.9) are also applied to the Iris problem as well as to the Wine data set (Ishibuchi, Murata, and Nakashima, 1997; Ishibuchi, Nakashima, and Nii, 2000).

Finally, Shi, Eberhart, and Chen (1999) proposed a GFRBCS which is also based on the Pittsburgh approach coding membership function parameters and types and using an FLC to control the GA mutation and crossover rates. An FRBCS with four rules was evolved for the Iris problem by this learning system.

For a short review on the use of GAs to design FRBCSs, focused on genetic feature selection and GFRBCSs, the interested reader can refer to (Cordón, del Jesús, and Herrera, 1999a).

11.1.2 *Diagnosis of myocardial infarction*

The exact diagnosis in cases of suspected myocardial infarction is a practical problem in forensic pathology. The combination of morphological methods and biochemical analysis of both myocardial tissue and pericardial fluid are accepted as the best markers for post-mortem diagnosis of myocardial infarction (Hougen et al., 1992; Valenzuela et al., 1994).

González, Pérez, and Valenzuela (1995) applied SLAVE to solve this classification problem. A FRBCS was obtained from a database containing the attribute values of each patient together with the classification of a positive or negative diagnosis of myocardial infarction.

The infarction database contains 14 attributes (2 nominal and 12 continuous) to determine the nominal diagnostic variable. The attributes are divided into two different types of tests, morphological methods and biochemical analysis.

The database was partitioned into four training sets and SLAVE was invoked on each partition. Table 11.1 compares the average classification rate on the four test sets obtained by the four FRBCSs designed by SLAVE, CART (Breiman et al., 1984) and back-propagation NN (BP) (White, 1992) algorithms. SLAVE clearly outperforms the CART and BP algorithms.

Table 11.1 Average test set classification results for myocardial infarction

SLAVE	CART	BP
91.86	81.39	83.73

A statistical analysis based on a logistic regression method was applied to the same data, obtaining a classification accuracy of 87.32%. SLAVE achieves a better accuracy because of the consistency parameters (see Sec. 8.2.1.1). In addition, the set of linguistic rules generated by SLAVE are more understandable and therefore more useful to clinical experts than a black-box NN.

11.1.3 *Breast cancer diagnosis*

Breast cancer is one of the most common forms of cancer among females. The presence of a breast mass is an alert sign, but it does not always indicate a malignant cancer. Fine needle aspiration (FNA) of breast masses is a cost-effective, non-traumatic, and mostly non-invasive diagnostic test that obtains information needed to evaluate malignancy.

The Wisconsin breast cancer diagnosis (WBCD) database (Merz and Murphy, 1996) is the result of the efforts made at the University of Wisconsin Hospital for accurately diagnosing breast masses based solely on an FNA test (Mangasarian, Setiono, and Goldberg, 1990). Nine visually assessed characteristics of an FNA sample considered relevant for diagnosis were identified, and assigned an integer value between 1 and 10 that indicates the appearance of the feature. The measured variables are as follows:

(1) Clump Thickness (v_1);
(2) Uniformity of Cell Size (v_2);
(3) Uniformity of Cell Shape (v_3);
(4) Marginal Adhesion (v_4);
(5) Single Epithelial Cell Size (v_5);
(6) Bare Nuclei (v_6);
(7) Bland Chromatin (v_7);
(8) Normal Nucleoli (v_8);
(9) Mitosis (v_9).

Specialists in the field furnished the diagnostics in the WBCD database, which contains 683 instances, with each entry representing the classification *benign* or *malignant* for a certain ensemble of attributes. There are 444 *benign* and 239 *malignant* cases. The diagnosis does not provide any information about the degree of benignity or malignancy.

Bennett and Mangasarian (1992) achieved a classification rate of 99.6% on the reduced data set of 487 instances available in 1992 using linear programming techniques. However, the obtained classifier was difficult to understand, i.e., diagnostic decisions are essentially black boxes, with no explanation as to how they are attained. As interpretability is a primary objective in medical diagnosis tasks, some approaches extract Boolean rules from NNs for the problem (Setiono, 1996; Setiono and Liu, 1996; Taha and Ghosh, 1997). The results are encouraging, as they exhibit good performance as well as reduce the number of rules and selected input variables. Nevertheless, these systems still employ crisp Boolean rules and are unable to furnish the user with a measure of confidence for the decision made. The first approach on evolution of fuzzy rules developed by Peña-Reyes and Sipper (1998) demonstrated that it is possible to generate an accurate classifier, whose decisions are not only interpretable but also indicate the confidence in the diagnosis given by the classifier.

The following subsections describe the FRBCS parameters, the genetic learning approach followed and the experimental results.

11.1.3.1 *Fuzzy rule-based classification system parameters*

The method proposed by Peña-Reyes and Sipper (1999a, 1999b) consists of an FRBS and a threshold unit. The FRBS computes a continuous appraisal value of the malignancy of a case, based on the input values. The threshold unit outputs a *benign* or *malignant* diagnosis according to the FRBS output.

Peña-Reyes and Sipper (1999a) chose the fuzzy parameters based on prior knowledge about the WBCD problem and the existing rule-based models, which lead to the following observations:

(1) Small number of rules: Systems with no more than four rules demonstrate a high classification rate (Setiono, 1996; Peña-Reyes and Sipper, 1998).

(2) Small number of variables: Rules with no more than four antecedents have proven to be appropriate (Setiono and Liu, 1996; Taha and Ghosh, 1997; Peña-Reyes and Sipper, 1998).

(3) Monotonicity of the input variables: Observation of the relationship between input and diagnosis shows that higher-valued variables are associated with malignancy (Peña-Reyes and Sipper, 1998).

Some fuzzy models sacrifice interpretability for improved performance. In the case of medical diagnosis, interpretability constitutes the major advantage of FRBSs compared to other classification algorithms. This observation motivated the authors to take the following five semantic criteria into account which impose additional constraints on the fuzzy parameters: (i) distinguishability; (ii) justifiable number of elements; (iii) coverage; (iv) normalisation; and (v) orthogonality (see (Peña-Reyes and Sipper, 1999b) for a more detailed description).

Taking into account these criteria, the following FRBS elements are considered:

- *Inference and defuzzification:* min-max fuzzy operators; orthogonality (which refers to the use of strong fuzzy partitions where the sum of the membership values for each element of the universe of discourse is equal to one); simplified TSK fuzzy rules with trapezoidal membership functions in the inputs; weighted-average defuzzification.

  For the output singletons, two reference values (2 and 4) are fixed by the authors to represent the values *benign* and *malignant*, respectively.

- *Structural parameters:* fuzzy partitions composed of two membership functions (*Low* and *High*) for the inputs; two output singletons; number of rules limited to be between 1 and 5.

- *Operational parameters:* the GA learns the composition of the rule antecedents by both selecting the relevant input variables and the linguistic terms associated to them. Moreover, it simultaneously adapts the input membership function parameters.

Every fuzzy rule identified by the GFRBCS is associated to the *benign* diagnosis. The *malignant* diagnosis is considered as an **else** rule condition that is given as rule output when the input to the system does not match the rule antecedent at all. The Output of a fired rule has a weight of 1 in the final decision, whilst the output obtained from the **else** condition has a weight of 0.25.

11.1.3.2 *Genetic learning approach*

The GFRBS is based on the Pittsburgh approach. The chromosome encodes the relevant input variables, the linguistic labels they take as a value and the membership function parameters associated to those labels:

- *Membership function parameters:* The fuzzy partitions of the nine input variables $(v_1, \ldots, v_9)$ —each one composed of two trapezoidal membership functions— are parameterised by a pair of values (P,d), where P is the rightmost point in the core of the fuzzy set *Low* and d is the relative distance from that point to the leftmost point in the core of the fuzzy set *High*.

- *Antecedents:* The i-th rule has the form:

$$IF\ v_1\ is\ A_1^i\ and\ \ldots\ and\ v_9\ is\ A_9^i\ THEN\ output\ is\ benign$$

where A_j^i represents the linguistic term associated to variable v_j. In the coding scheme, four possible values $\{0, 1, 2, 3\}$ are considered, with 1 and 2 standing for *Low* and *High*, respectively, and the other two values, 0 and 3, representing the absence (irrelevance) of the corresponding variable in the antecedent of the current rule. This way, the algorithm can implicitly discard irrelevant variables, by assigning the "absence" linguistic terms as their antecedent values.

Table 11.2 depicts the binary coding scheme considered for the complete KB representation.

The fitness function is given by

$$F = F_c - \alpha \cdot F_v - \beta \cdot F_e$$

Table 11.2 Parameter encoding

Parameter	Values	Bits	Quantity	Total bits
P	[1-8]	3	9	27
d	[1-8]	3	9	27
A	[0-3]	2	$9 \cdot N_r$	$18 \cdot N_r$

where $\alpha = 0.05$ and $\beta = 0.01$ (the values for α and β are derived empirically). The function combines three criteria:

(1) F_c: classification performance, computed as the percentage of correctly classified instances,
(2) F_v: the average number of variables per rule.
(3) F_e: the quadratic error between the continuous appraisal value (in the range [2,4]) and the correct discrete target diagnosis in the WBCD database (either 2 or 4), and

F_c, the ratio of correctly classified instances, is the most important performance measure. F_v measures linguistic integrity (interpretability) of the FRBS penalising rules that contain a large number of variables.

11.1.3.3 *Results*

The data is partitioned into a training set and a test set in three different ways: (1) the training set contains all 683 instances of the WBCD database, the test set is empty; (2) the training set contains a 75% of the WBCD instances, and the test set contains the remaining 25%; (3) the training set contains 50% of the WBCD instances, and the test set contains the remaining 50%. A total of 120 evolutionary runs were invoked on the three data set partitions.

Table 11.3 Experimental results

Train./test ratio	Training set	Test set	Overall	Number of vars.
100% / 0%	—	—	96.97	3.32
75% / 25%	97.00	96.02	96.76	3.46
50% /50%	97.71	94.73	96.23	3.41

Table 11.3 lists the average performance of the best individual over all

runs. Table 11.4 compares the best FRBS with the best solutions obtained by three other rule-based diagnostic approaches. The three approaches (Setiono, 1996; Setiono and Liu, 1996; Taha and Ghosh, 1997) involve Boolean rule sets extracted from trained NNs. The average number of variables per rule are given in parentheses.

Table 11.4   Comparison of results

Rules per system	Setiono-96	Setiono and Liu-96	Taha and Ghosh-97	GFRBCS
1	95.42(2)			97.07(4)
2				97.36(3)
3	97.14(4)	97.21(4)		97.80(4.7)
4				97.80(4.8)
5			96.19(1.8)	97.51(3.4)

In view of the results shown in Table 11.4, the GFRBCS obtains the best performance for all five RB sizes, i.e., from a single rule FRBS all up to a FRBS composed of five rules.

The following rule, together with its parameters, is the best single rule FRBS identified by the GFRBCS:

> R_1 : *IF v_1 is Low and v_2 is Low and v_6 is Low and v_8 is Low*
> *THEN output is benign*
> *Default* : *ELSE output is malignant*

with parameters values $P = (4, 4, -, -, -, 2, -, 2, -)$ and $d = (3, 1, -, -, -, 5, -, 7, -)$. It achieves a 97.07% correct classification rate both over the *benign* and the *malignant* cases, thus giving an overall classification rate of 97.07%.

Other examples of RBs obtained by the GFRBCS and additional experimental results can be found in (Peña-Reyes and Sipper, 1999b).

11.2   System Modelling

In *Fuzzy system modelling* (Bardossy and Duckstein, 1995; Pedrycz, 1996; Sugeno and Yasukawa, 1993), the underlying behaviour of the system is described by a set of fuzzy rules and variables. The term *linguistic mod-*

elling has been coined for fuzzy models based on descriptive Mamdani-type FRBSs, since the linguistic description provides an interpretable, comprehensible model of the system behaviour.

On the other hand, the term *fuzzy modelling* refers to the case in which TSK or approximate Mamdani-type FRBSs are used to model the behaviour of an underlying system. In fuzzy modelling, the accuracy of the model is more important than its mere interpretability and descriptive power.

The trade-off between interpretability and accuracy of the model usually depends on the requirements of the specific problem. GFRBSs have been applied to both fields, linguistic and fuzzy modelling. The following subsections present four applications from engineering as well as non-technical domains: two electrical engineering problems, the rice taste evaluation problem and the dental development age prediction.

11.2.1 *Power distribution problems in Spain*

The situation of Spanish electrical power market changes continuously as it evolves towards increased competition, although it is not completely deregulated yet. Four major power companies (Iberdrola, Endesa, Unión Fenosa, and Hidroeléctrica del Cantábrico) own almost all power generation plants and distribution networks in Spain and share the ownership of a company called *R.E.E.* (*Red Eléctrica Española*). Due to the characteristics of the Spanish market, any of these companies has a great influence on the price of electrical energy, which has an obvious impact on the economical development of the country.

Hence, the Spanish government decided some years ago to nationalise high voltage lines and to separate distribution and generation markets, thus forcing the said companies to act as two different entities each. This way, generation plants sell the power they produce in a partially regulated market and the distribution companies buy the energy in this same market (bilateral contracts between suppliers of energy and consumers are also allowed in certain cases). The energy price paid by the consumer is not completely received by the companies, but the payments are redistributed according to some criteria. Nowadays, this redistribution depends on the global amount of power that each company provides, which is previously contracted by the government according to the country needs from a pooling system where each company offers a certain energy amount at a specific price at each time of the day.

However, until January 1999, the payment redistribution was made according to several very complex criteria with the objective to achieve and maintain market equilibrium. These criteria among others included not only the amount of power generation of every company, but also the number of customers and the maintenance costs of the network owned by each of them (Sánchez, 1997). Due to this regulation, every company needed to estimate the maintenance costs of their global network, which constituted a difficult task due to two main reasons:

- Maintenance costs depend —among other factors— on the total length of electrical line each company owns, and on the type of this line (high, medium, urban low and rural low voltage). High and medium voltage lines can be easily measured, but low voltage lines are spread across cities and villages, and it is very difficult and expensive to measure them. This kind of line uses to be very convoluted and, in some cases —as in Asturias, the Spanish region tackled in this case—, companies can serve as much as 10,000 small villages.
- Moreover, some of the payment distribution criteria were based on the optimal network and not on the actual network. It is argued that more and more low and medium voltage lines have been installed incrementally over the years and thus the actual distribution is far from the optimal one. If the government rewards lengthy networks, there would be no incentive for the companies to modernise obsolete distribution networks.

Thus, there is an obvious need for a model of the length of low voltage line and of the maintenance costs of the optimal medium voltage network. All companies developed their own models and the government also used some of these models to determine the share of the payment that the companies receive.

In this context, two different problems were solved in (Cordón, Herrera, and Sánchez, 1999; Sánchez, 2000)*, to relate some characteristics of a certain village with the actual length of low voltage line contained in it,

*Within contract CN-96-055-B1 between Hidroeléctrica del Cantábrico and the University of Oviedo, and research project TIC96-0778 developed by University of Granada and supported by CICYT.

Table 11.5 Notation considered for the low voltage line estimation problem variables

Symbol	Meaning
A_i	Number of clients in population
R_i	Radius of i population in the sample
n	Number of populations in the sample
l_i	Line length, population i
$\tilde{l}_i$	Estimation of l_i
s_i	Number of sectors in town i

and to relate the maintenance cost of the optimal network that must be installed in certain towns with some other characteristics of these towns. In both cases, it was preferable that the solutions obtained are not only numerically accurate, but are also able to explain how a specific value is computed for a certain village or town. It is of major relevance that *these solutions are interpretable by human beings to some degree.* Thus, linguistic modelling seems an appropriate approach that simultaneously addresses the problem specific needs for accuracy and interpretability.

In the following two subsections, both modelling problems are solved with different types of GFRBSs, and compared with classical regression, NNs and other linguistic and fuzzy modelling techniques not based on EAs. Although one of the problem requirements is to obtain human interpretable models, some of the non-linguistic GFRBSs introduced in Chapters 7 and 8 to design TSK and approximate Mamdani-type fuzzy models are considered as well.

11.2.1.1 *Computing the length of low voltage lines*

The first problem is to find a model that relates the total length of low voltage line installed in a rural town with the number of inhabitants in the town and the mean of the distances from the centre of the town to the three furthest clients in it (Cordón, Herrera, and Sánchez, 1999; Sánchez, 2000). This model is then used to estimate the total length of line being maintained by one of the companies, Hidroeléctrica del Cantábrico. The data set contained instances of 495 towns (Sánchez, 1997) in which the length of the line was actually measured. The company used the model to predict the line length of more than 10,000 towns according to the set of attributes listed in Table 11.5.

The system model relates the target attribute l_i (line length) with the two attributes A_i and R_i (population and village radius). Different approaches such as regression techniques, and evolutionary and non-evolutionary processes to design linguistic models, have been investigated. Moreover, although one of the problem requirements is the interpretability of the model, neural modelling and evolutionary and non-evolutionary fuzzy modelling were also taken into account to obtain a performance threshold for the descriptive techniques.

To compare the performance of the different modelling techniques, the data set was randomly divided into two sets comprising 396 and 99 samples —80 and 20 percent of the whole data set— to be used as training and test sets. The prediction accuracy of the different approaches is measured by the mean square error (SE) (see Sec. 4.3.3.1), which in this case is defined as

$$\frac{1}{2 \cdot N} \sum_{i=1}^{N} (\tilde{l}_i - l_i)^2$$

and is computed over both sets, SE_{tra} and SE_{tst}.

The following subsections briefly describe the application of the said techniques. The next subsection compares the numerical results.

Application of classical regression methods

In order to apply classical methods to this problem, it becomes necessary to make some assumptions about the network structure (Cordón, Herrera, and Sánchez, 1999; Sánchez, 2000). In the villages studied, electrical networks are star-shaped and arranged in sectors. Consumers are connected by short power segments to a main line that provides power to the population (see Fig. 11.1).

The theoretical model is based on the following simplifying assumptions:

- Village i comprises s_i sectors. Each sector in the same village covers a radial segment of size $2 \cdot \theta_i$. Main lines depart from the centre of the village.
- The density of consumers is constant within a sector.
- All sectors in a village have the same radius, R_i, and contain a main line of length R_i and as many branches as consumers.

If we assume that consumers are uniformly distributed, the total length can be approximated by the product of the mean distance between con-

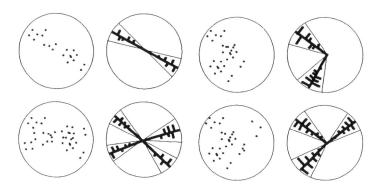

Fig. 11.1   Models of possible sub-nets

sumers and the main line by the number of inhabitants. As the sectors size is $2 \cdot \theta_i$, this mean distance d_i is computed as for each population i:

$$d_i = \frac{2 \cdot (1 - cos\theta_i)}{3 \cdot \theta_i} \cdot R_i$$

such that the power line length becomes

$$\tilde{l}_i = s_i \cdot (R_i + \frac{A_i}{s_i} \cdot d_i) = s_i \cdot R_i + A_i \cdot \frac{2 \cdot (1 - cos\theta_i)}{3 \cdot \theta_i} \cdot R_i$$

If the angles θ_i and the numbers s_i are similar across different areas, they can be regarded as constants and estimated by the parameters $\bar{\theta}_i = \theta$ y $\bar{s}_i = s$ using least squares linear regression

$$\tilde{l}_i / R_i = s + k(\theta) \cdot A_i$$

developed over a set of pairs $(x, y) = (A_i, l_i / R_i)$.

This estimate can be improved if the number of sectors, their angles and the number of consumers are no longer regarded as constants but depend on each other. This can be done by partitioning the data set into classes or by changing the regression variables. Both cases were studied, and the best estimate was obtained with the parameterised non-linear model

$$\frac{\tilde{l}_i}{R_i} = k_1 \cdot A_i^{k_2}$$

In the experiments, the parameters of the polynomial models were fitted by non-linear least squares (Levenberg-Marquardt method), whilst exponential and linear models were fitted by linear least squares.

Neural Modelling
The power line length was also modelled by a multi-layer perceptron with two input and a single output neurons, and a single hidden layer (Cordón, Herrera, and Sánchez, 1999; Sánchez, 2000). The network was trained with the *quick-propagation* algorithm. Several NNs with a different number of neurons in the hidden layer were tested, taking the one best performing in the approximation of the test set, which made use of 25 neurons.

Non-evolutionary Linguistic and Fuzzy Modelling
A variety of non-evolutionary linguistic and fuzzy modelling techniques have also been considered to solve the problem (Cordón, Herrera, and Sánchez, 1999; Cordón and Herrera, 2000). These *ad hoc data-driven linguistic and fuzzy rule learning methods* directly generate new rules from the training data based on a covering criterion (Casillas, Cordón, and Herrera, 2000), such as the widely known *Wang and Mendel* method (WM) (Wang and Mendel, 1992b), the one proposed by Nozaki, Ishibuchi, and Tanaka (1997) (NIT) or the grid-based method proposed by Cordón and Herrera (2000) (CH) for descriptive Mamdani-type FRBSs and the *Weighted Counting Algorithm* (WCA) (Bardossy and Duckstein, 1995) for approximate Mamdani-type FRBSs.

GFRBS Linguistic and Fuzzy Modelling
Finally, several of the GFRBSs introduced in this book have been considered to design different linguistic and fuzzy models to solve the problem. They are shown as follows:

- Two GFRBSs to design descriptive Mamdani-type FRBSs: the one proposed by Thrift (1991) (see Sec. 7.3.1) —Pittsburgh approach— and the descriptive variant of MOGUL —IRL approach— introduced in Chapter 8 (Cordón and Herrera, 2001; Cordón, del Jesús, Herrera, and Lozano, 1999; Cordón, Herrera, and Sánchez, 1999).
- Two different two-stage genetic learning processes to design linguistic FRBSs based on first generating the RB —starting from a previous DB definition which remains fixed during the RB learning— and then obtaining an appropriate DB definition by means of the descriptive genetic tuning process presented in Sec. 4.3.3.3. The two RB learning processes considered for the first stage are the WM *ad hoc* data-driven method and Thrift's GFRBS (Cordón and Herrera, 2001; Cordón, del Jesús, Herrera, and Lozano, 1999;

Cordón, Herrera, and Sánchez, 1997).

- Two genetic learning processes under the ALM methodology, introduced in Sec. 9.2.3, in which an RB with a large number of single and double-consequent rules is generated and a genetic selection process identifies the subset of rules that demonstrate the best cooperation (Cordón and Herrera, 2000). The two variants are the WM-based one introduced in Sec. 9.2.3 and another one that uses a different RB generation method (CH-based ALM) (Cordón and Herrera, 2000).

- An adaptation of the genetic learning process proposed by Ishibuchi, Nozaki, Yamamoto, and Tanaka (1995), that generates multiple RBs from fuzzy partitions of different granularity by means of the CH ad-hoc data driven learning method, joining them all in a global RB and applying a genetic selection process to obtain the best subset of rules (Cordón, Herrera, and Zwir, 1999), which is described in Sec. 9.2.4.2.

- The hard and soft constrained MOGUL variants for approximate Mamdani-type FRBSs introduced in Chapter 8 (Cordón and Herrera, 2001; Cordón, del Jesús, Herrera, and Lozano, 1999).

- A two-stage genetic learning process to design approximate Mamdani-type FRBSs that first generates the RB by means of the WCA and then adjusts the membership function definitions by means of the approximate genetic tuning process presented in Sec. 4.3.3.2 (Cordón and Herrera, 2001; Cordón, del Jesús, Herrera, and Lozano, 1999).

- The GFRBS based on the IRL approach to design TSK FRBSs introduced in Chapter 8 under the MOGUL paradigm (Cordón, Herrera, and Sánchez (1997, 1999)).

Comparison between modelling techniques

Table 11.6 compares the training and test set error achieved by the different modelling techniques for the low voltage line estimation problem. The column *Complexity* contains the number of parameters and the number of nodes in the parse tree of the expression, as well as the number of rules in the KB or FRB of the generated linguistic or fuzzy model. The fuzzy partitions in the linguistic modelling processes considered contained five triangular-shaped membership functions.

The results demonstrate that linguistic and fuzzy models clearly out-

Table 11.6   Results obtained in the low voltage line estimation problem

Method	Train.	Test	Complexity
Regression techniques			
Linear	287775	209656	7 nodes, 2 par.
Exponential	232743	197004	7 nodes, 2 par.
2th order polynomial	235948	203232	25 nodes, 6 par.
3rd order polynomial	235934	202991	49 nodes, 10 par.
Neural modelling			
3 layer perceptron 2-25-1	169399	167092	102 par.
Non-evol. Ling. Modelling			
WM linguistic model	298446	282058	13 rules
NIT linguistic model	229104	206636	40 rules
CH linguistic model	310308	286775	20 rules
Non-evol. Fuzzy Modelling			
WCA	356434	311195	20 rules
GFRBS Ling. Modelling			
WM + Tuning	175337	180102	13 rules
Thrift	218591	204426	25 rules
Thrift + Tuning	154314	199551	25 rules
descriptive MOGUL	150559	166669	19 rules
WM-based ALM	178571	180847	13 rules
CH-based ALM	179383	181284	22 rules
Ishibuchi et al.	177735	180721	15 rules
GFRBS Fuzzy Modelling			
WCA + Tuning	175887	180211	20 rules
soft const. approx. MOGUL	142108	166578	19 rules
hard const. approx. MOGUL	108203	166186	33 rules
TSK MOGUL (without ref.)	162609	148514	20 rules

perform classical non-linear regression methods and are equal or superior to NNs. In fact, the descriptive MOGUL GFRBS generates a very simple linguistic model —the RB is composed of only 19 rules— which shows a better performance than the NN in both training and test set errors. This result is significant, as it shows that a highly descriptive fuzzy model is able to achieve the same accuracy as a black-box NN.

Notice that better accuracy is obtained with non-linguistic, and therefore less descriptive fuzzy models, such as approximate Mamdani-type and TSK FRBSs. The three fuzzy models designed by means of the GFRBSs significantly outperform the NN both in abstraction (training set error) and generalisation (test set error) power. Moreover, although these fuzzy models do not lend themselves to an intuitive linguistic interpretation, they are still more descriptive than neural models, as the behaviour of the system can be understood and analysed by means of local fuzzy rules.

The overall best generalisation results correspond to the TSK fuzzy model obtained after the first learning stage since the refinement phase tends to over-fit the training data, as it further reduces the training set error but at the same time increases the test set error. In case accuracy becomes more important than interpretability, the TSK fuzzy model is the best option as it is only slightly more complex than the descriptive MOGUL-based model (20 rules).

11.2.1.2 *Computing the maintenance costs of medium voltage line*

Modelling the maintenance costs of the medium voltage electrical network is more complicated than estimating the low voltage line length, as maintenance costs must be related to the optimal network structure rather than to the actual network installed in the city. The estimated minimal maintenance cost is obtained from a model of the optimal electrical network for a town (Cordón, Herrera, and Sánchez, 1998, 1999). These estimates of the minimum costs are somewhat lower than the actual costs. As the true maintenance costs are anyhow exactly accounted, a model that relates these costs to the specific characteristics of a town is of little practical merit.

The 1059 instances in the data set are characterised by four attributes (see Table 11.7) and the associated minimum maintenance cost in that town. The objective is to design a model that relates the maintenance costs with the other four attributes. The following modelling techniques were considered for this purpose: classical regression, neural modelling and GFRBSs and non-evolutionary processes for fuzzy and linguistic modelling.

In order to compare the said techniques, the data set was partitioned into a training and a test set, comprised of 847 and 212 instances —80 and 20 percent of the whole data set—.

The following techniques have been applied to solve the medium voltage line maintenance cost problem:

Table 11.7   Notation considered for the medium voltage line maintenance cost estimation problem variables

Symbol	Meaning
x_1	Sum of the lengths of all streets in the town
x_2	Total area of the town
x_3	Area that is occupied by buildings
x_4	Energy supply to the town
y	Maintenance costs of medium voltage line

Classical methods
The linear and polynomial models investigated by Cordón, Herrera, and Sánchez (1998, 1999). The parameters of the polynomial models were fitted using the Levenberg-Marquardt method.

Neural Modelling
The neural model (three-layer perceptron) was trained with the conjugate gradient algorithm (Cordón, Herrera, and Sánchez, 1998, 1999). Again, the number of neurons in the hidden layer was determined by minimising the test set error. The final NN was composed of 4 input nodes, 5 hidden nodes, and 1 output node.

Non-evolutionary Linguistic Modelling
Two ad hoc data-driven linguistic rule learning methods have been applied (Cordón, Herrera, Magdalena, and Villar, 2000): WM and CH.

GFRBS Linguistic Modelling
The following GFRBSs introduced in this book to learn linguistic KBs have been considered to solve the problem:

- The three-stage descriptive MOGUL GFRBS introduced in Chapter 8 (Cordón, Herrera, and Sánchez, 1998, 1999).
- A two-stage genetic learning process to design linguistic FRBSs that generates the RB by means of the WM learning process and adjusts the membership functions using the descriptive genetic tuning process (see Sec. 4.3.3.3) (Cordón, Herrera, and Sánchez, 1998, 1999).
- One of the hybrid learning processes introduced in Sec. 9.3 that

Table 11.8 Results obtained in the medium voltage line maintenance cost estimation problem

Method	Training	Test	Complexity
Regression techniques			
Linear	164662	36819	17 nodes, 5 par.
2th order polynomial	103032	45332	77 nodes, 15 par.
Neural modelling			
3 layer perceptron 4-5-1	86469	33105	35 par.
Non-evol. Ling. Modelling			
WM linguistic model	24867	26974	95 rules
CH linguistic model	27698	26134	589 rules
GFRBS Ling. Modelling			
descriptive MOGUL	19679	22591	63 rules
WM + Tuning	20318	27615	66 rules
genetic DB learning + WM	9859	9559	71 rules
genetic DB learning + CH	10972	10232	348 rules
GFRBS Fuzzy Modelling			
TSK MOGUL	11074	11836	268 rules

generates the DB by means of a genetic learning process and invokes a non-evolutionary learning process (WM and CH) to derive the RB (Cordón, Herrera, Magdalena, and Villar, 2000).

GFRBS Fuzzy Modelling
The GFRBS to learn TSK KBs based on the IRL approach introduced in Chapter 8 was also used to design fuzzy models for this application (Cordón and Herrera, 1999b).

Comparison between modelling techniques
Table 11.8 compares the training and test set error achieved by the various approaches.

The conclusions are similar to the ones drawn in the low voltage line length estimation problem. Linguistic and fuzzy models outperform classical non-linear regression methods and NNs.

The main difference between both applications is that in this case we found two linguistic models that achieve a better accuracy than the TSK fuzzy model. Moreover, although the TSK model outperforms the linguistic

Table 11.9   Notation considered for the rice taste evaluation problem variables

Symbol	Meaning
x_1	Flavor
x_2	Appearance
x_3	Taste
x_4	Stickiness
x_5	Toughness
y	Rice evaluation

model generated by the descriptive variant of MOGUL, it is more difficult to interpret as it requires a substantially larger number of rules (268 versus 63).

The two hybrid processes based on a genetic learning of the DB and a non-evolutionary derivation of the RB achieve the highest accuracy. This is of significant importance as it demonstrates that an accurate linguistic model comprised of only 71 rules —which makes it easy to interpret by human beings— can be obtained from a simple ad hoc data-driven RB learning method for a properly defined DB. This linguistic model is twice as accurate on both training and test data set as the MOGUL one.

11.2.2   *The rice taste evaluation problem*

Subjective qualification of food taste constitutes an important but complex problem. The quality of rice is evaluated by a group of usually 24 experts that performs a subjective *sensory test* (Ishibuchi, Nozaki, Tanaka, Hosaka, and Matsuda, 1994; Nozaki, Ishibuchi, and Tanaka, 1997). The attributes that affect the rice quality are shown in Table 11.9.

Modelling the rice quality evaluation becomes a complex problem due to the large number of relevant attributes and the highly non-linear dependency of the classification on these attributes. Moreover, the objective is not only to obtain an accurate model, but also a user-interpretable model that is able to provide some insight into the reasoning process of the human experts. Hence, we are dealing again with a linguistic modelling problem.

This section reviews the performance of different techniques applied to solve the rice evaluation problem. All of them employ the data set presented in (Nozaki, Ishibuchi, and Tanaka, 1997), which is composed of 105 instances of different rice sorts grown in Japan (for example, Sasanishiki, Akita-Komachi, etc.) and contains the subjective expert evaluations of the five input and the output attributes, which are normalised in $[0, 1]$.

In order not to bias the learning process, the data set was partitioned into ten pairs of training sets of 75 and test sets of 30 instances. Each method is evaluated on the respective partitions and the results are averaged.

As the interpretability is a key issue, less complex fuzzy models with a small number of rules are preferred. Therefore, each input variable is characterised by only two linguistic labels, that refer to normalised triangular fuzzy sets. Fuzzy models of higher granularity are ignored as they drastically increase the number of rules in the generated linguistic model RBs.

Non-evolutionary Linguistic Modelling
Two modelling techniques different from GFRBSs have been applied to the problem, ad hoc data-driven algorithms (Casillas et al., 2000) and neuro-fuzzy systems (Nauck and Kruse, 1997). Within the first group, Cordón and Herrera (2000) applied the ad hoc data-driven processes WM, NIT and CH. The neuro-fuzzy processes considered by Cordón and Herrera (1999a) and Alcalá et al. (2000) are NefProx (Nauck, Klawoon, and Kruse, 1997), ANFIS (Jang, 1993) and the one proposed by Shann and Fu (1995).

GFRBS Linguistic Modelling
The GFRBSs considered are shown as follows:

- MOGUL and SLAVE as descriptive Mamdani-type GFRBSs based on the IRL approach introduced in Chapter 8 (Cordón, Herrera, González, and Pérez, 1998).
- The GFRBSs based on the Pittsburgh approach proposed by Thrift (1991) —introduced in Sec. 7.3.1— and by Liska and Melsheimer (1994) (Alcalá et al., 2000).
- The two variants of the ALM methodology previously considered in Sec. 11.2.1.1: WM-based and CH-based ALM (Cordón and Herrera, 2000).

Notice that, in addition, the descriptive genetic tuning process introduced in Sec. 4.3.3.3 was also applied by Alcalá et al. (2000) to refine the DBs of the generated linguistic models. The results obtained from the refined linguistic models are reported as well.

Comparison between modelling techniques

Table 11.10 compares the results of different modelling techniques. The numbers shown in columns SE_{tra}, SE_{tst}, and $\#R$ are the average results over the ten data set partitions. The table is comprised of two parts, one that refers to the pure linguistic modelling techniques, and the other to the refined linguistic models obtained by means of the descriptive genetic tuning. The first column contains the results obtained from the generation and selection processes, whereas the second column reports the results obtained after tuning. A straight line denotes unavailable results.

Table 11.10   Results for the rice taste evaluation

Method	#R	Generation SE_{tra}	SE_{tst}	Tuning SE_{tra}	SE_{tst}
Ad hoc data-driven					
WM	15	0.01328	0.01311	0.00111	0.00214
NIT	64	0.00862	0.00985	—	—
CH	32	0.00910	0.01012	—	—
Neuro-fuzzy					
NefProx	15	0.00633	0.00568	—	—
ANFIS	32	0.00503	0.00563	0.00277	0.00387
Shann-Fu	32	0.01940	0.02137	0.00183	0.00331
GFRBSs					
MOGUL	6	0.00486	0.00370	0.00108	0.00233
SLAVE	2 *	0.00411	0.00728	—	—
Thrift	15.9	0.00495	0.00600	0.00115	0.00293
Liska-Mels.	30.6	0.00128	0.00236	0.00081	0.00234
WM-based ALM	5	0.00341	0.00398	0.00103	0.00274
CH-based ALM	7.9	0.00359	0.00470	—	—

The results demonstrate that most GFRBSs clearly outperform ad hoc data-driven learning methods and neuro-fuzzy systems, both in accuracy and RB complexity. The best overall results without the subsequent tuning process are obtained by Liska-Melsheimer's GFRBS, although the generated models are on average composed of a large number of rules, 30.6. Significantly simpler models with comparable performance can be obtained from other GFRBSs. SLAVE tends to generate the most simple models, which are comprised by only 2 DNF rules in average. The following RB corresponds to one of these models:

R_1 : *IF Taste is Bad THEN Evaluation is Low*
also
R_2 : *IF Appearance is Good AND Taste is Good*
        THEN Evaluation is Good

In regard to the genetic tuning process results, all models demonstrate a good performance independent of the method that generated the original RB. Both training and test set errors are significantly reduced during the second learning stage. The Liska-Melsheimer's GFRBS still demonstrates the best abstraction capability but due to the large number of rules it over-fits the training data, as can be seen from the relatively small improvement in the generalisation error. Surprisingly, the simple WM learning process achieves the smallest test set error after the tuning process. MOGUL provides a good compromise solution between accuracy and complexity, as on average it employs only six rules but is still second best in classification error.

11.2.3 *Dental development age prediction*

Lee and Takagi (1996) proposed TSK FRBSs to estimate human dental age from tooth eruption status and patient chronological age. The authors compared an FRBS derived from expert knowledge with another automatically generated from a set of data collected by a Japanese researcher by means of their GFRBS based on the Pittsburgh approach (Lee and Takagi, 1993b, 1993c). A description of this GFRBS can be found in Sec. 7.3.3.

The prediction is based on human body tissues, which matures at different rates, thus causing the different subsystems composing the tissue to have different physiological ages. Due to this reason, accurate dental development age estimates are especially valuable to orthodontic and paediatrics dentists, which often use it to determine if dental treatment is mandatory to insure proper oral function. The dental age estimates are useful for identifying eruption trends in an individual, allowing the possibility of avoiding future dental problems by preventive measures.

X-ray images give dentists the most accurate assessment of the dental

---

*The fuzzy rules used in SLAVE are in DNF form

Table 11.11   Notation considered for the dental development age prediction problem
variables

Symbol	Meaning
x_1	State of tooth eruption (STE)
x_2	Deviation from the average time of tooth eruption (DATTE)
y	Evaluation results of dental development level (EDL)

development. The size and location of teeth below the gumline can be seen
and the state of development is easily assessed. In the absence of this information, the dentist must rely on averages and standard deviation measures
of the tooth eruption (The Japanese Society of Pedontics, 1988). However,
the large values for standard deviation of the premolars and canines suggest that modelling the data as normal distributions may be inadequate
and calls for a more powerful technique.

Linguistic and fuzzy modelling seem to be a suitable tool to solve the
problem since expert knowledge is readily available —dentists usually have
some rules of thumb about the dental development age estimation based on
visual inspection and patient chronological age— and the final rule-based
structure of the model allow dentists to better understand the relationship
between dental and chronological age.

To solve the problem, the authors start from a previously designed linguistic model. The original Mamdani-type FRBS is comprised of 15 linguistic rules relating the input and output variables shown in Table 11.11.

The three variables are partitioned into fuzzy sets based on expert
knowledge. STE, which represents the dentists visual assessment of the
tooth eruption state, is defined in [-1.0,1.0] and has a term set associated
comprised by three linguistic labels: *pre eruption*, *tooth emerging* and *post
eruption*. DATTE is defined in [-2.0,2.0] and represents the number of
standard deviations the chronological age of the patient differs from the
chronological age at which the tooth erupts (which is different for each
tooth). The following term set, comprised by five labels, is considered for
this variable: { *far behind, behind, on time, ahead, far ahead* }. Finally,
the EDL gives the dental development level, is defined in [-3.0,3.0] and has
seven terms associated: *very delayed, delayed, slightly delayed, standard,
slightly premature, premature*, and *very premature*.

Using a data set of 200 instances of male patients at the age of 3 to

13, Lee and Takagi designed a TSK fuzzy model to estimate the dental development age of the permanent lower first molar of males. They first modified the fuzzy input variables by directly considering the patient age instead of the previously described variable DATTE.

The square error $(\sum_{i=1}^{N}(y - y')^2)$ was used as optimisation criterion. In order not to bias the learning process, the authors invoked four-fold cross-validation, i.e., the available data set was randomly divided into four subsets, with three of them used for training and the remaining one as test data. Table 11.12 compares the average results of the automatically generated TSK fuzzy model with the manually designed linguistic model. The parameter values of the GFRBS are reported in the original paper (Lee and Takagi, 1996).

Table 11.12 Results for the dental development age estimation

Data set	Linguistic model		TSK fuzzy model	
	Training	Test	Training	Test
1	1.422497	1.333736	0.295571	0.464555
2	1.520893	1.058851	0.381718	0.531591
3	1.389938	1.437610	0.383921	0.694926
4	1.277530	1.775366	0.310246	0.336980

Analysing the results, it is clear that the TSK fuzzy model designed from the GFRBS outperforms the hand-made linguistic model by a wide margin. As the authors note in the paper, the differences between both models were particularly obvious in regions for which no data was available in the training data set, i.e., the TSK fuzzy model abstracted and generalised better than the linguistic model. Moreover, the GFRBS is able to generate simpler models with a fewer number of rules (for example, in the case of the fourth data set, a model with only six rules was obtained).

11.3 Control Systems

Control problems involve the design of a system (the controller) in such a way that the overall behaviour of a compound structure comprising the controller and the controlled system (or plant) shows some specific properties, usually, minimising or maximising some performance measure, such as fuel consumption, tracking deviation, wasted material, etc.

The classical approaches to control system design usually involve two important characteristics:

- The system to be controlled (plant) has to be known in such a way that given a specific input, it is possible to predict the output generated by the system.
- The goal of the control protocol must be described in terms of a mathematical expression (referred as the performance index) involving the variables of the system.

Expert control, knowledge-based control, and fuzzy control overcome these requirements as they implicitly include the necessary information in the generation of the control protocol. However, in case of a genetic approach to FLC design, at least the second issue (performance index) has to be made explicit in the genetic process.

Previous sections considered modelling and classification problems, in which the main goal of the learning process is to reduce the identification error, namely the distance between the output generated by the model or classifier for a given input, and the target output (the output of the system to be modelled or the class of the given element).

From the point of view of GFRBS design, the main difference between identification (modelling and classification) and control problems is the definition of the evaluation functions employed by the EA. This disparity not only centres on the fact of minimising (optimising) the consumption, or the deviation instead of the error, but in particular on the critical role of time in the evaluation of the effect of a specific control action. In the general case, the current state of a system depends not only on the previous control action, but also on the entire history of control actions invoked in the past. As a result of these questions, one of the main problems in real-world control applications of GFRBSs is the definition of a suitable performance index or evaluation function.

This section considers two different types of FLCs (from the point of view of the application):

(1) *Direct controllers.* The output of the controller is directly applied to the plant to close the control loop.
(2) *Supervisory controllers.* The output of the system serves as the reference input to a low-level controller, or as a recommendation to

a human operator, which translates it to low-level commands to be applied directly to the plant.

Three different applications covering direct control (the cart-pole balancing system) and supervisory control (biped walking machine gait synthesis, and supervision of fossil power plant operation) will be described.

11.3.1   *The cart-pole balancing system*

The cart-pole balancing system is a classical control problem that the literature has established as a benchmark for learning control systems. Several papers cited along the book use this problem to test their proposals. The objective of such a system is the control of the translational force applied to a cart, to position it at the centre of a finite track, while balancing an inverted pendulum hinged on the cart.

The problem is represented through a set of non-linear differential equations simulating the dynamics of the cart-pole system. The non-linear differential equations are:

$$\ddot{x} = \frac{F - \mu_c \cdot sign(\dot{x}) + m \cdot l \cdot \dot{\theta}^2 \cdot \sin\theta + \frac{3}{4} \cdot m \cdot \cos\theta \cdot (\frac{\mu_p \cdot \dot{\theta}}{m \cdot l} - g \cdot \sin\theta)}{M + m \cdot (1 - \frac{3}{4} \cdot \cos^2\theta)}$$

and

$$\ddot{\theta} = -\frac{3}{4 \cdot l} \cdot (\ddot{x} \cdot \cos\theta - g \cdot \sin\theta + \frac{\mu_p \cdot \dot{\theta}}{m \cdot l})$$

where $\ddot{x}$ and $\ddot{\theta}$ are functions of F, $\dot{x}$, θ and $\dot{\theta}$.

A set of dimensions and parameters of the system, commonly used in the literature, is presented in Table 11.13.

Jamshidi (1997) and Wang (1994) report different solutions to this problem using fuzzy control.

11.3.1.1   *Goal and fitness function*

The evaluation of a control system should be based on the analysis of the evolution of the system state over time. A KB is evaluated by testing it for an appropriate time period[†] starting from different initial states. A partial

---

[†]The time to reach the desired state from any initial condition, that in most papers is fixed to 60 seconds (6000 control cycles).

Table 11.13   Model Dimensions and Parameters

Symbol	Description	Value
x	Cart position	[-1.0,1.0] m
θ	Pole angle from vertical	[-0.26,0.26] rad
F	Force applied to cart	[-10,10] N
g	Force of gravity	9.8 m/s^2
l	Half-length of pole	0.5 m
M	Mass of the cart	1.0 kg
m	Mass of the pole	0.1 kg
μ_c	Friction of cart on track	0.0005 N
μ_p	Friction of pole's hinge	0.000002 kg m^2

simulation is terminated if any of the state variables exceeds the physical limits ($x = [-1, 1]$ m, $\theta = [-\theta_l, \theta_l]$ rad). Different starting positions are considered, usually arranged symmetrically around $\theta = 0, x = 0$. A performance index (KB evaluation) based on the cart position is computed upon termination of a trial. The experiments below employ the following performance index:

$$P = \frac{1}{20 \cdot 6000} \sum_{20 \text{ init.}} \sum_{6000 \text{ cycles}} \exp(-|x|)$$

The performance index depends on the variable x alone, rather than on (x, θ) because preliminary experiments demonstrated that learning based on θ usually focuses on only balancing the pole, whereas the cart position is ignored.

The evaluation of the learning process, which is illustrated in Fig. 11.2, is similar to that described by Cooper and Vidal (1993).

11.3.1.2   *Some results*

In the following, the learning method described in Sec. 7.3.4 is applied to the cart-pole problem, and results from a set of experiments with different initial population and learning parameters are reported (Magdalena and Monasterio, 1997). The experiments use twenty initial states generated from five pairs of (θ, x): $\theta(0) = 0.01$ rad with $x(0) = 0.01, 0.1$ and 0.5 m, and $\theta(0) = 0.15$ rad with $x(0) = 0.1$ and 0.5 m; and the symmetrical values

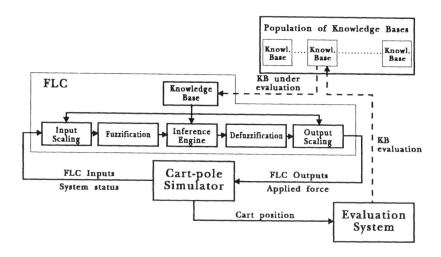

Fig. 11.2   Evaluation Process

$(\theta, -x)$, $(-\theta, x)$ and $(-\theta, -x)$. All the initial states have null derivatives $(\dot{\theta}(0) = 0$ and $\dot{x}(0) = 0)$. The EA employs a population size of two hundred, due to elitism the fifty best individuals are directly copied to the next generation.

Three groups of experiments are described:

(1) Learning based on the use of linear scaling, working only with rules and gain.
(2) Learning based on the rules and the non-linearity of the scaling, but maintaining the range of the variables.
(3) Learning based on the use of non-linear scaling (with variable range) and rules.

Experiments evolving rules and gain
Different values of reordering rate, rule mutation rate, scaling functions mutation rate and maximum gain variation have been applied. Linear scaling has been imposed by using a fixed initial value of a equal to 1 (linear scaling), and no sensitivity mutation ($\alpha = 1$).

A typical set of learning parameters for the evolutionary process is: a rule mutation rate of 0.05, a range mutation rate of 0.1 and a maximum gain variation (K) of 0.2 or 0.3 (see Eq. 7.10). Figure 11.3 shows the evolution of the best performance (average of four experiments with different,

randomly generated, initial population), over sixty generations. The solid
line represents the evolution using $K = 0.2$, and the dotted line corresponds
to $K = 0.3$. By analysing these results, it is possible to assert that the genetic process has a certain degree of robustness in relation to the variation
of the parameter K. The results obtained with $K = 0.2$ and $K = 0.3$ are
not qualitatively different.

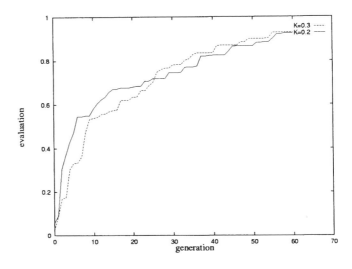

Fig. 11.3   Average evolution of higher evaluation (evolving rules and gain)

Experiments evolving rules and sensitivity

A set of experiments has been carried out for the following set of learning
parameters: a rule mutation rate of 0.05, a range mutation rate of 0.1 and
a maximum variation of sensitivity parameter (α) of 2 or 4 (see Eq. 7.11).
Figure 11.4 shows the evolution of the maximum performance (average of
four experiments with different, randomly generated, initial population),
over sixty generations. The solid line represents the evolution using $\alpha = 2$,
and the dotted line corresponds to $\alpha = 4$. As in the previous experiments,
it is possible to assert that the genetic process has a certain degree of
robustness in relation to the variation of the parameter α. The results
obtained with $\alpha = 2$ and $\alpha = 4$ are not qualitatively different.

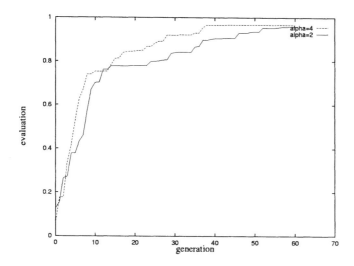

Fig. 11.4   Average evolution of higher evaluation (evolving rules and sensitivity)

Experiments evolving rules, gain and sensitivity

As in previous experiments, different values of the evolution parameters have been applied. A typical set of learning parameters for the evolutionary process is: a rule mutation rate of 0.1, a range mutation rate of 0.1, a maximum gain variation (K) of 0.2 and a sensitivity (α) of 2. Figure 11.5 shows the evolution of the maximum performance (average of four experiments with different, randomly generated, initial population), over sixty generations. The solid line represents the evolution using $K = 0.2$ and $\alpha = 2$, and the dotted lines represent the average maximum performance over experiments producing Fig. 11.3 (fixed sensitivity) and Fig. 11.4 (fixed gain). Figure 11.6 shows a trajectory of the cart position controlled by an FLC generated by the evolutionary process with $K = 0.2$ and $\alpha = 2$.

11.3.2   *A diversification problem*

This section show a similar control problem as the previous one, based on the same genetic approach (working with linear scaling functions), but for a completely different learning purpose.

The controlled system (supervisory control) is a simulated anthropomorphic ($1.75m$, $70kg$) biped walking machine (Magdalena, 1994; Magdalena

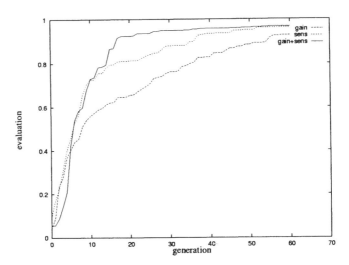

Fig. 11.5   Average evolution of higher evaluation by evolving gain and/or sensitivity

and Monasterio, 1995, 1997). The model is a six-link, 2-D structure, with two legs, a hip and a trunk (see Fig. 11.7). Each leg has a punctual foot, a knee with a degree of freedom and a hip with another degree of freedom. The contact of the stance-foot with the ground defines an unpowered degree of freedom (α_1) whose behaviour is not directly controlled, but controlled through the appropriate movements of the joints ($\alpha_2 - \alpha_6$). The complete mathematical description of the model is obtained by Magdalena (1994).

The goal of the FLC is to define joint trajectories in such a way that the biped system describes a regular walk without falling or stopping. In this case, the objective of the GAs is diversification, rather than optimisation that is usually the objective of genetic learning. The idea is to use bio-mechanical studies to obtain the KB of an FLC controlling a regular walk with a certain speed and stride length, and then apply GAs to create other KBs capable of controlling regular walks at different speeds and stride lengths.

The FLC generates a sequence of movements of the biped model according to the information in the KB. This sequence of movements is evaluated generating a value in [0,1], based on the *stability* and *regularity* of the walk, over a ten-second simulation (a thousand control cycles). A walk is considered stable while the biped is still walking forward (without stopping,

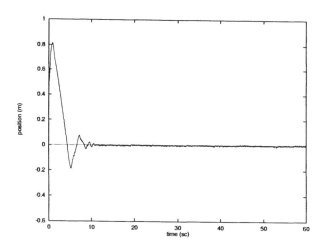

Fig. 11.6 Trace of cart position applying FLCs obtained by evolving rules, gain and sensitivity

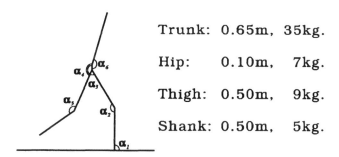

Fig. 11.7 Variables and dimensions of the model

falling, or stepping back). The stability is evaluated from 0 to 0.8, according to the part of the simulation period that is considered as stable walk. Only for those walks receiving 0.8 evaluation for stability, i.e., situations generating a complete stable simulation of 10 sc., the regularity of walk is evaluated and added to the previous value. The regularity will evaluate up to 0.2, and is computed considering the standard deviation of the stride period (time between two successive contacts). Regularity will evaluate to 0.2 when deviation is 0, and will decrease linearly to be 0 when the deviation equals to 20% of the mean period of the evaluated walk.

Experimental studies demonstrated the ability of the learning system to generate proper gaits (Magdalena and Monasterio, 1995, 1997). Figure 11.8 shows four walking sequences under the control of genetically generated FLCs. All KBs are obtained from a single learning process with a reproduction rate of 0.8, a reordering rate of 0.5, a rule mutation rate of 0.01, a range mutation rate of 0.05 and a mutation constant (the parameter K on Eq. 7.10) equal to 0.5. The maximum population size was 500 individuals, the number of generations was 40 and the initial population contained five expert KBs that generated proper gaits[‡], whereas the remaining KBs caused the biped system to tumble. The five expert KBs were extracted from bio-mechanical studies and the generated gaits showed speeds in the $[1.05,1.15]m/sc$ interval and stride lengths in the interval $[0.67,0.68]m$. The sequences (Fig. 11.8) present simulation results obtained with the biped model, with a temporal resolution of twenty images per second, except for sequence 4 which displays ten images per second. Each sequence represents a particular case: the best, the shortest, the longest and fastest, and the slowest gait.

The main characteristics of each sequence are described below: the average speed of the biped (S), the stride length (L) and the duration of the sequence (T):

(1) The most regular (best): $S = 1.21m/sc$, $L = 0.68m$ and $T = 2sc$.
(2) The shortest: $S = 0.76m/sc$, $L = 0.64m$ and $T = 3sc$.
(3) The longest and fastest: $S = 1.29m/sc$, $L = 0.78m$ and $T = 2sc$.
(4) The slowest[§]: $S = 0.54m/sc$, $L = 0.67m$ and $T = 4.5sc$.

The genetic learning process improved the stride length and speed from the initial intervals $[0.67,0.68]m$ and $[1.05,1.15]m/sc$, to $[0.64,0.78]m$ and $[0.54,1.29]m/sc$, generating a total of more than hundred different gaits within this parameter range. Fifty of these gaits achieve a better performance than the best individual in the initial population (0.9491). The overall best individual within forty generations of the previously described experiment achieved a performance of 0.9817, which corresponds to a standard deviation of stride length of less than 2%.

[‡]Four of these gaits had an evaluation 0.8 (meaning that the walking sequence was quite irregular), and the fifth had an evaluation of 0.9491.

[§]This sequence contains only ten images per second.

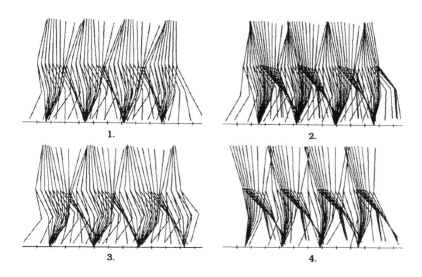

Fig. 11.8   Different genetically generated gaits

11.3.3   *Supervision of fossil power plant operation*

11.3.3.1   *Problem statement*

Power plants are complex processes, in which several low-level automatic controllers stabilise the plant on a short-term horizon. However, human operators are in charge of the supervisory control in order to optimise the long-term performance of the power plant. In such a process, a supervisor control system may assist the human operator to achieve an improved overall plant operation. The task at hand constitutes a modelling problem similar to the power line length prediction problem discussed earlier in this chapter. In this case, the problem is the improvement of the plant efficiency and the reduction of emissions caused by its operation. Emission reduction is a key issue in particular for fossil power plants which are only operated during on peak hours in order to complement the production of other power plants (nuclear, water, ...). The assignment of these peaks is mainly based on the previously mentioned criteria (efficiency and emissions). Consequently, these power plants have to improve their power generation process in order to justify their operation. This was the situation in late eighties when UITESA (Unión Iberoamericana de Tecnología Eléctrica SA) and Universidad Politécnica de Madrid started a first research project (Magdalena et al., 1993; Velasco et al., 1992) with the result of a prototype system installed (December 1990) in the second group of Térmicas del Besós SA, which was a company of Enher and Hidroeléctrica de Cataluña (50%, 50%), at Sant Adriá del Besós (Barcelona, Spain).

As a result, a second project (CORAGE) founded by the European Commission and the Spanish CDTI (Centre for Technological and Industrial Development) took place from 94 to 96, generating a supervisory control system that was installed in the second group of Velilla, a coal plant of Iberdrola, located at Velilla del Río Carrión (Palencia, Spain).

11.3.3.2 *Genetic fuzzy rule-based system*

The control architecture was composed of an FLC that receives its input from an acquisition module in charge of filtering and pre-processing 23 variables that reflect the current status of the plant and its environment. These variables were selected by operators as the minimum set for properly defining the plant state. The supervisory FS suggested eleven set points for the operation variables. These set points were presented to the operator through a graphical interface module and the operators decided whether to follow or reject the recommendation made by the FLC.

It has been decided to use a continuous learning system for on-line optimisation, due to the complexity and time variability of the system (broken or dirty pipes, slow pumps, air temperature, humidity, ... , are time-varying parameters that can not be captured within the control system). New data is acquired at a rate of one sample per every 10 minutes[¶], upon which the FS responds with a recommendation for the next control action.

An approximate FRBS working with trapezoidal fuzzy sets and following the Michigan approach was applied. The system is described in detail in Sec. 6.3.2. The fuzzy rules of the form

$$IF\ V_1 \in [0.1, 0.2, 0.4, 0.5]\ and\ V_3 \in [0.3, 0.3, 0.4, 0.6]$$
$$THEN\ V_7 \in [0.2, 0.3, 0.3, 0.4]$$

refer to a variable number of input variables.

[¶]The first project used a cycle of two minutes that was extended after analysing the comments of the operators and their control protocols.

The objective of the system was to improve the efficiency of the process (measured as a reduction of heat rate). The performance of the controller was evaluated based on the current heat rate ten minutes after the control action was applied according to the measure introduced in Sec. 6.3.2:

$$SP = \begin{cases} 1 & \text{if } HR_t \leq HR_{min} \\ \frac{HR - HR_{t-1}}{HR_{min} - HR_{t-1}} & \text{if } HR_t > HR_{min} \wedge HR_t \leq HR_{t-1} \\ \frac{HR_{t-1} - HR}{HR_{max} - HR_{t-1}} & \text{if } HR_t < HR_{max} \wedge HR_t > HR_{t-1} \\ -1 & \text{if } HR_t > HR_{max} \end{cases}$$

where HR denotes the heat rate, and the maximum and minimum values are evaluated over the past N values of the variable.

The final rule evaluation is obtained by multiplying the rule influence estimation (as described in Sec. 6.3.2) and the system performance. A further problem resulted from the fact that the instantaneous, true heat rate was not directly observable on the real plant. In fact, only an overall heat rate measured on a daily basis was available, and the instantaneous heat rate was variable considering that the caloric properties of the used coal varied as it originates from different mines in the area. An estimated instantaneous heat rate was computed as a function of measurable values (such as the air consumption or the volume and composition of emission).

The *Limbo* (Sec. 6.3.2) is one of the key aspects of the system design. The *Limbo* is an intermediate storage for rules generated by the GA, that are tested and evaluated without directly affecting the output of the controller. The role of the Limbo is different when dealing with direct control systems or with supervisory systems that interface with human operators. In the first case, the Limbo reduces the effect of the integration of new (unevaluated) rules in the system. In the second case, its role is mostly psychological, as it avoids that randomly generated new rules that obviously propose incorrect control action are presented to the human operator who might loose confidence in the supervisory system and tends to ignore future correct suggestions as well.

11.3.3.3 *Application results*

As a result of a preliminary analysis, the following parameters have been used:

- Maximum number of rules in the KB: 1000.

- Learning rate K (for strength update): 0.25.
- Age limit for the limbo: 300.
- Minimum rule activations: 10.
- Minimum equivalent evaluation: 0.1.

The on-site test of the FLC took place in June 1996. In the validation phase, two periods of ten days of continuous operation were evaluated. The two periods were qualitatively equivalent in terms of generated power and environmental conditions. During the first period, the system was run in the background (to allow learning) and the output was hidden from the operators. Consequently, the human operators applied their own control decisions as in the usual operation mode. During the second period, the operators received the recommendations of the system and which they were supposed to follow, as close as possible.

Two important properties were tested, first the validity of the estimated instantaneous heat rate as an approximation of the actual heat rate, and second, the improvement of the process (through the reduction of heat rate).

The results were convincing since at the end of the test, the heat rate was reduced by 1.12%. In fact, the reduction on the estimated heat rate was even larger (2.10%).

The main ideas of CORAGE were incorporated to the European project TOPGEN‖. The objective of that project were "to recommend on-line adjustments to the operators of Ballylumford Power Plant (North Ireland), property of Premier Power Plc (PREMIER), in order to achieve performance improvements during steady-state and in fault-free conditions, given a set for exploitation constraints. At the same time, continuous performance monitoring will be provided. Thus, the expected result is a software system, fully installed at Ballylumford Power Plant, that by means of operation recommendations and performance monitoring, will help improve the performance of the plant, reducing the amount of fuel necessary to obtain the electrical power required. This system will work on-line, continuously receiving data from the process".

‖Esprit 22945, Application of Advanced Software Technologies to the Optimization of Power Generation Plants.

11.4 Robotics

The last section of this chapter reviews the application of GFRBSs to mobile robotics. The first part introduces the basic concepts and necessary background information on behaviour-based robotics and evolutionary robotics. The second part describes the application of a GFRBS to learn an obstacle avoidance behaviour for a mobile robot.

11.4.1 *Behaviour-based robotics*

Behaviour-based robotics is a new field that takes animal behaviour as a guideline to build robotic systems that navigate, plan and operate in the real world. Therefore, the study of neuroscience, psychology and ethology provides valuable insight into the principles that underlie intelligent behaviour. Whereas some roboticists attempt to imitate the biological systems as accurate as possible, most researchers tend to abstract their details and rather use them as a mere inspiration for the design of intelligent robots.

The traditional artificial intelligence approach to robotics assumes a central, usually symbolic representation of the world as the basis for planning, reasoning and action. The common sense-plan-act scheme builds a world model from its sensory input, generates a plan based on the model which is then executed. This approach suffers from the drawback that perceptual information is subject to imprecision and uncertainty, which in all but the most predictable worlds makes it impossible to obtain an accurate, complete model of the environment. Even though prior knowledge about the environment can be beneficial for planning, behaviours that entirely rely on it to decide upon the robot actions are prone to fail as metric maps are approximate and incomplete and the location of objects and landmarks might change over time. In addition, the external environmental conditions, such as lighting, occlusion and sensor noise, might lead to incorrect interpretation of the sensor data.

Another problem is to achieve a timely robotic response in case the execution fails and a new plan has to be generated. As an alternative, Brooks (1986) proposed a behaviour-based approach, the so-called subsumption architecture, in which perception and action are tightly coupled. The two key aspects of the approach, are *situatedness*, which means to avoid abstract, symbolic representations, and *embodiment*, which refers to the fact that robots experience the world as physical entities. Behavioural responses result from the interaction between perception and action rather than through deliberation based on apparent goals and plans. The motor commands issued by a reactive control scheme are based on the external perception of features in the environment, rather than an internally represented, pre-computed path.

One of the major challenges in behaviour-based robotics is the question of how to coordinate among a collection of basic behaviours. Behaviour *arbitration* decides which behaviours should be activated in a particular context. *Command fusion* is the method to combine the different control actions proposed by individual behaviours into a coherent actuator command. The subsumption architecture uses inhibition and suppression as the mechanism to coordinate among multiple concurrent, primitive behaviours that directly ground on sensors and actuators. Complex behavioural sequences emerge as the currently active behaviour alters the state of the robot which in turn gives rise to new perceptual inputs that subsequently trigger the next appropriate behaviour.

In summary, a reactive behaviour is a map from a perceived stimulus to a reflex-type of motor response, that avoids the need for an abstract world representation. Behaviours constitute the basic building blocks for robotic actions as they are inherently modular and operate in parallel. Animal behaviours provide a source of inspiration for the design and operation of artificial robotic systems. The objective of behaviour-based robotics is to design robust control systems that enable a robot to reliably execute complex tasks in an at least partially unknown and unstructured environment.

Fuzzy control provides a valuable method for the design and implementation of robotic behaviours, as the information provided by the sensory apparatus usually lacks precision and completeness. The mode of approximate reasoning employed by FSs is particularly suitable to cope with the type of uncertainty and vagueness inherent to real-world situations. FLCs smoothly interpolate the control output across the input domain resulting in a steady, regular movement of the mobile robot despite the presence of sensor and environment noise. FLCs are based on qualitative knowledge rather than a precise model of the process, a feature that makes them less susceptible to variations in system parameters and external conditions. This type of robustness is particularly beneficial when fuzzy rule-based behaviours adapted in simulation are transferred to physical robots or are supposed to operate on different robotic platforms.

Fuzzy control provides an intuitive way to formulate the type of stimulus-response rules that constitute a distinguishable feature of behaviour-based robotics. The local scope of fuzzy rules and the linguistic representation of knowledge facilitate the specification and analysis of behaviours.

In the past, FLCs have been successfully implemented on mobile robots to perform path tracking, obstacle avoidance and navigation (Driankov and Saffiotti, 2000; Hoffmann, 2000; Takeuchi, Nagai, and Enomoto, 1988). In addition, roboticists used FL for behaviour coordination in a way that allows a smooth transition among different behaviours as an alternative to strictly prioritised arbitration scheme. The arbitration policy is formulated by fuzzy context rules that define the degree to which a behaviour is activated in a particular situation (Saffiotti, Konolige, and Ruspini, 1995).

11.4.2 *Evolutionary robotics*

As an alternative to the tedious manual design of robotic systems, evolutionary robotics is concerned with the automatic design of robotic behaviours by imitating the processes that occur in natural evolution. The genotype encodes parameters of the behavioural controller, for example the gains associated to the input stimuli. The designer implicitly specifies the desired behavioural properties by means of a scalar objective function that describes how well the robot accomplishes the task at hand. In case of a collision avoidance behaviour for example, the objective function punishes the robot for bumping into an obstacle. In all but a few cases, evolutionary behaviour design becomes a trial and error process, in which the human designer refines the original objective function upon observing the robotic behaviour evolved by means of the EA.

As EAs typically require a large number of fitness evaluations, most of the adaptation schemes for robotic behaviours utilise a simulation rather than learning in real-time on the robot itself. However, evolving in simulation bears the risk of over-adapting the robotic behaviour to peculiar features of the simulation scheme. This kind of brittleness is avoided if noise is added to the simulated sensors and actuators and by grounding the behaviour directly on the perception itself rather than some abstract representation of the world.

11.4.3 *Robotic platform*

The mobile robot used in the experiments is a Nomad Scout shown in
Fig. 11.9. It is equipped with ultrasonic, tactile and odometry sensors. A
low-level controller governs the sensor data acquisition, motion commands
and communications. The high level behavioural control either runs on an
on-board computer or a remote workstation that communicates with the
robot via radio modem. The robot possesses a two wheel differential drive
located at the geometric centre which allows omni-directional steering at
zero turn radius. The platform has a diameter of 41 *cm* and moves at a
speed of up to 1 *m/s*. Six bumpers for collision detection are arranged
around the outside perimeter.

Fig. 11.9   Nomad Scout

A ring of sixteen uniformly distributed sonar sensors measures distances
to objects in a range from 15 *cm* to 650 *cm* with a typical error margin
of about ±1%. Objects that are larger than the sonar wavelength of ≈
7 *mm* reflect the incoming sound wave, whereas objects smaller than the
wavelength act as diffractors. Objects with smooth surfaces such as walls,
doors or plates act as mirrors with respect to sound waves. Due to this
mirror-like property, a sonar echo might undergo multiple reflections before
reaching the receiver, which results in a potential over-estimation of the
distance to the object. Convex sharp edges such as door frames and table

legs produce a specular echo. Leonard and Durrant-Whyte (1992) provide a detailed discussion of sonar sensing for mobile robot navigation.

The basic motion command for the 2-DOF differential drive specifies the left and right wheel velocities for the 2-DOF differential drive. However, for navigation purposes, the desired robot motion is more intuitively described through the commanded translational $v \in [-50 \ cm/s, 50 \ cm/s]$ and turn rate $\omega \in [-45 \ °/s, 45 \ °/s]$ from which the wheel velocities are computed. The robot state is given by its position x, y and orientation ϕ and evolves according to

$$\begin{aligned} \dot{x} &= v \cdot \cos\phi \\ \dot{y} &= v \cdot \sin\phi \\ \dot{\phi} &= \omega \end{aligned}$$

11.4.4 *Perception of the environment*

The obstacle avoidance behaviour only takes distance measurements of the nine front sonar sensors into account. The translational velocity v and turn rate ω are controlled independent of each other. The translational velocity control is based on the concept of an inner and outer safety zone around the robot shown in Fig. 11.10. The white circle at the origin depicts the robot base. The plot is relative to the robot frame with the current direction of motion pointing downwards along the y-axis. The grey level indicates the danger of objects perceived under a certain angle and distance. Dark shades indicate a substantial danger of a collision, whereas a light shade corresponds to an at least currently harmless object.

Notice that the safety zone of the front sonar extends farther then those pointing to the sides. The overall safety index s is computed as the minimum over one complete cycle of sonar distance measurements $r_1, \ldots, r_9$ as

$$s = \min_i s_i(r_i)$$

where $s_i(r_i)$ is the safety index at a radius r_i along the axis of sonar i. The commanded velocity v assumes a value in the interval $[v_{min}, v_{max}] = [1cm/s, 25cm/s]$ in a way that the robot moves slower as the safety index decreases.

$$v = v_{min} + s \cdot (v_{max} - v_{min})$$

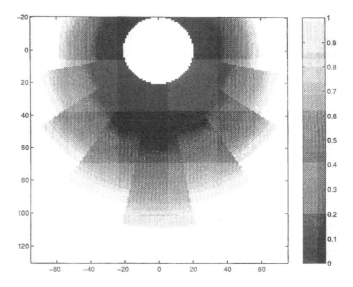

Fig. 11.10   Robot safety zone

The low angular resolution of sonar distance measurements makes it difficult to obtain exact, unambiguous geometric information based on a single sonar scan. In order to generate a map of the environment, the robot has to actively explore its environment and collect sonar data from multiple perspectives. However, reactive robotic behaviours, such as obstacle avoidance, do not require a complex representation but rather operate in a stimulus-response mode which establishes the control action directly on the perception rather than a geometric model of the environment.

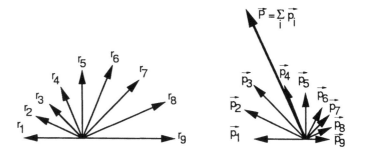

Fig. 11.11   Perception vector

The relevant perception in an obstacle avoidance behaviour has to capture somehow the proximity and location of the nearest object. Therefore, we aggregate a complete sonar scan $\{r_1, \ldots, r_9\}$ into a planar perception vector, whose direction vaguely indicates the location of obstacles and whose magnitude reflects the proximity of the nearest obstacle. A local perception vector $\vec{p}_i$ is calculated for each sonar sensor as shown in Fig. 11.11. Its direction coincides with the sensor axis while its length is inversely proportional to the sonar reading $r_i \in [0, d_{max}]$:

$$|\vec{p}_i(t)| = \begin{cases} 0 & : r_i > d_{max} \\ \frac{d_{max} - r_i}{d_{max}} & : r_i \le d_{max} \end{cases}$$

where $d_{max} = 2 \cdot m$ specifies an upper range of reaction above which obstacles are ignored. The global perception vector $\vec{P}$ is computed as the vector addition

$$\vec{P} = \sum_{i=1}^{9} \vec{p}_i$$

of the local vectors $\vec{p}_i$.

11.4.5 *Fuzzy logic controller*

The magnitude $|\vec{P}|$ and orientation $\angle \vec{P}$ of the perception vector $\vec{P}$ constitute the two input variables to the obstacle avoidance FLC that regulates the turn rate ω. The input space is partitioned (Fig. 11.12) into four fuzzy sets *very small, small, medium, large* abbreviated by *VS, SM, ME, LG* for the magnitude and seven fuzzy sets *very left, medium left, small left, zero, small right, medium right, very right* abbreviated by *VL, ML, SL, ZE, SR, MR, VR* for the orientation of the perception vector.

The domain of the turn rate ω is partitioned into nine output fuzzy sets *positive big, positive medium, positive small, positive zero, zero, negative zero, negative small, negative medium, negative big* denoted by *PB, PM, PS, PZ, ZE, NZ, NS, NM, NB* in Fig. 11.13. *PB* corresponds to a sharp left turn at near maximum rate of $40°/s$. The terms *PS, PZ, ZE, NZ, NS* correspond to relatively smooth turns with $|\omega| \le 10°/s$.

The obstacle avoidance behaviour is implemented by a standard descriptive Mamdani-type FLC whose RB contains 28 rules. The decision table of a manually designed controller is depicted in Table 11.14.

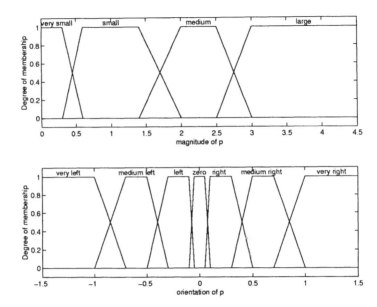

Fig. 11.12   Membership function for input variables magnitude and orientation of the perception vector $\vec{P}$

Table 11.14   Decision table of manually designed FLC

ω		$\angle\vec{P}$						
		VL	ML	SL	ZE	SR	MR	VR
	VS	ZE	ZE	ZE	NZ	ZE	ZE	ZE
$\|\vec{P}\|$	SM	ZE	NZ	NZ	NZ	PZ	PZ	ZE
	ME	ZE	NZ	NS	NM	PS	PZ	ZE
	LG	NZ	NS	NM	NB	PM	PS	PZ

11.4.6   *Genetic fuzzy rule-based system*

The GFRBS that is utilised for designing the obstacle avoidance behaviour is based on the Pittsburgh approach described in Chapter 7 in which the chromosome codes an entire RB. Chromosomes $\{s_1, \ldots, s_{28}\}$ are of fixed length and represent the 28 entries of the decision table coded as integers. Each gene $s_i \in [1, 2, \ldots, 9]$ encodes a possible output fuzzy label *PB, PM, PS, PZ, ZE, NZ, NS, NM, NB* for a particular fuzzy rule. The mutation

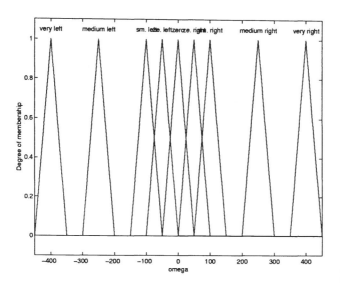

Fig. 11.13   Membership function for output variable ω

operator either shifts the integer either up or down, such that the offspring output fuzzy label for a particular rule is always a neighbour of the parent label. For example, mutation changes the label *PM* either to *PB* or *PS*. Despite its positional bias, one-point crossover is used for recombination as the influence on the behaviour of fuzzy rules that are neighbours in the decision table is likely to be correlated.

In the first experiment, the initial population is seeded with the manually designed RB. Each controller is evaluated on the Nomadic simulator starting from different initial positions. The training environment contains cluttered obstacles of various dimensions. A single run either stops if the robot collides with an obstacle as indicated by the bumper state or until a maximum number t_{max} of control steps elapses. In order to demonstrate a meaningful obstacle avoidance behaviour, the robot is only supposed to turn when an obstacle is present and otherwise to move as straight as feasible. In order to prevent the evolution of a behaviour that simply gyrates the robot on the spot, the fitness increases as the average turn rate ω decreases. The fitness attributed to the behaviour for a single control step t_i is

$$f(t) = \frac{v(t)}{v_{max}} \cdot \frac{\omega_{max} - |\omega(t)|}{\omega_{max}}$$

Table 11.15   Decision table of the evolutionary designed FLC

ω		$\angle\vec{P}$						
		VL	ML	SL	ZE	SR	MR	VR
	VS	ZE	PS	NM	NB	NS	NB	PM
$|\vec{P}|$	SM	NZ	NZ	NB	PZ	NB	NM	NZ
	ME	NZ	PS	NS	PZ	NM	ZE	NB
	LG	PS	PZ	NZ	NZ	ZE	NB	PZ

The fitness $f(t) \in [0,1]$ increases with decreasing absolute turn rate $|\omega(t)|$ and increasing velocity $v(t)$. Notice that the FLC only governs ω, whereas it has no direct control over $v(t)$ which is regulated according to the safety index. However, by timely steering the robot away from obstacles, the safety index and therefore the translational velocity remain large. The objective of the adaptation process is to find a suitable compromise that minimises the steering action while at the same maintaining a safe distance to obstacles.

In addition, the fitness function penalises collisions with obstacles. The fitness values $f(t)$ of individual control steps are accumulated over time:

$$f = \sum_{t=0}^{\min\{t_{coll}, t_{max}\}} f(t)$$

Fitness accumulation either terminates upon collision t_{coll} or after the maximum allowed time t_{max} in case no collision occurs. Therefore, behaviours that reliably avoid a collision achieve a better overall fitness even though they might gather less fitness per individual control step. Finally, the fitness values f_i achieved over multiple runs i are aggregated into a scalar fitness F. In order to emphasise the evolution of a robust behaviour, the total fitness F used in the selection step becomes the mean of the average fitness f_i and the worst fitness f_i:

$$F = \frac{\min_i f_i + <f_i>}{2}$$

The population size is 30 individuals, the initial generation is seeded randomly in that the nine output labels occur with equal probability. Table 11.15 shows the best decision table that emerged in the course of 50 generations. In general, the evolved FLC tends to steer the robot to the

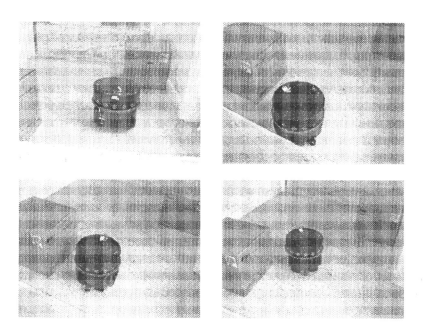

Fig. 11.14   Robot during experiment

right ($\omega < 0$) in order to avoid an obstacle. In those cases in which the robot turns left (*PZ, PS, PM, PB*), the magnitude of ω is usually smaller than for left turns (*NZ, NS, NM, NB*), as the decision table nine entries for sharp left turns (*NM, NB*) but only one case in which it turns right sharply(*PM, PB*). Demonstrating a preference for one turning direction is a good strategy as a symmetric behaviour bears the risk of cornering the robot by alternating left-right turns.

Figure 11.14 shows images from a robot experiment in a real environment. Figure 11.15 depicts the path described by the robot. A snapshot of the robot location and orientation is plotted every ten control steps based on the odometry data obtained from the wheel encoders. Notice that the distance between two snapshots indicates the velocity at which the robot is moving. A wide gap corresponds to a large velocity, densely spaced snapshots indicate a slow translational motion. For reasons of clarity, the figure only shows the first 40 snapshots of the entire trajectory.

The robot starts in the top right corner facing east (see top, left image Fig. 11.14). It moves straight ahead until it encounters the right wall and

Fig. 11.15   Robot path for obstacle avoidance behaviour with the fuzzy decision table in Table 11.15

performs a 180° turn (see top, right image and bottom left Fig. 11.14). As the robot finds more space after passing the rectangular obstacle it moves diagonally across the room at a large velocity (see bottom, right image Fig. 11.14). It again slows down as it reaches the lower left corner and performs another sharp right turn. In the following, the robot continues along a loop in which it performs right turns similar to the ones shown in the figure.

Bibliography

Abe, S. and R. Thawonmas (1997). A fuzzy classifier with ellipsoidal regions. *IEEE Transactions on Fuzzy Systems 5*, 358–368.

Alba, E., C. Cotta, and J. M. Troya (1999). Evolutionary design of fuzzy logic controllers using strongly-typed GP. *Mathware & Soft Computing 6*(1), 109–124.

Alcalá, R., J. Casillas, O. Cordón, and F. Herrera (1999). Fuzzy graphs: features and taxonomy of learning methods for non-grid oriented fuzzy rule-based systems. Technical Report DECSAI-99117, Dept. of Computer Science and Artificial Intelligence, University of Granada. Spain.

Alcalá, R., J. Casillas, O. Cordón, F. Herrera, and I. Zwir (2000). Techniques for learning and tuning fuzzy rule-based systems for linguistic modeling and their application. In C. T. Leondes (Ed.), *Knowledge-Based Systems. Techniques and Applications*, Volume 3, pp. 889–941. Academic Press.

Anderberg, M. R. (1973). *Cluster Analysis for Applications*. Kluwer Academic Press.

Angeline, P. J., G. M. Saunders, and J. B. Pollack (1994). An evolutionary algorithm that constructs recurrent neural networks. *IEEE Transactions on Neural Networks 5*(1), 54–64.

Babu, G. P. and M. N. Murty (1994). Clustering with evolution strategies. *Pattern Recognition 27*(2), 321–329.

Babuska, R. (1998). *Fuzzy Modeling for Control*. Kluwer Academic Press.

Bäck, T. (1996). *Evolutionary Algorithms in Theory and Practice*. Oxford University Press.

Baglio, S., L. Fortuna, S. Graziana, and G. Muscato (1993). Membership function shape and the dynamical behavior of a fuzzy system. In *Proc. First European Congress on Fuzzy and Intelligent Technologies (EUFIT'93)*, Aachen, Germany, pp. 645–650.

Baker, J. E. (1987). Reducing bias and inefficiency in the selection algorithm. In *Proc. Second International Conference on Genetic Algorithms (ICGA'87)*,

Hillsdale, USA, pp. 14–21.

Banzhaf, W., P. Nordin, R. Keller, and F. D. Francone (1998). *Genetic Programming: An Introduction.* Morgan Kaufmann.

Bardossy, A. and L. Duckstein (1995). *Fuzzy Rule-Based Modeling with Application to Geophysical, Biological and Engineering Systems.* CRC Press.

Bastian, A. (1994). How to handle the flexibility of linguistic variables with applications. *International Journal of Uncertainty, Fuzziness and Knowledge-Based Systems 3*(4), 463–484.

Bastian, A. (2000). Identifying fuzzy models utilizing genetic programming. *Fuzzy Sets and Systems 113*, 333–350.

Battle, D., A. Homaifar, E. Tunstel, and G. Dozier (1999). Genetic programming of full knowledge bases for fuzzy logic controllers. In *Proc. Genetic and Evolutionary Computation Conference (GECCO'99)*, Volume 2, Orlando, Florida, pp. 1463–1468.

Beasly, D., D. R. Bull, and R. R. Martin (1993). A sequential niche technique for multimodal function optimization. *Evolutionary Computation 1*(2), 101–125.

Benachenhou, D. (1994). Smart trading with FRET. In G. J. Deboeck (Ed.), *Trading on the Edge: Neural, Genetic, and Fuzzy Systems for Chaotic and Financial Markets*, pp. 215–242. John Willey & Sons.

Benítez, J. M., J. L. Castro, and I. Requena (2001). FRUTSA: Fuzzy rule tuning by simulated annealing. To appear in International Journal of Approximate Reasoning.

Bennett, K. P. and O. L. Mangasarian (1992). Neural network training via linear programming. In P. M. Pardalos (Ed.), *Advances in Optimization and Parallel Computing*, pp. 56–57. Elsevier Science.

Berenji, H. (1992). Fuzzy logic controllers. In R. R. Yager and L. A. Zadeh (Eds.), *An Introduction to Fuzzy Logic Applications in Intelligent Systems*, pp. 69–96. Kluwer Academic Press.

Bezdek, J. C. (1981). *Pattern Recognition with Fuzzy Objective Function Algorithms.* Plenum Press.

Bezdek, J. C. (1992). Guest editorial. *IEEE Transactions on Neural Networks 3*(5), 641.

Bezdek, J. C., C. Coray, R. Gunderson, and J. Watson (1981). Detection and characterization of cluster substructure. Part I: linear structure. *SIAM Journal on Applied Mathematics 40*(2), 339–357.

Bezdek, J. C. and R. J. Hathaway (1994). Optimization of fuzzy clustering criteria using genetic algorithms. In *Proc. First IEEE Conference on Evolutionary Computation (ICEC'94)*, Volume 2, Orlando, FL, USA, pp. 589–594.

Bezdek, J. C. and S. K. Pal (Eds.) (1992). *Fuzzy Models for Pattern Recognition. Methods that Search for Structures in Data.* IEEE Press.

Bezdek, J. C., E. C.-K. Tsao, and N. R. Pal (1992). Fuzzy Kohonen clustering networks. In *Proc. First IEEE International Conference on Fuzzy Systems (FUZZ-IEEE'92)*, San Diego, USA, pp. 1035–1043.

Binaghi, E., P. A. Brivio, and A. Rampini (Eds.) (1996). *Soft Computing in Remote Sensing Data Analysis.* World Scientific.

Bonarini, A. (1993). ELF: learning incomplete fuzzy rule sets for an autonomous robot. In *Proc. of the First European Congress on Intelligent Technologies and Soft Computing (EUFIT '93)*, Aachen, Germany, pp. 69–75.

Bonarini, A. (1994). Learning to coordinate fuzzy behaviors for autonomous agents. In *Proc. Second European Congress on Intelligent Technologies and Soft Computing (EUFIT'94)*, Aachen, Germany, pp. 475–479.

Bonarini, A. (1996a). Delayed reinforcement, fuzzy Q-learning and fuzzy logic controllers. In F. Herrera and J. L. Verdegay (Eds.), *Genetic Algorithms and Soft Computing*, Number 8 in Studies in Fuzziness and Soft Computing, pp. 447–466. Physica-Verlag.

Bonarini, A. (1996b). Evolutionary learning of fuzzy rules: competition and co-operation. In W. Pedrycz (Ed.), *Fuzzy Modelling: Paradigms and Practice*, pp. 265–284. Norwell, MA: Kluwer Academic Press.

Bonarini, A. (1996c). Learning dynamic fuzzy behaviors from easy missions. In *Proc. Sixth International Conference on Information Processing and Management of Uncertainty in Knowledge-Based Systems (IPMU'96)*, Granada, Spain, pp. 1223–1228.

Bonarini, A. (1997). Anytime learning and adaptation of hierarchical fuzzy logic behaviors. *Adaptive Behavior Journal 5*, 281–315.

Bonarini, A. and F. Basso (1997). Learning to compose fuzzy behaviours for automatic agents. *International Journal of Approximate Reasoning 17*(4), 409–432.

Bonarini, A. and V. Trianni (2001). Learning fuzzy classifier systems for multi-agent coordination. To appear in Information Sciences.

Bonissone, P. P. (1994). Fuzzy logic controllers: an industrial reality. In J. M. Zurada, R. J. Marks, and C. J. Robinson (Eds.), *Computational Intelligence: Imitating Life*, pp. 316–327. IEEE Press.

Bonissone, P. P., P. S. Khedkar, and Y. Chen (1996). Genetic algorithms for automated tuning of fuzzy controllers: A transportation application. In *Proc. Fifth IEEE International Conference on Fuzzy Systems (FUZZ-IEEE'96)*, New Orleans, USA, pp. 674–680.

Booker, L. B. (1982). *Intelligent behavior as an adaptation to the task environment.* Ph. D. thesis, Department of Computer and Communication Science. University of Michigan.

Booker, L. B. (1990). Representing attribute-based concepts in a classifier system. In *Proc. First Workshop on Foundations of Genetic Algorithms*, pp. 115–127. Morgan Kaufmann.

Booker, L. B., D. E. Goldberg, and J. H. Holland (1989). Classifier systems and genetic algorithms. *Artificial Intelligence 40*, 235–282.

Breiman, L., J. H. Friedman, R. A. Olshen, and C. J. Stone (1984). *Classification and regression trees.* Wadsworth.

Brooks, R. (1986). A robust layered control system for a mobile robot. *IEEE*

Journal of Robotics and Automation RA-2(1), 14–23.

Brown, H. and R. E. Smith (1996). Classifier systems renaissance: New analogies, new directions. In J. R. Koza, D. E. Goldberg, and R. L. Riolo (Eds.), *Proc. First International Conference on Genetic Programming (GP'96)*, pp. 547–552. MIT Press.

Buckles, B. P., F. E. Petry, D. Prabhu, R. George, and R. Srikanth (1994). Fuzzy clustering with genetic search. In *Proc. First IEEE Conference on Evolutionary Computation (ICEC'94)*, Orlando, FL, USA, pp. 46–50.

Buckley, J. J. (1993). Sugeno type controllers are universal approximators. *Fuzzy Sets and Systems 53*, 299–304.

Burdsall, B. and C. Giraud-Carrier (1997). Evolving fuzzy prototypes for efficient data clustering. In *Proc. Second International ICSC Symposium on Fuzzy Logic and Applications (ISFL'97)*, pp. 217–223.

Burkhardt, D. and P. P. Bonissone (1992). Automated fuzzy knowledge base generation and tuning. In *Proc. First IEEE International Conference on Fuzzy Systems (FUZZ-IEEE'92)*, San Diego, USA, pp. 179–188.

Butz, M. and W. Stolzmann (1999). Action-planning in anticipatory classifier systems. In *Proc. Second International Workshop on Learning Classifier Systems (2-IWLCS) on the Genetic and Evolutionary Computation Conference (GECCO'99)*, Orlando, Florida, pp. 242–249.

Butz, M. V., D. E. Goldberg, and W. Stolzmann (1999). New challenges for an anticipatory classifier system: Hard problems and possible solutions. Technical Report IlliGAL Report No. 99019, Illinois Genetic Algorithms Laboratory, University of Illinois at Urbana-Champaign.

Butz, M. V., D. E. Goldberg, and W. Stolzmann (2000a). Introducing a genetic generalization pressure to the anticipatory classifier system. Part 1. Theoretical approach. Technical Report IlliGAL Report No. 2000005, Illinois Genetic Algorithms Laboratory, University of Illinois at Urbana-Champaign.

Butz, M. V., D. E. Goldberg, and W. Stolzmann (2000b). Introducing a genetic generalization pressure to the anticipatory classifier system. Part 2. Performance analysis. Technical Report IlliGAL Report No. 2000006, Illinois Genetic Algorithms Laboratory, University of Illinois at Urbana-Champaign.

Butz, M. V., D. E. Goldberg, and W. Stolzmann (2000c). Investigating generalization in the anticipatory classifier system. Technical Report IlliGAL Report No. 2000014, Illinois Genetic Algorithms Laboratory, University of Illinois at Urbana-Champaign.

Butz, M. V., D. E. Goldberg, and W. Stolzmann (2000d). Probability-enhanced predictions in the anticipatory classifier system. Technical Report IlliGAL Report No. 2000016, Illinois Genetic Algorithms Laboratory, University of Illinois at Urbana-Champaign.

Butz, M. V. and S. W. Wilson (2000). An algorithmic description of XCS. Technical Report IlliGAL Report No. 2000017, Illinois Genetic Algorithms Laboratory, University of Illinois at Urbana-Champaign.

Cao, Z. and A. Kandel (1989). Applicability of some fuzzy implication operators.

Fuzzy Sets and Systems 31, 151–186.

Carpenter, G. A., S. Grossberg, N. Markuzon, J. H. Reynolds, and D. B. Rosen (1992). Fuzzy ARTMAP: A neural network architecture for incremental supervised learning of analog multidimensional maps. *IEEE Transactions on Neural Networks 3*(5), 698–713.

Carpenter, G. A., S. Grossberg, and D. B. Rosen (1991). Fuzzy ART: fast stable learning and categorization of analog patterns by an adaptive resonance system. *Neural Networks 4*, 759–771.

Carse, B., T. C. Fogarty, and A. Munro (1996). Evolving fuzzy rule based controllers using genetic algorithms. *Fuzzy Sets and Systems 80*, 273–294.

Casillas, J., O. Cordón, and F. Herrera (2000). A methodology to improve ad hoc data-driven linguistic rule learning methods by inducing cooperation among rules. Technical Report DECSAI-00101, Dept. of Computer Science and Artificial Intelligence, University of Granada. Spain.

Castillo, L., A. González, and R. Pérez (2001). Including a simplicity criterion in the selection of the best rule in a fuzzy genetic learning algorithm. To appear in Fuzzy Sets and Systems.

Castro, J. L. (1995). Fuzzy logic controllers are universal approximators. *IEEE Transactions on Systems, Man, and Cybernetics 25*(4), 629–635.

Castro, J. L. and M. Delgado (1996). Fuzzy systems with defuzzification are universal approximators. *IEEE Transactions on Systems, Man, and Cybernetics 26*(1), 149–152.

Chan, P. T., W. F. Xie, and A. B. Rad (2000). Tuning of fuzzy controller for an open-loop unstable system: A genetic approach. *Fuzzy Sets and Systems 111*(2), 137–152.

Chang, T. C., K. Hasegawa, and C. W. Ibbs (1991). The effects of membership functions in fuzzy reasoning. *Fuzzy Sets and Systems 44*, 169–186.

Cheong, F. and R. Lai (2000). Constraining the optimization of a fuzzy logic controller using an enhanced genetic algorithm. *IEEE Transactions on Systems, Man and Cybernetics - Part B: Cybernetics 30*, 31–46.

Chi, Z., H. Yan, and T. Pham (1996). *Fuzzy Algorithms: With Applications to Image Processing and Pattern Recognition*. World Scientific.

Chin, T. C. and X. M. Qi (1998). Genetic algorithms for learning the rule base of fuzzy logic controller. *Fuzzy Sets and Systems 97*(1), 1–7.

Chiu, S. (1994). Fuzzy model identification based on cluster estimation. *Journal of Intelligent Fuzzy Systems 2*, 267–278.

Chung, I.-F., C. J. Lin, and C. T. Lin (2000). A GA-based fuzzy adaptive learning control network. *Fuzzy Sets and Systems 112*(1), 65–84.

Clark, P. and T. Niblett (1986). Learning if-then rules in noisy environments. Technical Report TIRM 86-019, The Turing Institute, Glasgow. U.K.

Combs, W. E. and J. E. Andrews (1998). Combinatorial rule explosion eliminated by a fuzzy rule configuration. *IEEE Transactions on Fuzzy Systems 6*(1), 1–11.

Cooper, M. G. and J. J. Vidal (1993). Genetic design of fuzzy controllers. In

Proc. Second International Conference on Fuzzy Theory and Technology (FTT'93).

Cordón, O., M. J. del Jesús, and F. Herrera (1998). Genetic learning of fuzzy rule-based classification systems cooperating with fuzzy reasoning methods. *International Journal of Intelligent Systems 13*(10-11), 1025–1053.

Cordón, O., M. J. del Jesús, and F. Herrera (1999a). Evolutionary approaches to the learning of fuzzy rule-based classification systems. In L. C. Jain (Ed.), *Evolution of Engineering and Information Systems and Their Applications*, pp. 107–160. CRC Press.

Cordón, O., M. J. del Jesús, and F. Herrera (1999b). A proposal on reasoning methods in fuzzy rule-based classification systems. *International Journal of Approximate Reasoning 20*, 21–45.

Cordón, O., M. J. del Jesús, F. Herrera, and M. Lozano (1999). MOGUL: A methodology to obtain genetic fuzzy rule-based systems under the iterative rule learning approach. *International Journal of Intelligent Systems 14*(11), 1123–1153.

Cordón, O. and F. Herrera (1995). A general study on genetic fuzzy systems. In J. Periaux, G. Winter, M. Galán, and P. Cuesta (Eds.), *Genetic Algorithms in Engineering and Computer Science*, pp. 33–57. John Wiley and Sons.

Cordón, O. and F. Herrera (1996a). Generating and selecting fuzzy control rules using evolution strategies and genetic algorithms. In *Proc. Sixth International Conference on Information Processing and Management of Uncertainty in Knowledge-Based Systems (IPMU'96)*, Volume 2, Granada, Spain, pp. 733–738.

Cordón, O. and F. Herrera (1996b). A hybrid genetic algorithm-evolution strategy process for learning fuzzy logic controller knowledge bases. In F. Herrera and J. L. Verdegay (Eds.), *Genetic Algorithms and Soft Computing*, Number 8 in Studies in Fuzziness and Soft Computing, pp. 251–278. Physica-Verlag.

Cordón, O. and F. Herrera (1997a). Evolutionary design of TSK fuzzy rule-based systems using (μ, λ)-evolution strategies. In *Proc. Sixth IEEE International Conference on Fuzzy Systems (FUZZ-IEEE'97)*, Volume 1, Barcelona, Spain, pp. 509–514.

Cordón, O. and F. Herrera (1997b). Identification of linguistic fuzzy models by means of genetic algorithms. In D. Driankov and H. Hellendoorn (Eds.), *Fuzzy Model Identification. Selected Approaches*, pp. 215–250. Springer-Verlag.

Cordón, O. and F. Herrera (1997c). A three-stage evolutionary process for learning descriptive and approximate fuzzy logic controller knowledge bases from examples. *International Journal of Approximate Reasoning 17*(4), 369–407.

Cordón, O. and F. Herrera (1999a). ALM: A methodology for designing accurate linguistic models for intelligent data analysis. In D. J. Hand, J. N. Kok, and M. R. Berthold (Eds.), *Proc. Third International Conference on Intelligent Data Analysis (IDA'99)*., Number 1642 in Lecture Notes in Computer

Science, pp. 15–26. Berlin: Springer-Verlag.

Cordón, O. and F. Herrera (1999b). A two-stage evolutionary process for designing TSK fuzzy rule-based systems. *IEEE Transactions on Systems, Man, and Cybernetics 29*(6), 703–715.

Cordón, O. and F. Herrera (2000). A proposal for improving the accuracy of linguistic modeling. *IEEE Transactions on Fuzzy Systems 8*(3), 335–344.

Cordón, O. and F. Herrera (2001). Hybridizing genetic algorithms with sharing scheme and evolution strategies for designing approximate fuzzy rule-based systems. *Fuzzy Sets and Systems 118*(2), 235–255.

Cordón, O., F. Herrera, A. González, and R. Pérez (1998). Encouraging cooperation in the genetic iterative rule learning approach for qualitative modeling. In J. Kacprzyk and L. A. Zadeh (Eds.), *Computing with Words in Intelligent/Information Systems 2*, pp. 95–117. Physica-Verlag.

Cordón, O., F. Herrera, and M. Lozano (1996). A three-stage method for designing genetic fuzzy systems by learning from examples. In H. M. Voight, W. Ebeling, E. Rechemberg, and H. P. Schwefel (Eds.), *Proc. Fourth International Conference on Parallel Problem Solving from Nature - PPSN IV*, Number 1141 in Lecture Notes in Computer Science, pp. 720–729. Berlin: Springer-Verlag.

Cordón, O., F. Herrera, and M. Lozano (1997). A classified review on the combination fuzzy logic-genetic algorithms bibliography: 1989-1995. In E. Sanchez, T. Shibata, and L. Zadeh (Eds.), *Genetic Algorithms and Fuzzy Logic Systems. Soft Computing Perspectives*, pp. 209–241. World Scientific.

Cordón, O., F. Herrera, L. Magdalena, and P. Villar (2000). A genetic learning process for the scaling factors, granularity and contexts of the fuzzy rule-based system data base. In *Proc. Eighth International Conference on Information Processing and Management of Uncertainty in Knowledge-Based Systems (IPMU'00)*, Volume I, Madrid, Spain, pp. 430–437.

Cordón, O., F. Herrera, L. Magdalena, and P. Villar (2001). A genetic learning process for the scaling factors, granularity and contexts of the fuzzy rule-based system data base. To appear in Information Sciences.

Cordón, O., F. Herrera, and A. Peregrín (1997). Applicability of the fuzzy operators in the design of fuzzy logic controllers. *Fuzzy Sets and Systems 86*, 15–41.

Cordón, O., F. Herrera, and A. Peregrín (2000). Searching for basic properties obtaining robust implication operators in fuzzy control. *Fuzzy Sets and Systems 111*(2), 237–251.

Cordón, O., F. Herrera, and L. Sánchez (1997). Evolutionary learning processes for data analysis in electrical engineering applications. In D. Quagliarelli, J. Periaux, C. Poloni, and G. Winter (Eds.), *Genetic Algorithms in Engineering and Computer Science*, pp. 205–224. John Wiley and Sons.

Cordón, O., F. Herrera, and L. Sánchez (1998). Computing the spanish medium electrical line maintenance costs by means of evolution-based learning processes. In M. Ali, A. P. del Pobil, and J. Mira (Eds.), *Proc. Eleventh*

International Conference on Industrial & Engineering Applications of Artificial Intelligence & Expert Systems - IEA-98-AIE., Number 1415 in Lecture Notes in Artificial Intelligence, pp. 478–486. Berlin: Springer-Verlag.

Cordón, O., F. Herrera, and L. Sánchez (1999). Solving electrical distribution problems using hybrid evolutionary data analysis techniques. *Applied Intelligence 10*(1), 5–24.

Cordón, O., F. Herrera, and P. Villar (2000). Analysis and guidelines to obtain a good uniform fuzzy partition granularity for fuzzy rule-based systems using simulated annealing. *International Journal of Approximate Reasoning 25*(3), 187–215.

Cordón, O., F. Herrera, and P. Villar (2001). Generating the knowledge base of a fuzzy rule-based system by the genetic learning of the data base. To appear in IEEE Trans. on Fuzzy Systems.

Cordón, O., F. Herrera, and I. Zwir (1999). Linguistic modeling by hierarchical systems of linguistic rules. Technical Report DECSAI-99114, Dept. of Computer Science and Artificial Intelligence, University of Granada. Spain.

Cordón, O., F. Herrera, and I. Zwir (2000a). Hierarchical knowledge bases for fuzzy rule-based systems. In *Proc. Eighth International Conference on Information Processing and Management of Uncertainty in Knowledge-Based Systems (IPMU'00)*, Volume III, Madrid, Spain, pp. 1770–1777.

Cordón, O., F. Herrera, and I. Zwir (2000b). An iterative learning methodology for hierarchical systems of linguistic rules introducing rule reinforcement. Technical Report DECSAI-00106, Dept. of Computer Science and Artificial Intelligence, University of Granada. Spain.

Corcoran, A. L. and S. Sen (1994). Using real-valued genetic algorithms to evolve rule sets for classification. In *Proc. First IEEE Conference on Evolutionary Computation (ICEC'94)*, Volume 1, Orlando, FL, USA, pp. 120–124.

Cox, E. (1995). A fuzzy system for detecting anomalous behaviors in healthcare provider claims. In S. Goonatilake and P. Treleaven (Eds.), *Intelligent Systems for Finance and Business*, pp. 111–134. John Willey & Sons.

Daugherity, W. C., B. Rathakrishnan, and J. Yen (1992). Performance evaluation of a self-tuning fuzzy controller. In *Proc. First IEEE International Conference on Fuzzy Systems (FUZZ-IEEE'92)*, San Diego, USA, pp. 389–397.

Davé, R. N. (1990). Fuzzy shell clustering and application to circle detection in digital images. *International Journal on General Systems 16*(4), 343–355.

Davidor, Y. (1991). A genetic algorithm applied to robot trajectory generation. In L. Davis (Ed.), *Handbook of Genetic Algorithms*. Van Nostrand-Reinhold.

de Sousa, M. and M. K. Madrid (2000). Optimization of Takagi-Sugeno fuzzy controllers using a genetic algorithm. In *Proc. Ninth IEEE International Conference on Fuzzy Systems (FUZZ-IEEE'2000)*, Volume 1, San Antonio, Texas, USA, pp. 30–35.

Deb, K. and D. E. Goldberg (1989). An investigation of niche and species formation in genetic function optimization. In *Proc. Third International Conference on Genetic Algorithms (ICGA'89)*, Hillsdale, USA, pp. 42–50. Morgan

Kaufmann.

DeJong, K. (1988). Learning with genetic algorithms: An overview. *Machine Learning 3*(3), 121–138.

DeJong, K. A., W. M. Spears, and D. F. Gordon (1993). Using genetic algorithms for concept learning. *Machine Learning 13*, 161–188.

Delgado, M., A. F. Gómez-Skarmeta, and F. Martín (1997). A fuzzy clustering based rapid-prototyping for fuzzy rule-based modeling. *IEEE Transactions on Fuzzy Systems 5*(2), 223–233.

Delgado, M., M. A. Vila, and W. Voxman (1998). A fuzziness measure for fuzzy numbers: applications. *Fuzzy Sets and Systems 94*(2), 205–216.

Dorigo, M. (1991). New perspectives about default hier formation in learning classifier systems. Technical Report 91-002, Politecnico di Milano, Dipartimento di Elettronica. Politecnico di Milano. Italy.

Dorigo, M. and M. Colombetti (1997). *Robot shaping: An experimental in behavior engineering.* Cambridge, MA: MIT Press/Bradford.

Dorigo, M. and E. Sirtori (1991). Alecsys: a parallel laboratory for learning classifier systems. In *Proc. Fourth International Conference on Genetic Algorithms (ICGA'91)*, San Diego, USA, pp. 296–302. Morgan Kaufmann.

Driankov, D., H. Hellendoorn, and M. Reinfrank (1993). *An Introduction to Fuzzy Control.* Springer-Verlag.

Driankov, D. and A. Saffiotti (2000). *Fuzzy logic techniques for autonomous vehicle navigation.* Studies in Fuzziness. Springer Verlag.

Dujet, C. and N. Vincent (1995). Force implication: a new approach to human reasoning. *Fuzzy Sets and Systems 69*, 53–63.

Dunn, J. C. (1974). A fuzzy relative of the ISODATA process and its use in detecting compact well-separated clusters. *Journal of Cybernetics 3*, 32–57.

Egan, M. A., M. Krishnamoorthy, and K. Rajan (1998). Comparative study of a genetic fuzzy c-means algorithm and a validity guided fuzzy c-means algorithm for locating clusters in noisy data. In *Proc. 1998 IEEE Conference on Evolutionary Computation (ICEC'98)*, Anchorage, Alaska, USA, pp. 440–445.

Eshelman, L. and J. Schaffer (1993). Real-coded genetic algorithms and interval-schemata. In L. D. Whitley (Ed.), *Foundation of Genetic Algorithms 2*, pp. 187–202. Morgan Kaufmann.

Eshelman, L. J. (1991). The CHC adaptive search algorithm: How to have safe search when engaging in non traditional genetic recombination. In G. J. E. Rawlins (Ed.), *Foundations of Genetic Algorithms I*, pp. 265–283. Kaufmann Pub.

Farag, W. A., V. H. Quintana, and G. Lambert-Torres (1998). A genetic-based neuro-fuzzy approach for modeling and control of dynamical systems. *IEEE Transactions on Neural Network 9*(5), 756–767.

Fathi-Torbaghan, M. and L. Hildebrand (1997). Model-free optimization of fuzzy rule based systems using evolution strategies. *IEEE Transactions on Sys-*

tems, Man and Cybernetics-Part B: Cybernetics 27(2), 270–277.

Filipič, B. and D. Juričić (1996). A genetic algorithm to support learning fuzzy control rules from examples. In F. Herrera and J. L. Verdegay (Eds.), *Genetic Algorithms and Soft Computing*, Number 8 in Studies in Fuzziness and Soft Computing, pp. 403–418. Physica-Verlag.

Fisher, F. A. (1936). The use of multiple measurements in taxonomic problems. *Annals of Eugenics 7*, 179–188.

Fogel, D. B. (1991). *System Identification through Simulated Evolution: A Machine Learning Approach to Modeling*. Ginn Press.

Fogel, D. B. (1992). *Evolving artificial intelligence*. Ph. D. thesis, University of California, San Diego.

Fogel, D. B. (1993). Evolving fuzzy clusters. In *Proc. Second IEEE International Conference on Fuzzy Systems (FUZZ-IEEE'93)*, San Francisco, USA, pp. 1829–1834.

Fogel, D. B. (1998). *Evolutionary Computation. The Fossil Record*. IEEE Press.

Fogel, L. J. (1962). Autonomous automata. *Industrial Research 4*, 14–19.

Forrest, S. (1991). *Parallelism and Programming in Classifier Systems*. Morgan Kaufmann Publishers.

Furuhashi, T. (1997). Developement of if-then rules with the use of DNA coding. In W. Pedrycz (Ed.), *Fuzzy Evolutionary Computation*, pp. 107–125. Kluwer Academic Publishers.

Furuhashi, T., Y. Miyata, and Y. Uchikawa (1997). A new approach to genetic based machine learning for efficient local improvement and its application to a graphic problem. *Information Science 103*, 87–100.

Furuhashi, T., K. Nakaoka, K. Morikawa, H. Maeda, and Y. Uchikawa (1995). A study of knowledge finding using fuzzy classifier systems. *Journal of Japan Society for Fuzzy Theory and Systems 7*, 555–567.

Furuhashi, T., K. Nakaoka, K. Morikawa, and Y. Uchikawa (1994). An acquisition of control knowledge using multiple fuzzy classifier systems. *Journal of Japan Society for Fuzzy Theory and Systems 6*, 603–609.

Furuhashi, T., K. Nakaoka, and Y. Uchikawa (1994a). A new approach to genetic based machine learning and an efficient finding of fuzzy rules. In *Proc. 1994 IEEE/Nagoya University WWW on Fuzzy Logic and Neural Networks/Genetic Algorithms*, Nagoya, Japan, pp. 114–121.

Furuhashi, T., K. Nakaoka, and Y. Uchikawa (1994b). Suppression of excessive fuzziness using multiple fuzzy classifier. In *Proc. Third IEEE International Conference on Fuzzy Systems (FUZZ-IEEE'94)*, Volume 1, Orlando, FL, USA, pp. 411–414.

Furuhashi, T., K. Nakaoka, and Y. Uchikawa (1995). An efficient finding of fuzzy rules using a new approach to genetic based machine learning. In *Proc. Fourth IEEE International Conference on Fuzzy Systems (FUZZ-IEEE'95)*, Yokohama, Japan, pp. 715–722.

Furuhashi, T., K. Nakaoka, and Y. Uchikawa (1996). A study on fuzzy classifier systems for finding control knowledge of multi-input systems. In F. Herrera and J. L. Verdegay (Eds.), *Genetic Algorithms and Soft Computing*, Number 8 in Studies in Fuzziness and Soft Computing, pp. 489–502. Physica-Verlag.

Garibaldi, J. M. and E. C. Ifeator (1999). Application of simulated annealing fuzzy model tuning to umbilical cord acid-base interpretation. *IEEE Transactions on Fuzzy Systems 7*(1), 72–84.

Gath, I. and A. Geva (1989). Unsupervised optimal fuzzy clustering. *IEEE Transactions on Pattern Analysis and Machine Intelligence 7*, 773–781.

Geyer-Schulz, A. (1995). *Fuzzy Rule-Based Expert Systems and Genetic Machine Learning*. Heidelberg: Physica-Verlag.

Geyer-Schulz, A. (1996). The mit beer distribution game revisited: Genetic machine learning and managerial behavior in a dynamic decision making experiment. In F. Herrera and J. L. Verdegay (Eds.), *Genetic Algorithms and Soft Computing*, Number 8 in Studies in Fuzziness and Soft Computing, pp. 658–682. Physica-Verlag.

Giordana, A. and F. Neri (1996). Search intensive concept induction. *Evolutionary Computation 3*, 375–416.

Giordana, A. and L. Saitta (1993). REGAL: an integrated system for learning relations using genetic algorithms. In *Proc. Second International Workshop on Multistrategy Learning*, Harpers Ferry, VA, USA, pp. 234–249.

Glorennec, P. Y. (1991). Adaptive fuzzy control. In *Proc. Fourth International Fuzzy Systems Association World Congress (IFSA'91)*, Brussels, Belgium, pp. 33–36.

Glorennec, P. Y. (1996). Constrained optimization of FIS using an evolutionary method. In F. Herrera and J. L. Verdegay (Eds.), *Genetic Algorithms and Soft Computing*, Number 8 in Studies in Fuzziness and Soft Computing, pp. 349–368. Physica-Verlag.

Glorennec, P. Y. (1997). Coordination between autonomous robots. *International Journal of Approximate Reasoning 17*(4), 433–446.

Goldberg, D. E. (1989). *Genetic Algorithms in Search, Optimization, and Machine Learning*. Addison-Wesley.

Goldberg, D. E. (1991). Real-coded genetic algorithms, virtual alphabets, and blocking. *Complex Systems 5*, 139–167.

Gómez-Skarmeta, A. F. and F. Jiménez (1999). Fuzzy modeling with hybrid systems. *Fuzzy Sets and Systems 104*, 199–208.

González, A. and F. Herrera (1997). Multi-stage genetic fuzzy systems based on the iterative rule learning approach. *Mathware & Soft Computing 4*, 233–249.

González, A. and R. Pérez (1996a). Aplicación de un sistema de refinamiento de reglas a problemas de clasificación. In *Memoria del V Congreso Iberoamericano de Inteligencia Artificial (IBERAMIA'96)*, pp. 20–29. In spanish.

González, A. and R. Pérez (1996b). A learning system of fuzzy control rules based on genetic algorithms. In F. Herrera and J. L. Verdegay (Eds.), *Genetic Algorithms and Soft Computing*, Number 8 in Studies in Fuzziness and Soft

Computing, pp. 202–225. Physica-Verlag.

González, A. and R. Pérez (1997a). A two level genetic fuzzy learning algorithm for solving complex problems. In *Proc. Seventh International Fuzzy Systems Association World Congress (IFSA'97)*, Volume 1, Prague, Czech Republic, pp. 192–197.

González, A. and R. Pérez (1997b). Using information measures for determining the relevance of the predictive variables in learning problems. In *Proc. Sixth IEEE International Conference on Fuzzy Systems (FUZZ-IEEE'97)*, Volume 3, Barcelona, Spain, pp. 1423–1428.

González, A. and R. Pérez (1998a). Completeness and consistency conditions for learning fuzzy rules. *Fuzzy Sets and Systems 96*, 37–51.

González, A. and R. Pérez (1998b). A fuzzy theory refinement algorithm. *International Journal of Approximate Reasoning 14*(3-4), 193–220.

González, A. and R. Pérez (1998c). A simplicity criterion for learning fuzzy rules. In *Proc. Sixth European Congress on Intelligent Technologies and Soft Computing (EUFIT'98)*, Volume 1, Aachen, Germany, pp. 418–422.

González, A. and R. Pérez (1999a). SLAVE: a genetic learning system based on an iterative approach. *IEEE Transactions on Fuzzy Systems 7*(2), 176–191.

González, A. and R. Pérez (1999b). A study about the inclusion of linguistic hedges in a fuzzy rule learning algorithm. *International Journal of Uncertainty, Fuzziness and Knowledge-Based Systems 7*(3), 257–266.

González, A. and R. Pérez (2001). Selection of relevant features in a fuzzy genetic learning algorithm. To appear in IEEE Transactions on Systems, Man, and Cybernetics.

González, A., R. Pérez, and A. Valenzuela (1995). Diagnosis of myocardial infarction through fuzzy learning techniques. In *Proc. Sixth International Fuzzy Systems Association World Congress (IFSA'95)*, Sao Paulo, Brazil, pp. 273–276.

González, A., R. Pérez, and J. L. Verdegay (1993). Learning the structure of a fuzzy rule: a genetic approach. In *Proc. First European Congress on Fuzzy and Intelligent Technologies (EUFIT'93)*, Aachen, Germany, pp. 814–819. Also in Fuzzy System and Artificial Intelligence 3(1): 57-70 (1994).

Greene, D. P. and S. F. Smith (1992). COGIN: symbolic induction with genetic algorithms. In *Proc. Tenth National Conference on Artificial Intelligence*, San Mateo, CA, USA, pp. 111–116. Morgan Kaufmann.

Greene, D. P. and S. F. Smith (1993). Competition-based induction of decision models from examples. *Machine Learning 13*, 229–257.

Greene, D. P. and S. F. Smith (1994). Using coverage and a model building constraint in learning classifier systems. *Evolutionary Computation 2*, 67–91.

Grefenstette, J. J. (1988). Credit assignment in rule discovery systems based on genetic algorithms. *Machine Learning 8*, 225–246.

Grefenstette, J. J. (1991). Lamarckian learning in multi-agent environments. In *Proc. Fourth International Conference on Genetic Algorithms (ICGA'91)*,

San Diego, USA, pp. 303–310. Morgan Kaufmann.

Gudwin, R., F. Gomide, and W. Pedrycz (1997). Nonlinear context adaptation with genetic algorithms. In *Proc. Seventh International Fuzzy Systems Association World Congress (IFSA '97)*, Prague, Czech Republic.

Gudwin, R., F. Gomide, and W. Pedrycz (1998). Context adaptation in fuzzy processing and genetic algorithms. *International Journal of Intelligent Systems 13*(10/11), 929–948.

Guêly, F., R. La, and P. Siarry (1999). Fuzzy rule base learning through simulated annealing. *Fuzzy Sets and Systems 105*(3), 353–363.

Gupta, M. M. (1992). Fuzzy logic and neural networks. In *Proc. Tenth International Conference on Multiple Criteria Decision Making*, Taipei, Japan, pp. 281–294.

Gupta, M. M. and J. Qi (1991). Design of fuzzy logic controllers based on generalized T-operators. *Fuzzy Sets and Systems 40*, 473–489.

Gupta, M. M. and D. H. Rao (1993). Neuro-control systems: A tutorial. In M. M. Gupta and D. H. Rao (Eds.), *Neuro-Control Systems: Theory and Applications*, pp. 1–43. IEEE Press.

Gurocak, H. B. (1999). A genetic-algorithm-based method for tuning fuzzy logic controllers. *Fuzzy Sets and Systems 108*(1), 39–47.

Gustafson, D. E. and W. C. Kessel (1979). Fuzzy clustering with a fuzzy covariance matrix. In *Proc. IEEE Conference on Decision and Control*, San Diego, CA, USA, pp. 761–766.

Halgamuge, S. and M. Glesner (1994). Neural networks in designing fuzzy systems for real world applications. *Fuzzy Sets and Systems 65*(1), 1–12.

Hall, L. O., J. C. Bezdek, S. Boggavarapu, and A. Bensaid (1994). Genetic fuzzy clustering. In *Proc. NAFIPS'94*, San Antonio, Texas, USA, pp. 411–415.

Hall, L. O. and B. Ozyurt (1995). Scaling genetically guided fuzzy clustering. In *Proc. ISUMA-NAFIPS'95*, pp. 328–332.

Hall, L. O., B. Ozyurt, and J. C. Bezdek (1998). The case for genetic algorithms in fuzzy clustering. In *Proc. Seventh International Conference on Information Processing and Management of Uncertainty in Knowledge-Based Systems (IPMU'98)*, Paris, France, pp. 288–295.

Hall, L. O., B. Ozyurt, and J. C. Bezdek (1999). Clustering with a genetically optimized approach. *IEEE Transactions on Evolutionary Computation 3*(2), 103–112.

Hanebeck, U. D. and G. K. Schmidt (1996). Genetic optimization of fuzzy networks. *Fuzzy Sets and Systems 79*(1), 59–68.

Harp, S. A., T. Samad, and A. Guha (1989). Towards the genetic synthesis of neural networks. In J. D. Schaffer (Ed.), *Proc. Third International Conference on Genetic Algorithms (ICGA'89)*, Hillsdale, USA, pp. 360–369. Morgan Kaufmann.

Harris, C. J., C. G. Moore, and M. Brown (1993). *Intelligent Control. Aspects of Fuzzy Logic and Neural Networks*. World Scientific.

Hathaway, R. and J. C. Bezdek (1993). Switching regression models and fuzzy clustering. *IEEE Transactions on Fuzzy Systems 1*(3), 195–204.

Hayashi, Y., E. Czogala, and J. J. Buckley (1992). Fuzzy neural controller. In *Proc. First IEEE International Conference on Fuzzy Systems (FUZZ-IEEE'92)*, San Diego, USA, pp. 197–202.

Hekanaho, J. (1996). Background knowledge in GA-based concept learning. In *Proc. Thirteenth International Conference on Machine Learning*, pp. 234–242.

Hekanaho, J. (1997). GA-based rule enhancement concept learning. In *Proc. Third International Conference on Knowledge Discovery and Data Mining*, Newport Beach, CA, USA, pp. 183–186.

Hellendoorn, H. and C. Thomas (1993). Defuzzification in fuzzy controllers. *Journal of Intelligent & Fuzzy Systems 1*, 109–123.

Herrera, F. and M. Lozano (1996). Adaptation of genetic algorithm parameters based on fuzzy logic controllers. In F. Herrera and J. L. Verdegay (Eds.), *Genetic Algorithms and Soft Computing*, Number 8 in Studies in Fuzziness and Soft Computing, pp. 95–125. Physica-Verlag.

Herrera, F. and M. Lozano (1998). Fuzzy genetic algorithms: Issues and models. Technical Report DECSAI TR-98116, Dept. of Computer Science and Artificial Intelligence, University of Granada.

Herrera, F. and M. Lozano (2000). Gradual distributed real-coded genetic algorithms. *IEEE Transactions on Evolutionary Computation 4*(1), 43–63.

Herrera, F. and M. Lozano (2001). Adaptive genetic algorithms based on coevolution with fuzzy behaviours. To appear in IEEE Transactions on Evolutionary Computation.

Herrera, F., M. Lozano, and J. L. Verdegay (1995). Tuning fuzzy controllers by genetic algorithms. *International Journal of Approximate Reasoning 12*, 299–315.

Herrera, F., M. Lozano, and J. L. Verdegay (1996). Dynamic and heuristic fuzzy connectives-based crossover operators for controlling the diversity and convergence of real-coded genetic algorithms. *International Journal of Intelligent Systems 11*, 1018–1041.

Herrera, F., M. Lozano, and J. L. Verdegay (1997). Fuzzy connectives based crossover operators to model genetic algorithms population diversity. *Fuzzy Sets and Systems 92*(1), 21–30.

Herrera, F., M. Lozano, and J. L. Verdegay (1998a). A learning process for fuzzy control rules using genetic algorithms. *Fuzzy Sets and Systems 100*, 143–158.

Herrera, F., M. Lozano, and J. L. Verdegay (1998b). Tackling real-coded genetic algorithms: Operators and tools for behavioural analysis. *Artificial Intelligence Review 12*, 265–319.

Herrera, F. and L. Magdalena (1997). Genetic fuzzy systems. *Tatra Mountains Mathematical Publications 13*, 93–121. Lecture Notes of the Tutorial: Genetic Fuzzy Systems. Seventh IFSA World Congress (IFSA'97), Prage, June 1997.

Hewahi, N. M. and K. K. Bharadwaj (1996). Bucket brigade algorithm for hierarchical censored production rule-based system. *International Journal of Intelligent Systems 11*, 197–225.

Hirota, K. (Ed.) (1993). *Industrial Applications of Fuzzy Technology*. Springer-Verlag.

Hoffman, F. and O. Nelles (2000). Structure identification of TSK-fuzzy models using genetic programming. In *Proc. Eighth International Conference on Information Processing and Management of Uncertainty in Knowledge-Based Systems (IPMU'00)*, Volume I, Madrid, Spain, pp. 438–445.

Hoffman, F. and O. Nelles (2001). Genetic programming for model selection of tsk-fuzzy systems. To appear in Information Sciences.

Hoffmann, F. (2000). Soft computing techniques for the design of mobile robot behaviours. *Journal of Information Sciences 122*(2-4), 241–258.

Hoffmann, F. and G. Pfister (1996). Learning of a fuzzy control rule base using messy genetic algorithms. In F. Herrera and J. L. Verdegay (Eds.), *Genetic Algorithms and Soft Computing*, Number 8 in Studies in Fuzziness and Soft Computing, pp. 279–305. Physica-Verlag.

Hoffmann, F. and G. Pfister (1997). Evolutionary design of a fuzzy knowledge base for a mobile robot. *International Journal of Approximate Reasoning 17*(4), 447–469.

Holland, J. H. (1975). *Adaptation in Natural and Artificial Systems*. Ann Arbor: University of Michigan Press.

Holland, J. H. (1976). Adaptation. In R. Rosen and F. M. Snell (Eds.), *Progress in Theoretical Biology IV*, pp. 263–293. Academic Press.

Holland, J. H. (1985). Properties of the bucket brigade algorithm. In *Proc. First International Conference on Genetic Algorithms and Their Applications (ICGA'85)*, Pittsburgh, USA, pp. 1–7.

Holland, J. H. (1986). Escaping brittleness: the possibilities of general purpose learning algorithms applied to parallel rule-based systems. In R. S. Michalski, J. M. Carbonell, and T. M. Mitchell (Eds.), *Machine Learning: An AI Approach. Vol II*, pp. 593–623. Morgan-Kaufmann.

Holland, J. H. and J. S. Reitman (1978). Cognitive systems based on adaptive algorithms. In D. A. Waterman and F. Hayes-Roth (Eds.), *Pattern-Directed Inference Systems*. Academic Press.

Homaifar, A., D. Battle, E. Tunstel, and G. Dozier (2000). Genetic programming design of fuzzy logic controllers for mobile robot path tracking. *International Journal of Knowledge-Based Intelligent Engineering Systems 4*, 33–52.

Horikawa, S.-I., T. Furuhashi, and Y. Uchikawa (1992). On fuzzy modeling using fuzzy neural networks with the back-propagation algorithm. *IEEE Transactions on Neural Networks 3*(5), 801–806.

Horn, J. and D. E. Goldberg (1996). Natural niching for evolving cooperative classifiers. In J. R. Koza, D. E. Goldberg, and R. L. Riolo (Eds.), *Proc. First International Conference on Genetic Programming (GP'96)*, pp. 553–

564. MIT Press.

Horn, J., D. E. Goldberg, and K. Deb (1994). Implicit niching in a learning classifier systems: Nature's way. *Evolutionary Computation 2*(1), 37–66.

Hougen, H. P., E. Villanueva, and A. Valenzuela (1992). Sudden cardiac death: a comparative study of morphological, histochemical and biochemical methods. *Forensic Science International 52*, 161–169.

Howard, L. and D. D'Angelo (1995). GA-P: a genetic algorithm and genetic programming hybrid. *IEEE Expert 10*(3), 11–15.

Ichihashi, H. and M. Tokunaga (1993). Neuro-fuzzy optimal control of backing up a trailer truck. In *Proc. IEEE International Conference on Neural Networks (ICNN'93)*, San Francisco, USA, pp. 306–311.

Ishibuchi, H., T. Morisawa, and T. Nakashima (1996). Voting schemes for fuzzy-rule-based classification systems. In *Proc. Fifth IEEE International Conference on Fuzzy Systems (FUZZ-IEEE'96)*, New Orleans, USA, pp. 614–620.

Ishibuchi, H. and T. Murata (1996). A genetic-algorithm-based fuzzy partition method for pattern classification problems. In F. Herrera and J. L. Verdegay (Eds.), *Genetic Algorithms and Soft Computing*, Number 8 in Studies in Fuzziness and Soft Computing, pp. 555–578. Physica-Verlag.

Ishibuchi, H., T. Murata, and T. Nakashima (1997). Genetic-algorithm-based approaches to classification problems. In W. Pedrycz (Ed.), *Fuzzy Evolutionary Computation*, pp. 127–153. Kluwer Academic Publishers.

Ishibuchi, H., T. Murata, and I. B. Türksen (1997). Single-objective and two-objective genetic algorithms for selecting linguistic rules for pattern classification problems. *Fuzzy Sets and Systems 89*, 135–150.

Ishibuchi, H., T. Nakashima, and T. Kuroda (1999). A hybrid fuzzy genetic based machine learning algorithm: Hybridization of Michigan approach and Pittsburgh approach. In *Proc. 1999 IEEE Systems, Man, and Cybernetics Conference*, Tokyo, Japan, pp. 296–301.

Ishibuchi, H., T. Nakashima, and T. Kuroda (2000a). A hybrid fuzzy GBML algorithm for designing compact fuzzy rule-based classification systems. In *Proc. Ninth IEEE International Conference on Fuzzy Systems (FUZZ-IEEE'2000)*, Volume 2, San Antonio, Texas, USA, pp. 706–711.

Ishibuchi, H., T. Nakashima, and T. Kuroda (2000b). Minimizing the number of fuzzy rules by fuzzy genetics-based machine learning for pattern classification problems. In *Proc. Eighth International Conference on Information Processing and Management of Uncertainty in Knowledge-Based Systems (IPMU'00)*, Volume I, Madrid, Spain, pp. 96–103.

Ishibuchi, H., T. Nakashima, and T. Murata (1999). Performance evaluation of fuzzy classifier systems for multidimensional pattern classification problems. *IEEE Transactions on Systems, Man and Cybernetics - Part B: Cybernetics 29*, 601–618.

Ishibuchi, H., T. Nakashima, and T. Murata (2001). Three-objective genetics-based machine learning for linguistic rule extraction. To appear in *Information Sciences*.

Ishibuchi, H., T. Nakashima, and M. Nii (2000). Fuzzy if-then rules for pattern classification. In D. Ruan and E. E. Kerre (Eds.), *Fuzzy If-Then Rules in Computational Intelligence. Theory and Applications*, pp. 267–295. Kluwer Academic Publishers.

Ishibuchi, H., K. Nozaki, and H. Tanaka (1992). Distributed representation of fuzzy rules and its application to pattern classification. *Fuzzy Sets and Systems 52*, 21–32.

Ishibuchi, H., K. Nozaki, and H. Tanaka (1993). Efficient fuzzy partition for pattern space for classification problems. *Fuzzy Sets and Systems 59*, 295–304.

Ishibuchi, H., K. Nozaki, H. Tanaka, Y. Hosaka, and M. Matsuda (1994). Empirical study on learning in fuzzy systems by rice taste analysis. *Fuzzy Sets and Systems 64*, 129–144.

Ishibuchi, H., K. Nozaki, N. Yamamoto, and H. Tanaka (1994). Construction of fuzzy classification systems with rectangular fuzzy rules using genetic algorithms. *Fuzzy Sets and Systems 65*, 237–253.

Ishibuchi, H., K. Nozaki, N. Yamamoto, and H. Tanaka (1995). Selecting fuzzy if-then rules for classification problems using genetic algorithms. *IEEE Transactions on Fuzzy Systems 3*(3), 260–270.

Ishigami, H., T. Fukuda, T. Shibata, and F. Arai (1995). Structure optimization of fuzzy neural network by genetic algorithm. *Fuzzy Sets and Systems 71*, 257–264.

Jamshidi, M. (1997). *Large-Scale Systems: Modeling, Control and Fuzzy Logic*. Prentice Hall PTR.

Jang, J.-S. R. (1992). Self-learning fuzzy controllers based on temporal back propagation. *IEEE Transactions on Neural Networks 3*(5), 714–723.

Jang, J.-S. R. (1993). ANFIS: adaptive-network-based fuzzy inference system. *IEEE Transactions on Systems, Man, and Cybernetics 23*(3), 665–684.

Jang, J.-S. R. and C.-T. Sun (1993). Functional equivalence between radial basis function networks and fuzzy inference systems. *IEEE Transactions on Neural Networks 4*(1), 156–159.

Janikow, C. Z. (1993). A knowledge intensive genetic algorithm for supervised learning. *Machine Learning 13*, 198–228.

Janikow, C. Z. (1996). A genetic algorithm method for optimizing fuzzy decision trees. *Information Sciences 89*, 275–296.

Janikow, C. Z. (1998). A genetic algorithm method for optimizing fuzzy decision trees. *IEEE Transactions on Systems, Man and Cybernetics, Part B 28*, 1–14.

Jin, Y. (2000). Fuzzy modeling of high-dimensional systems: Complexity reduction and interpretability improvement. *IEEE Transactions on Fuzzy Systems 8*(2), 212–220.

Jin, Y., W. von Seelen, and B. Sendhoff (1999). On generating FC^3 fuzzy rule systems from data using evolutionary strategies. *IEEE Transactions on Systems, Man and Cybernetics - Part B: Cybernetics 29*(6), 829–845.

Jog, P., J. Y. Suh, and D. V. Gutch (1989). The effects of population size, heuristic crossover and local improvement on a genetic algorithm. In *Proc. Third International Conference on Genetic Algorithms (ICGA'89)*, Hillsdale, USA, pp. 110–115. Morgan Kaufmann.

Kang, S.-J., C.-H. Woo, H.-S. Hwang, and K. B. Woo (2000). Evolutionary design of fuzzy rule base for nonlinear system modeling and control. *IEEE Transactions on Fuzzy Systems 8*(1), 37–45.

Karr, C. (1991). Genetic algorithms for fuzzy controllers. *AI Expert 6*(2), 26–33.

Karr, C. and E. J. Gentry (1993). Fuzzy control of pH using genetic algorithms. *IEEE Transactions on Fuzzy Systems 1*(1), 46–53.

Keller, J. M. and D. J. Hunt (1985). Incorporating fuzzy membership functions into the perceptron algorithm. *IEEE Transactions on Pattern Analysis and Machine Intelligence 7*(6), 693–699.

Kim, J. and B. P. Zeigler (1996a). Designing fuzzy logic controllers using a multiresolutional search paradigm. *IEEE Transactions on Fuzzy Systems 4*, 213–226.

Kim, J. and B. P. Zeigler (1996b). Hierarchical distributed genetic algorithm: a fuzzy logic controller design application. *IEEE Expert 11*(3), 76–83.

Kinzel, J., F. Klawoon, and R. Kruse (1994). Modifications of genetic algorithms for designing and optimizing fuzzy controllers. In *Proc. First IEEE Conference on Evolutionary Computation (ICEC'94)*, Orlando, FL, USA, pp. 28–33.

Kiszka, J., M. Kochanska, and D. Sliwinska (1985). The influence of some fuzzy implication operators on the accuracy of a fuzzy model - Parts I and II. *Fuzzy Sets and Systems 15*, 111–128, 223–240.

Kitano, H. (1990). Empirical studies of the speed of convergence of neural networks training using genetic algorithms. In *Proc. Eigth National Conference in Artificial Intelligence*, pp. 789–796.

Klawoon, F. (1997). Fuzzy clustering with evolutionary algorithms. In *Proc. Seventh International Fuzzy Systems Association World Congress (IFSA'97)*, Volume 1, Prague, Czech Republic, pp. 312–317.

Klawoon, F. (1998). Fuzzy clustering with evolutionary algorithms. *International Journal of Intelligent Systems 13*(10-11), 975–991.

Klir, G. J. and B. Yuan (1995). *Fuzzy Sets and Fuzzy Logic*. Prentice-Hall.

Koczy, L. (1996). Fuzzy if ... then rule models and their transformation into one another. *IEEE Transactions on Systems, Man, and Cybernetics 26*(5), 621–637.

Kosko, B. (1992). *Neural Networks and Fuzzy Systems*. Prentice-Hall.

Kosko, B. (1995). Optimal fuzzy rules cover extrema. *International Journal of Intelligent Systems 10*, 249–255.

Koza, J. R. (1992). *Genetic Programming*. Cambridge; MA: MIT Press.

Koza, J. R. (1994). *Genetic Programming II: Automatic Discovery of Reusable Programs*. MIT Press.

Koza, J. R. (1999). *Genetic Programming III; Darwinian Invention and Problem*

Solving. San Francisco, CA: Morgan Kaufman.

Krishnakumar, K. and A. Satyadas (1995). GA-optimized fuzzy controller for spacecraft attitude control. In J. Periaux, G. Winter, M. Galán, and P. Cuesta (Eds.), *Genetic Algorithms in Engineering and Computer Science*, pp. 305–320. John Wiley and Sons.

Krishnamraju, P. V., J. J. Buckley, K. D. Reilly, and Y. Hayashi (1994). Genetic learning algorithms for fuzzy neural nets. In *Proc. Third IEEE International Conference on Fuzzy Systems (FUZZ-IEEE'94)*, Orlando, FL, USA, pp. 1969–1974.

Krishnapuram, R., H. Frigui, and O. Nasraoui (1995). Fuzzy and possibilistic shell clustering algorithms and their application to boundary detection and surface approximation. Parts I and II. *IEEE Transactions on Fuzzy Systems 3*, 29–60.

Krishnapuram, R., O. Nasraoui, and H. Frigui (1992). The fuzzy c spherical shells algorithm: a new approach. *IEEE Transactions on Neural Networks 3*, 663–671.

Krone, A., P. Krause, and T. Slawinski (2000). A new rule reduction method for finding interpretable and small rule bases in high dimensional search spaces. In *Proc. Ninth IEEE International Conference on Fuzzy Systems (FUZZ-IEEE'2000)*, Volume 2, San Antonio, Texas, USA, pp. 694–699.

Kuncheva, L. I. (2000). *Fuzzy Classifier Design*. Number 49 in Studies in Fuzziness and Soft Computing. Physica-Verlag.

Kuo, Y.-H., C.-I. Kao, and J.-J. Chen (1993). A fuzzy neural network model and its hardware implementation. *IEEE Transactions on Fuzzy Systems 1*(3), 171–183.

Kwan, H. K. and Y. Cai (1994). A fuzzy neural network and its application to pattern recognition. *IEEE Transactions on Fuzzy Systems 2*(3), 185–193.

Lanzi, P. L. (1999a). An analysis of generalization in the XCS classifier systems. *Evolutionary Computation 7*, 125–149.

Lanzi, P. L. (1999b). Extending the representation of classifier conditions. Part I: From binary to messy coding. In *Proc. Genetic and Evolutionary Computation Conference (GECCO'99)*, Orlando, Florida.

Lanzi, P. L. and A. Perrucci (1999). Extending the representation of classifier conditions. Part II: From messy coding to S-expressions. In *Proc. Genetic and Evolutionary Computation Conference (GECCO'99)*, Orlando, Florida.

Lanzi, P. L., W. Stolzmann, and S. W. Wilson (2000). *Learning Classifier Systems: From Foundations to Applications*. Number 1813 in Lecture Notes in Computer Science. Springer Verlag.

Larrañaga, P., C. M. Kuijpers, R. H. Murga, and Y. Yurramendi (1996). Learning bayesian network structures by searching for the best ordering with genetic algorithms. *IEEE Transactions on Systems, Man and Cybernetics 26*(4), 487–493.

Larrañaga, P., M. Poza, Y. Yurramendi, R. H. Murga, and C. M. Kuijpers (1996). Structure learning of bayesian network structures by genetic algorithms: A performance analysis of control parameters. *IEEE Transactions on Pattern Analysis and Machine Intelligence 18*(9), 912–926.

Lee, C. C. (1990). Fuzzy logic in control systems: fuzzy logic controller – Parts I and II. *IEEE Transactions on Systems, Man, and Cybernetics 20*(2), 404–418, 419–435.

Lee, M. A. and H. Takagi (1993a). Dynamic control of genetic algorithms using fuzzy logic techniques. In *Proc. Fifth International Conference on Genetic Algorithms (ICGA'93)*, pp. 78–83. Morgan Kaufmann.

Lee, M. A. and H. Takagi (1993b). Embedding apriori knowledge into an integrated fuzzy system design method based on genetic algorithms. In *Proc. Fifth International Fuzzy Systems Association World Congress (IFSA'93)*, Seoul, Korea, pp. 1293–1296.

Lee, M. A. and H. Takagi (1993c). Integrating design stages of fuzzy systems using genetic algorithms. In *Proc. Second IEEE International Conference on Fuzzy Systems (FUZZ-IEEE'93)*, San Francisco, USA, pp. 613–617.

Lee, M. A. and H. Takagi (1994). A framework for studying the effects of dynamic crossover, mutation and population sizing in genetic algorithms. In T. Furuhashi (Ed.), *Advances in Fuzzy Logic, Neural Networks and Genetic Algorithms*, Number 1011 in Lecture Notes in Computer Science, pp. 111–126. Springer-Verlag.

Lee, M. A. and H. Takagi (1996). Hybrid genetic-fuzzy systems for intelligent systems design. In F. Herrera and J. L. Verdegay (Eds.), *Genetic Algorithms and Soft Computing*, Number 8 in Studies in Fuzziness and Soft Computing, pp. 226–250. Physica-Verlag.

Lee, S. C. and E. T. Lee (1975). Fuzzy neural networks. *Mathematical Biosciences 23*, 151–177.

Leonard, J. and H. Durrant-Whyte (1992). *Directed Sonar Sensing for Mobile Robot Navigation.* Kluwer Academic Publishers.

Leondes, C. T. (Ed.) (2000). *Knowledge Engineering. Systems, Techniques and Applications.* Academic Press.

Li, H., P. T. Chan, A. B. Rad, and Y. K. Wong (1997). Optimization of scaling factors of fuzzy logic controller by genetic algorithm. In *Proc. IFAC Symp. on Artificial Intelligence in Real Time Control*, pp. 397–402.

Linkens, D. A. and H. O. Nyongesa (1995). Evolutionary learning in fuzzy neural control systems. In *Proc. Third European Congress on Fuzzy and Intelligent Technologies (EUFIT'95)*, Aachen, Germany, pp. 990–995.

Liska, J. and S. S. Melsheimer (1994). Complete design of fuzzy logic system using genetic algorithms. In *Proc. Third IEEE International Conference on Fuzzy Systems (FUZZ-IEEE'94)*, Volume 2, Orlando, FL, USA, pp. 1377–1382.

Liu, J. and W. Xie (1995). A genetics-based approach to fuzzy clustering. In *Proc. Fourth IEEE International Conference on Fuzzy Systems (FUZZ-IEEE'95)*, Yokohama, Japan, pp. 2233–2237.

López, S., L. Magdalena, and J. R. Velasco (1998). Genetic fuzzy c-means algorithm for the automatic generation of fuzzy partitions. In *Proc. Seventh*

International Conference on Information Processing and Management of Uncertainty in Knowledge-Based Systems (IPMU'98), Paris, France, pp. 705–711.

López, S., L. Magdalena, and J. R. Velasco (2000). Non-euclidean genetic FCM clustering algorithm. In *Proc. Eighth International Conference on Information Processing and Management of Uncertainty in Knowledge-Based Systems (IPMU'00)*, Volume III, Madrid, Spain, pp. 1313–1319.

Magdalena, L. (1994). *Estudio de la coordinación inteligente en robots bípedos: aplicación de lógica borrosa y algoritmos genéticos.* Doctoral dissertation, Universidad Politécnica de Madrid (Spain). In spanish.

Magdalena, L. (1996). A first approach to a taxonomy of fuzzy-neural systems. In R. Sun and F. Alexandre (Eds.), *Connectionist Symbolic Integration*, Chapter 5, pp. 69–88. Lawrence Erlbaum Associates.

Magdalena, L. (1997). Adapting the gain of an FLC with genetic algorithms. *International Journal of Approximate Reasoning 17*(4), 327–349.

Magdalena, L. (1998). Crossing unordered sets of rules in evolutionary fuzzy controllers. *International Journal of Intelligent Systems 13*(10/11), 993–1010.

Magdalena, L. and F. Monasterio (1995). Evolutionary-based learning applied to fuzzy controllers. In *Proc. 4th IEEE International Conference on Fuzzy Systems, FUZZ-IEEE'95*, Volume 3, Yokohama, Japan, pp. 1111–1118.

Magdalena, L. and F. Monasterio (1997). A fuzzy logic controller with learning through the evolution of its knowledge base. *International Journal of Approximate Reasoning 16*(3/4), 335–358.

Magdalena, L. and J. R. Velasco (1997). Evolutionary based learning of fuzzy controllers. In W. Pedrycz (Ed.), *Fuzzy Evolutionary Computation*, pp. 249–268. Kluwer Academic Publishers.

Magdalena, L., J. R. Velasco, G. Fernández, and F. Monasterio (1993). A control architecture for optimal operation with inductive learning. In A. Ollero and E. F. Camacho (Eds.), *Intelligent Components and Instrument for Control Applications*, pp. 105–110. Pergamon Press.

Mamdani, E. H. (1974). Applications of fuzzy algorithm for control a simple dynamic plant. *Proceedings of the IEE 121*(12), 1585–1588.

Mamdani, E. H. and S. Assilian (1975). An experiment in linguistic synthesis with a fuzzy logic controller. *International Journal of Man-Machine Studies 7*, 1–13.

Mandal, D. P., C. A. Murthy, and S. K. Pal (1992). Formulation of a multivalued recognition system. *IEEE Transactions on Systems, Man, and Cybernetics 22*, 607–620.

Mangasarian, O. L., R. Setiono, and W. H. Goldberg (1990). Pattern recognition via linear programming: Theory and application to medical diagnosis. In T. F. Coleman and Y. Li (Eds.), *Large-Scale Numerical Optimization*, pp. 22–31. SIAM.

McInerney, M. and A. P. Dhawan (1993). Use of genetic algorithms with back-propagation in training of feedforward neural networks. In *Proc. IEEE International Conference on Neural Networks (ICNN'93)*, San Francisco, USA, pp. 203–208.

Mendel, J. M. (1995). Fuzzy logic systems for engineering: a tutorial. *Proceedings of the IEEE 83*(3), 345–377.

Merz, C. J. and P. M. Murphy (1996). UCI repository of machine learning databases. http://www.ics.uci.edu/~mlearn/MLRepository.html.

Michalewicz, Z. (1996). *Genetic Algorithms + Data Structures = Evolution Programs* (Third, Extended ed.). Springer-Verlag.

Michalski, R. S. (1983). Theory and methodology of inductive learning. *Artificial Intelligence 20*, 111–161.

Michalski, R. S., I. Mozetic, J. Hong, and N. Lavrae (1986). *The AQ15 inductive learning system: an overview and experiments*. University of Illinois, Urbana-Champain.

Miller, G., P. Todd, and S. Hedge (1989). Designing neural networks using genetic algorithms. In J. D. Schaffer (Ed.), *Proc. Third International Conference on Genetic Algorithms (ICGA'89)*, Hillsdale, USA, pp. 379–384. Morgan Kaufmann.

Mitchell, M. (1996). *An Introduction to Genetic Algorithms*. Cambridge, Mass. : MIT Press.

Mitra, S. and S. K. Pal (1996). Fuzzy self-organization, inferencing, and rule generation. *IEEE Transactions on Systems, Man, and Cybernetics 26*(5), 608–619.

Montana, D. J. and L. Davis (1989). Training feedforward neural networks using genetic algorithms. In *Proc. Eleventh International Joint Conference on Artificial Intelligence*, pp. 762–767. Morgan Kaufmann.

Mühlenbein, H. and D. Schlierkamp-Voosen (1993). Predictive models for the breeder genetic algorithm (I) Continuous parameter optimization. *Evolutionary Computation 1*, 25–49.

Nakaoka, K., T. Furuhashi, and Y. Uchikawa (1994). A study on apportionment of credit of fuzzy classifier system for knowledge acquisition of large scale systems. In *Proc. Third IEEE International Conference on Fuzzy Systems (FUZZ-IEEE'94)*, Volume 3, Orlando, FL, USA, pp. 1797–1800.

Nakaoka, K., T. Furuhashi, Y. Uchikawa, and H. Maeda (1996). A proposal on payoffs and apportionment of credits of fuzzy classifier systems - Finding of knowledge for large scale systems. *Journal of Japan Society for Fuzzy Theory and Systems 8*, 25–34.

Narazaki, H. and A. L. Ralescu (1993). An improved synthesis method for multilayered neural networks using qualitative knowledge. *IEEE Transactions on Fuzzy Systems 1*(2), 125–137.

Nascimiento, S. and F. Moura-Pires (1996). A genetic fuzzy c-means algorithm. In *Proc. Sixth International Conference on Information Processing and Management of Uncertainty in Knowledge-Based Systems (IPMU'96)*, Volume 2, Granada, Spain, pp. 745–750.

Nascimiento, S. and F. Moura-Pires (1997). A genetic approach for fuzzy clustering with a validity measure function. In X. Liu, P. Cohen, and M. R. Berthold (Eds.), *Proc. Second International Conference on Intelligent Data Analysis (IDA'97).*, Number 1280 in Lecture Notes in Computer Science, Berlin, pp. 325–335. Springer-Verlag.

Nauck, D., F. Klawoon, and R. Kruse (1997). *Foundations of Neuro-Fuzzy Systems.* John Willey & Sons.

Nauck, D. and R. Kruse (1992). A neural fuzzy controller learning by fuzzy error backpropagation. In *Proc. NAFIPS'92*, Puerto Vallarta, Mexico, pp. 388–397.

Nauck, D. and R. Kruse (1993). A fuzzy neural network learning fuzzy control rules and membership functions by fuzzy error backpropagation. In *Proc. IEEE International Conference on Neural Networks (ICNN'93)*, San Francisco, USA, pp. 1022–1027.

Nauck, D. and R. Kruse (1997). A neuro-fuzzy method to learn fuzzy classification rules from data. *Fuzzy Sets and Systems 89*, 377–388.

Nelles, O. (1999). *Nonlinear System Identification with Local Linear Neuro-Fuzzy Models.* Aachen, Germany: Shaker Verlag.

Neri, F. and A. Giordana (1995). A parallel genetic algorithm for concept learning. In L. Eshelman (Ed.), *Proc. Sixth International Conference on Genetic Algorithms (ICGA'95)*, pp. 436–443. Morgan Kaufmann.

Ng, K. C. and Y. Li (1994). Design of sophisticated fuzzy logic controllers using genetic algorithms. In *Proc. Third IEEE International Conference on Fuzzy Systems (FUZZ-IEEE'94)*, Volume 3, Orlando, FL, USA, pp. 1708–1712.

Nie, J. and D. A. Linkens (1993). Learning control using fuzzified self-organizing radial basis function network. *IEEE Transactions on Fuzzy Systems 1*(4), 280–287.

Nie, J. and D. A. Linkens (1995). *Fuzzy-Neural Control: principles, algorithms, and applications.* Prentice Hall International.

Nomura, H., H. Hayashi, and N. Wakami (1991). A self-tuning method of fuzzy control by descendent method. In *Proc. Fourth International Fuzzy Systems Association World Congress (IFSA'91)*, Brussels, Belgium, pp. 155–158.

Nomura, H., H. Hayashi, and N. Wakami (1992a). A learning method of fuzzy inference rules by descent method. In *Proc. First IEEE International Conference on Fuzzy Systems (FUZZ-IEEE'92)*, San Diego, USA, pp. 203–210.

Nomura, H., H. Hayashi, and N. Wakami (1992b). A self-tuning method of fuzzy reasoning by genetic algorithms. In *Proc. the 1992 International Fuzzy Systems and Intelligent Control*, Lousville, USA, pp. 236–245.

Nordin, P. (1994). A compiling genetic programming system that directly manipulates the machine code. In K. E. Kinnear (Ed.), *Advances in Genetic Programming*, pp. 311–331. MIT Press.

Nozaki, K., H. Ishibuchi, and H. Tanaka (1996). Adaptive fuzzy rule-based classification systems. *IEEE Transactions on Fuzzy Systems 4*, 238–250.

Nozaki, K., H. Ishibuchi, and H. Tanaka (1997). A simple but powerful heuristic method for generating fuzzy rules from numerical data. *Fuzzy Sets and Systems 86*, 251–270.

Palm, R. (1995). Scaling of fuzzy controllers using the cross-correlation. *IEEE Transactions on Fuzzy Systems 3*(1), 116–123.

Palm, R., D. Driankov, and H. Hellendoorn (1997). *Model Based Fuzzy Control.* Springer-Verlag.

Park, D., A. Kandel, and G. Langholz (1994). Genetic-based new fuzzy reasoning models with application to fuzzy control. *IEEE Transactions on Systems, Man, and Cybernetics 24*(1), 39–47.

Parodi, A. and P. Bonelli (1993). A new approach to fuzzy classifier systems. In *Proc. Fifth International Conference on Genetic Algorithms (ICGA'93)*, pp. 223–230. Morgan Kaufmann.

Pedrycz, W. (1989). *Fuzzy Control and Fuzzy Systems.* John Willey & Sons.

Pedrycz, W. (1991). Neurocomputations in relational systems. *IEEE Transactions on Pattern Analysis and Machine Intelligence 13*(3), 289–297.

Pedrycz, W. (1992). Fuzzy neural networks with reference neurons as pattern classifiers. *IEEE Transactions on Neural Networks 3*(5), 770–775.

Pedrycz, W. (Ed.) (1996). *Fuzzy Modelling: Paradigms and Practice.* Kluwer Academic Press.

Pedrycz, W., R. Gudwin, and F. Gomide (1997). Nonlinear context adaptation in the calibration of fuzzy sets. *Fuzzy Sets and Systems 88*(1), 91–97.

Pedrycz, W. and J. Valente de Oliveira (1995). An algorithmic framework for development and optimization of fuzzy models. *Fuzzy Sets and Systems 80*(1), 37–55.

Pelikan, M. and D. E. Goldberg (2000). Research on the bayesian optimization algorithm. Technical Report IlliGAL Report No. 2000010, Illinois Genetic Algorithms Laboratory, University of Illinois at Urbana-Champaign.

Pelikan, M., D. E. Goldberg, and E. Cantu-Paz (2000). Linkage problem, distribution estimation, and bayesian networks. *Evolutionary Computation 8*(3), 311–340.

Pelikan, M., D. E. Goldberg, and F. Lobo (2000). A survey of optimization by building and using probabilistic models. Technical Report IlliGAL Report No. 99018, Illinois Genetic Algorithms Laboratory, University of Illinois at Urbana-Champaign.

Peña-Reyes, C. A. and M. Sipper (1998). Evolving fuzzy rules for breast cancer diagnosis. In *Proc. 1998 International Symposium on Nonlinear Theory and Applications (NOLTA'98)*, Lausanne, pp. 369–372.

Peña-Reyes, C. A. and M. Sipper (1999a). Designing breast cancer diagnostic systems via a hybrid fuzzy-genetic methodology. In *Proc. Eight IEEE International Conference on Fuzzy Systems (FUZZ-IEEE'99)*, Seoul, Korea.

Peña-Reyes, C. A. and M. Sipper (1999b). A fuzzy-genetic approach to breast cancer diagnosis. *Artificial Intelligence in Medicine 17*(2), 155.

Pham, D. T. and D. Karaboga (1991). Optimum design of fuzzy logic controllers using genetic algorithms. *Journal of Systems Engineering 1*, 114–118.

Procyk, T. J. and E. H. Mamdani (1979). A linguistic self-organizing process controller. *Automatica 15*(1), 15–30.

Quinlan, J. R. (1986). Induction of decision trees. *Machine Learning 1*, 81–106.

Quinlan, J. R. (1993). *C4.5 : Programs for Machine Learning*. Morgan Kaufman, San Mateo.

Rechenberg, I. (1973). *Evolutionsstrategie: optimierung technischer systeme nach prinzipien der biologischen evolution*. Frommann-Holzboog. In german.

Regattieri-Delgado, M., F. Von-Zuben, and F. Gomide (1999). Modular and hierarchical evolutionary design of fuzzy systems. In *Proc. Genetic and Evolutionary Computation Conference (GECCO'99)*, Volume 1, Orlando, Florida, pp. 180–187.

Regattieri-Delgado, M., F. Von-Zuben, and F. Gomide (2000a). Evolutionary design of Takagi-Sugeno fuzzy systems: a modular and hierarchical approach. In *Proc. Ninth IEEE International Conference on Fuzzy Systems (FUZZ-IEEE'2000)*, Volume 1, San Antonio, Texas, USA, pp. 447–452.

Regattieri-Delgado, M., F. Von-Zuben, and F. Gomide (2000b). Optimal parameterization of evolutionary Takagi-Sugeno fuzzy systems. In *Proc. Eighth International Conference on Information Processing and Management of Uncertainty in Knowledge-Based Systems (IPMU'00)*, Volume II, Madrid, Spain, pp. 738–745.

Regattieri-Delgado, M., F. Von-Zuben, and F. Gomide (2001). Hierarchical genetic fuzzy systems. To appear in Information Sciences.

Reilly, K. D. F., J. J. Buckley, and P. V. Krisnamraju (1996). Joint backpropagation and genetic algorithms for training neural nets with applications to the "robokid" problem. In *Proc. Sixth International Conference on Information Processing and Management of Uncertainty in Knowledge-Based Systems (IPMU'96)*, Volume 1, Granada, Spain, pp. 187–192.

Roubos, H. and M. Setnes (2000). Compact fuzzy models through complexity reduction and evolutionary optimization. In *Proc. Ninth IEEE International Conference on Fuzzy Systems (FUZZ-IEEE'2000)*, Volume 2, San Antonio, Texas, USA, pp. 762–767.

Rovatti, R., R. Guerrieri, and G. Baccarani (1993). Fuzzy rules optimization and logic synthesis. In *Proc. Second IEEE International Conference on Fuzzy Systems (FUZZ-IEEE'93)*, Volume 2, San Francisco, USA, pp. 1247–1252.

Saffiotti, A., K. Konolige, and E. H. Ruspini (1995). A multivalued logic approach to integrating planning and control. *Artificial Intelligence 76*(1-2), 481–526.

Sánchez, L. (1997). Study of the Asturias rural and urban low voltage network. Technical report, Hidroeléctrica del Cantábrico Research and Development Department, Asturias, Spain. In spanish.

Sánchez, L. (2000). Interval-valued GA-P algorithms. *IEEE Transactions on Evolutionary Computation 4*(1), 64–72.

Sánchez, L., I. Couso, and J. A. Corrales (2001). Combining GP operators with SA search to evolve fuzzy rule-based classifiers. To appear in Information Sciences.

Sánchez, L. and S. García (1999). Fuzzy classifier induction with GA-P algorithms. In *Proc. ESTYLF-EUSFLAT Joint Conference*, Palma de Mallorca, Spain.

Satyadas, A. and K. KrishnaKumar (1996). EFM-based controllers for space attitude control: applications and analysis. In F. Herrera and J. L. Verdegay (Eds.), *Genetic Algorithms and Soft Computing*, Number 8 in Studies in Fuzziness and Soft Computing, pp. 152–171. Physica-Verlag.

Schaffer, J. D. and A. Morishima (1987). An adaptive crossover distribution mechanism for genetic algorithms. In *Proc. Second International Conference on Genetic Algorithms (ICGA'87)*, Hillsdale, USA, pp. 36–40. Morgan Kaufmann.

Schulte, C. M. (1994). Genetic algorithms for prototype based fuzzy clustering. In *Proc. Second European Congress on Fuzzy and Intelligent Technologies (EUFIT'94)*, Aachen, Germany, pp. 913–921.

Schwefel, H.-P. (1995). *Evolution and Optimum Seeking*. John Wiley and Sons.

Seng, T. L., M. B. Khalid, and R. Yusof (1999). Tuning of a neuro-fuzzy controller by genetic algorithm. *IEEE Transactions on Systems, Man, and Cybernetics-Part B: Cybernetics 29*(2), 226–236.

Setiono, R. (1996). Extracting rules from pruned neural networks for breast cancer diagnosis. *Artificial Intelligence in Medicine*, 37–51.

Setiono, R. and H. Liu (1996). Symbolic representation of neural networks. *Computer 29*(3), 71–77.

Setnes, M., R. Babuska, U. Kaymak, and H. R. van Nauta-Lemke (1998). Similarity measures in fuzzy rule base simplification. *IEEE Transactions on Systems, Man and Cybernetics - Part B: Cybernetics 28*, 376–386.

Setness, M. and H. Hellendoorn (2000). Orthogonal transforms for ordering and reduction of fuzzy rules. In *Proc. Ninth IEEE International Conference on Fuzzy Systems (FUZZ-IEEE'2000)*, Volume 2, San Antonio, Texas, USA, pp. 700–705.

Setness, M. and H. Roubos (2000). GA-fuzzy modeling and classification: complexity and performance. *IEEE Transactions on Fuzzy Systems 8*(5), 509–522.

Shann, J. J. and H. C. Fu (1995). A fuzzy neural network for rule acquiring on fuzzy control systems. *Fuzzy Sets and Systems 71*, 345–357.

Shi, Y., R. Eberhart, and Y. Chen (1999). Implementation of evolutionary fuzzy systems. *IEEE Transactions on Fuzzy Systems 7*(2), 109–119.

Shi, Y. and M. Mizumoto (2000). A new approach of neuro-fuzzy learning algorithm for tuning fuzzy rules. *Fuzzy Sets and Systems 112*, 99–116.

Shimojima, K., N. Kubota, and T. Fukuda (1996). Virus-evolutionary genetic algorithm for fuzzy controller optimization. In F. Herrera and J. L. Verdegay (Eds.), *Genetic Algorithms and Soft Computing*, Number 8 in Studies in Fuzziness and Soft Computing, pp. 369–388. Physica-Verlag.

Siarry, P. and F. Guely (1998). A genetic algorithm for optimizing Takagi-Sugeno fuzzy rule bases. *Fuzzy Sets and Systems 99*, 37–47.

Simpson, P. K. (1992). Fuzzy min-max neural networks - Part 1: Classification. *IEEE Transactions on Neural Networks 3*(5), 776–786.

Simpson, P. K. (1993). Fuzzy min-max neural networks - Part 2: clustering. *IEEE Transactions on Fuzzy Systems 1*(1), 32–45.

Smith, R. E. and D. E. Goldberg (1992). Reinforcement learning with classifier systems: adaptive default hierarchy formation. *Applied Artificial Intelligence 6*, 79–102.

Smith, S. F. (1980). *A learning system based on genetic adaptive algorithms*. Doctoral dissertation, Department of Computer Science. University of Pittsburgh.

Smith, S. F. (1983). Flexible learning of problem solving heuristics through adaptive search. In *Proc. Eighth International Joint Conference on Artificial Intelligence*, pp. 422–425. Morgan Kaufmann.

Srikanth, R., R. George, N. Warsi, D. Prabhu, F. E. Petry, and B. P. Buckles (1995). A variable -length genetic algorithm for clustering and classification. *Pattern Recognition Letters 16*, 789–800.

Stolzmann, W. (1997). *Anticipative classifier systems [Anticipatory Classifier Systems]*. Cambridge, MA: MIT Press/Bradford.

Stolzmann, W. (1998). Anticipatory classifier systems. In J. R. Koza et al. (Eds.), *Proc. Third Int. Conference on Genetic Programming (GP'98)*, pp. 658–664. Morgan Kaufmann.

Stolzmann, W. (1999). Latent learning in khepera robots with anticipatory classifier systems. In *Proc. Second International Workshop on Learning Classifier Systems (2-IWLCS) on the Genetic and Evolutionary Computation Conference (GECCO'99)*, Orlando, Florida, pp. 290–297.

Stolzmann, W., M. V. Butz, J. Hoffmann, and D. E. Goldberg (2000). First cognitive capabilities in the anticipatory classifier system. Technical Report IlliGAL Report No. 2000008, Illinois Genetic Algorithms Laboratory, University of Illinois at Urbana-Champaign.

Streifel, R. J., R. J. Marks, R. Red, J. J. Choi, and M. Healy (1999). Dynamic fuzzy control of genetic algorithms parameter coding. *IEEE Transactions on Systems, Man and Cybernetics. Part B: Cybernetics 29*(3), 426–433.

Sugeno, M. and G. T. Kang (1988). Structure identification of fuzzy model. *Fuzzy Sets and Systems 28*(1), 15–33.

Sugeno, M. and T. Yasukawa (1993). A fuzzy-logic-based approach to qualitative modeling. *IEEE Transactions on Fuzzy Systems 1*(1), 7–31.

Suh, J. Y. and D. V. Gutch (1987). Incorporating heuristic information on genetic search. In *Proc. Second International Conference on Genetic Algorithms (ICGA'87)*, Hillsdale, USA, pp. 100–107. Morgan Kaufmann.

Surmann, H. (1996). Genetic optimization of fuzzy rule-based systems. In F. Herrera and J. L. Verdegay (Eds.), *Genetic Algorithms and Soft Computing*, Number 8 in Studies in Fuzziness and Soft Computing, pp. 389–402. Physica-Verlag.

Surmann, H., A. Kanstein, and K. Goser (1993). Self-organizing and genetic algorithms for an automatic design of fuzzy control and decision systems. In *Proc. First European Congress on Fuzzy and Intelligent Technologies (EUFIT'93)*, Aachen, Germany, pp. 1097–1104.

Taha, I. and J. Ghosh (1997). Evaluation and ordering or rules extracted from feedforward networks. In *Proc. IEEE International Conference Neural Networks*, pp. 221–226.

Takagi, H. (1994). Application of neural networks and fuzzy logic to consumer products. In R. J. Marks II (Ed.), *Fuzzy Logic Technology and Applications*, IEEE Technology Update Series, pp. 8–12. IEEE Press.

Takagi, H. and I. Hayashi (1991). NN-driven fuzzy reasoning. *International Journal of Approximate Reasoning 5*(3), 191–212.

Takagi, H., N. Suzuki, T. Koda, and Y. Kojima (1992). Neural networks designed on approximate reasoning architecture and their applications. *IEEE Transactions on Neural Networks 3*(5), 752–760.

Takagi, T. and M. Sugeno (1985). Fuzzy identification of systems and its application to modeling and control. *IEEE Transactions on Systems, Man, and Cybernetics 15*(1), 116–132.

Takeuchi, T., Y. Nagai, and N. Enomoto (1988). Fuzzy control of a mobile robot for obstacle avoidance. *Journal of Information Sciences 43*, 231–248.

Tano, S. (1995). Fuzzy logic for financial trading. In S. Goonatilake and P. Treleaven (Eds.), *Intelligent Systems for Finance and Bussines*, pp. 209–224. John Willey & Sons.

The Japanese Society of Pedontics (1988). The chronology of deciduous and permanent dentition in japanese children. *Japanese Journal of Pediatric Dentistry 26*, 1–18.

Thrift, P. (1991). Fuzzy logic synthesis with genetic algorithms. In *Proc. Fourth International Conference on Genetic Algorithms (ICGA'91)*, San Diego, USA, pp. 509–513. Morgan Kaufmann.

Trillas, E. and L. Valverde (1985). On implication and indistinguishability in the setting of fuzzy logic. In J. Kacpryzk and R. R. Yager (Eds.), *Management Decision Support Systems Using Fuzzy Logic and Possibility Theory*, pp. 198–212. Verlag TUV Rheinland.

Tunstel, E. and M. Jamshidi (1996). On genetic programming of fuzzy rule-based systems for intelligent control. *International Journal of Intelligent Automation and Soft Computing 2*, 271–284.

Turhan, M. (1997). Genetic fuzzy clustering by means of discovering membership functions. In X. Liu, P. Cohen, and M. R. Berthold (Eds.), *Proc. Second International Conference on Intelligent Data Analysis (IDA'97).*, Number 1280 in Lecture Notes in Computer Science, Berlin, pp. 383–393. Springer-Verlag.

Umbers, I. G. and P. J. King (1980). An analysis of human-decision making in cement kiln control and the implications for automation. *International Journal of Man-Machine Studies 12*, 11–23.

Valente de Oliveira, J. (1999). Semantic constraints for membership functions optimization. *IEEE Transactions on Systems, Man and Cybernetics-Part A: Systems and Humans 29*, 128–138.

Valenzuela, A., H. P. Hougen, and E. Villanueva (1994). Lipoproteins and opolipoproteins in pericardial fluid: new postmortem markers for coronary arteriosclerosis. *Forensic Science International 66*, 81–88.

Valenzuela-Rendón, M. (1991a). The fuzzy classifier system: A classifier system for continuously varying variables. In *Proc. Fourth International Conference on Genetic Algorithms (ICGA'91)*, San Diego, USA, pp. 346–353. Morgan Kaufmann.

Valenzuela-Rendón, M. (1991b). The fuzzy classifier system: Motivations and first results. In H. P. Schwefel and R. Männer (Eds.), *Proc. First International Conference on Parallel Problem Solving from Nature - PPSN I*, pp. 330–334. Berlin: Springer-Verlag.

Valenzuela-Rendón, M. (1998). Reinforcement learning in the fuzzy classifier system. *Expert Systems with Applications 14*, 237–247.

Van Le, T. (1995). Evolutionary fuzzy clustering. In *Proc. Second IEEE Conference on Evolutionary Computation (ICEC'95)*, Volume 2, Perth, Australia, pp. 753–758.

Velasco, J. R. (1998). Genetic-based on-line learning for fuzzy process control. *International Journal of Intelligent Systems 13*(10-11), 891–903.

Velasco, J. R., G. Fernández, and L. Magdalena (1992). Inductive learning applied to fossil power plants control optimization. In E. Welfonder, G. K. Lausterer, and H. Weber (Eds.), *Control of Power Plants and Power Systems*, pp. 205–210. Pergamon Press.

Velasco, J. R., S. López, and L. Magdalena (1997). Genetic fuzzy clustering for fuzzy sets definition. In *Proc. Sixth IEEE International Conference on Fuzzy Systems (FUZZ-IEEE'97)*, Volume 1, Barcelona, Spain, pp. 1665–1670.

Velasco, J. R. and L. Magdalena (1995). Genetic algorithms in fuzzy control systems. In J. Periaux, G. Winter, M. Galán, and P. Cuesta (Eds.), *Genetic Algorithms in Engineering and Computer Science*, pp. 141–165. John Wiley and Sons.

Venturini, G. (1993). SIA: a supervised inductive algorithm with genetic search for learning attribute based concepts. In *Proc. European Conference on Machine Learning*, Viena, pp. 280–296.

Vishnupad, P. S. and Y. C. Shin (1999). Adaptive tuning of fuzzy membership functions for non-linear optimization using gradient descent method. *Journal of Intelligent and Fuzzy Systems 7*, 13–25.

Voget, S. and M. Kolonko (1998). Multidimensional optimisation with a fuzzy genetic algorithm. *Journal of Heuristics 4*(3), 221–244.

Voigt, H. M., H. Mühlenbein, and H. Cvetković (1995). Fuzzy recombination for the breeder genetic algorithm. In L. Eshelman (Ed.), *Proc. Sixth International Conference on Genetic Algorithms (ICGA'95)*, pp. 104–111. Morgan Kaufmann.

Wang, C. H., T. P. Hong, and S. S. Tseng (1998). Integrating fuzzy knowledge by genetic algorithms. *IEEE Transactions on Evolutionary Computation* 2(4), 138–149.

Wang, L. X. (1992). Fuzzy systems are universal approximators. In *Proc. First IEEE International Conference on Fuzzy Systems (FUZZ-IEEE'92)*, San Diego, USA, pp. 1163–1170.

Wang, L. X. (1994). *Adaptive Fuzzy Systems and Control: Design and Analysis.* Prentice-Hall.

Wang, L. X. and J. M. Mendel (1992a). Fuzzy basis functions, universal approximation, and orthogonal least-squares learning. *IEEE Transactions on Neural Networks* 3(5), 807–814.

Wang, L. X. and J. M. Mendel (1992b). Generating fuzzy rules by learning from examples. *IEEE Transactions on Systems, Man, and Cybernetics* 22(6), 1414–1427.

Weiß, G. (1991). The action-oriented bucket brigade. Technical Report FKI-156-91, Technische Universität München, Institut für Informatik, Technische Universität München.

Weiß, G. (1992). Learning the goal relevance of actions in classifier systems. In *Proc. Tenth European Conference on Artificial Intelligence (ECAI'92)*, pp. 430–434.

Weiß, G. (1994). The locality/globality dilemma in classifier systems and an approach to its solution. Technical Report FKI-187-94, Technische Universität München, Institut für Informatik, Technische Universität München.

Weiland, A. P. (1991). Evolving neural network controllers for unstable systems. In *Proc. IEEE International Joint Conference on Neural Networks*, Volume II, pp. 667–673.

White, H. (1992). *Artificial Neural Networks: Approximation and Learning Theory.* Blackwell Publishers.

Whitley, D., S. Dominic, and R. Das (1991). Genetic reinforcement learning with multilayered neural networks. In *Proc. Fourth International Conference on Genetic Algorithms (ICGA'91)*, San Diego, USA. Morgan Kaufmann.

Whitley, D., S. Dominic, R. Das, and C. Anderson (1993). Genetic reinforcement learning for neurocontrol problems. *Machine Learning 13*, 259–284.

Whitley, D., T. Starkweather, and C. Bogart (1990). Genetic algorithms and neural networks: Optimizing connections and connectivity. *Parallel Computing 14*, 347–361.

Wilson, S. W. (1987). Hierarchical credit allocation in a classifier system. In L. Davis (Ed.), *Genetic Algorithm and Simulated Annealing*, pp. 104–105. Morgan Kaufmann.

Wilson, S. W. (1994). ZCS: A zeroth order classifier system. *Evolutionary Computation* 2(1), 1–18.

Wilson, S. W. (1995). Classifier fitness based on accuracy. *Evolutionary Computation* 3(2), 149–175.

Wilson, S. W. (1997). Generalization in the XCS classifier system. In J. R. Koza et al. (Eds.), *Proc. Second International Conference on Genetic Program-*

ming (GP'97), pp. 665–674. Morgan Kaufmann.

Withley, D. (1995). Genetic algorithms and neural networks. In G. Winter, J. Periaux, M. Galan, and P. Cuesta (Eds.), *Genetic Algorithms in Engineering and Computer Science*, Chapter 11, pp. 203–216. John Wiley and Sons.

Wright, A. (1991). Genetic algorithms for real parameter optimization. In G. J. E. Rawlin (Ed.), *Foundations of Genetic Algorithms 1*, pp. 205–218. Morgan Kaufmann.

Wu, B. and X. Yu (2000). Fuzzy modelling and identification with genetic algorithm based learning. *Fuzzy Sets and Systems 113*, 351–365.

Xu, H. Y. and G. Vukovich (1993). A fuzzy genetic algorithm with effective search and optimization. In *Proc. 1993 International joint Conference on Neural Networks*, pp. 2967–2970.

Yager, R. R. and L. A. Zadeh (Eds.) (1992). *An Introduction to Fuzzy Logic Applications in Intelligent Systems*. Kluwer Academic Press.

Yam, Y., P. Baranyi, and C.-T. Yang (1999). Reduction of fuzzy rule base via singular value decomposition. *IEEE Transactions on Fuzzy Systems 7*, 120–132.

Yang, M. S. (1993). A survey on fuzzy clustering. *Mathematical and Computer Modeling 11*(14), 1–16.

Ye, Z. and L. Gu (1994). A fuzzy system for trading the Shangai stock market. In G. J. Deboeck (Ed.), *Trading on the Edge: Neural, Genetic, and Fuzzy Systems for Chaotic and Financial Markets*, pp. 207–214. John Willey & Sons.

Yen, J. and W. Gillespie (1995). Integrating global and local evaluations for fuzzy model identification using genetic algorithms. In *Proc. Sixth International Fuzzy Systems Association World Congress (IFSA'95)*, Sao Paulo, Brazil, pp. 121–124.

Yen, J. and L. Wang (1999). Simplifying fuzzy rule-based models using orthogonal transformation methods. *IEEE Transactions on Systems, Man and Cybernetics - Part B: Cybernetics 29*, 13–24.

Yoshinari, Y., W. Pedrycz, and K. Hirota (1993). Construction of fuzzy models through clustering techniques. *Fuzzy Sets and Systems 54*, 157–165.

Yosikawa, T., T. Furuhashi, and Y. Uchikawa (1995). Acquisition of fuzzy rules for constructing intelligent systems using genetic algorithms based on DNA coding method. In *Proc. International Joint Conference of CFSA/IFIS/SOFT'95 on Fuzzy Theory and Applications*, pp. 447–448.

Yosikawa, T., T. Furuhashi, and Y. Uchikawa (1996a). *Acquisition of Fuzzy rules from DNA coding method*. Number 1152 in Lecture Notes in Artificial Intelligence. Springer-Verlag.

Yosikawa, T., T. Furuhashi, and Y. Uchikawa (1996b). DNA coding method and a mechanism of development for acquisition of fuzzy control rules. In *Proc. Fifth IEEE International Conference on Fuzzy Systems (FUZZ-IEEE'96)*, New Orleans, USA, pp. 2194–2200.

Yuan, B., G. J. Klir, and J. F. Swan-Stone (1995). Evolutionary fuzzy c-means clustering algorithm. In *Proc. Fourth IEEE International Conference on Fuzzy Systems (FUZZ-IEEE'95)*, Yokohama, Japan, pp. 2221–2226.

Yuize, H., T. Yagyu, M. Yoneda, Y. Katoh, S. Tano, M. Grabisch, and S. Fukami (1991). Decision support system for foreign exchange trading - practical implementation. In *Proc. International Fuzzy Engineering Symposium (IFES'91)*, Yokohama, pp. 971–982.

Zadeh, L. A. (1965). Fuzzy sets. *Information and Control 8*, 338–353.

Zadeh, L. A. (1973). Outline of a new approach to the analysis of complex systems and decision processes. *IEEE Transactions on Systems, Man, and Cybernetics 3*, 28–44.

Zadeh, L. A. (1975). The concept of a linguistic variable and its applications to approximate reasoning - Parts I, II and III. *Information Sciences 8-9*, 199–249, 301–357, 43–80.

Zadeh, L. A. (1977). Fuzzy sets and their applications to classification and clustering. In *Classification and Clustering*, pp. 251–299. Academic Press.

Zadeh, L. A. (1997). What is soft computing? *Soft Computing 1*(1), 1.

Zhang, J., A. J. Morris, and G. A. Montague (1994). Fault diagnosis of a CSTR using fuzzy neural networks. In *Proc. IFAC Artificial Intelligence in Real-Time Control (AIRTC'94)*, Valencia, Spain, pp. 191–196.

Zhao, L., Y. Tsujimura, and M. Gen (1996). Genetic algorithms for fuzzy clustering. In *Proc. Third IEEE Conference on Evolutionary Computation (ICEC'96)*, Nagoya, Japan, pp. 716–719.

Zheng, L. (1992). A practical guide to tune of proportional and integral (PI) like fuzzy controllers. In *Proc. First IEEE International Conference on Fuzzy Systems (FUZZ-IEEE'92)*, San Diego, USA, pp. 633–640.

Zimmermann, H. J. (1996). *Fuzzy Sets Theory and its Applications*. Kluwer Academic Press.

Acronyms

CA:	Credit assignment
CCP:	Cooperation versus Competition Problem
CG:	Centre of Gravity
CR:	Conflict-resolution
CS:	Classifier System
DB:	Data Base
EA:	Evolutionary Algorithm
EC:	Evolutionary Computation
ES:	Evolution Strategy
EP:	Evolutionary Programming
FCS:	Fuzzy Classifier System
FL:	Fuzzy Logic
FLC:	Fuzzy Logic Controller
FNN:	Fuzzy Neural Network
FRB:	Fuzzy Rule Base
FRBCS:	Fuzzy Rule-Based Classification System
FRBS:	Fuzzy Rule-Based System
FRM:	Fuzzy Reasoning Method
FS:	Fuzzy System
GA:	Genetic Algorithm
GBML:	Genetic-Based Machine Learning
GFC:	Genetic Fuzzy Clustering
GFRBCS:	Genetic Fuzzy Rule-Based Classification System
GFRBS:	Genetic Fuzzy Rule-Based System
GFS:	Genetic Fuzzy System

GP:	Genetic Programming
IRL:	Iterative Rule Learning
KB:	Knowledge Base
MV:	Maximum Value
NN:	Neural Network
PR:	Probabilistic Reasoning
RB:	Rule Base
RCGA:	Real-Coded GA

Index